MANUAL

OF

MINERALOGY,

INCLUDING

OBSERVATIONS ON MINES, ROCKS,

REDUCTION OF ORES,

AND THE

APPLICATIONS OF THE SCIENCE TO THE ARTS,

WITH 260 ILLUSTRATIONS.

DESIGNED FOR THE USE OF SCHOOLS AND COLLEGES.

BY JAMES D. DANA, A. M.,

Member of the Soc. Cæs. Nat Cur. of Moscow, the Soc Philomathique et Paris, the American Academy of Arts and Sciences at Boston, etc.; Author of a "System of Mineralogy."

NEW EDITION, REVISED AND ENLARGED.

NEW HAVEN:

PUBLISHED BY DURRIE & PECK.

PHILADELPHIA: PECK & BLISS.

1857.

PREFACE TO THE FIRST EDITION.

In the preparation of this Manual, the author has endeavored to meet a demand often urged, by making it, as far as possible, practical and American in character.

Prominence has been given to the more common species, while others are but briefly noticed in smaller type, or are mentioned only by name. The uses of minerals and their modes of application in the arts have been especially dwelt upon. The value of ores in mining, their modes of reduction, the yield of mines in different countries, and the various applications of the metals, have been described as minutely as was consistent with the extent of the work. The various rocks are in like manner included.

At the same time, the subject has been presented with all the strictness of a scientific system. The classification adopted throws together ores of the same metal, and associates the earthy species as far as possible in natural groups. This order is preferred by very many teachers of the science, and has advantages which for many purposes counterbalance those of a more perfectly natural system. The account of the ores of each metal is preceded by a brief statement of their distinctive characters; and after the descriptions, there follow general remarks on mines, metallurgical processes, and other useful information.

As the rarer mineral species are not altogether excluded, but are briefly mentioned each in its proper place in the system, the student, should he meet with them, will be guided by the Manual to some knowledge of their general characters, and aided in arranging them in his cabinet.

The list of American localities appended to the work, the descriptions of mineralogical implements, and the notice of foreign weights, measures and coins, will be found convenient to the student.

The author must refer to his larger work for more minute information on the localities of minerals and the associations of species—for full lists of synonyms—a more complete account of crystallography and its details, and more numerous analyses, with their authorities. * * * * * *

PREFACE TO THE NEW EDITION.

In bringing this Manual up to the present state of the science, numerous changes and additions have been required. The arrangement, however, remains unaltered, except in the order of some of the metals. The Table of American localities has been nearly doubled in length, giving it the completeness it has in the author's "Treatise on Mineralogy." A chapter has also been added on the chemical composition and formulas of minerals, in which the subject is explained with simple illustrations, and a list of the more prominent species with their chemical formulas is given, following the order of the descriptive part of the work.

New Haven, 1857.

TABLE OF CONTENTS.

1*

GLOSSARY AND INDEX OF TERMS.*

ACICULAR, [Lat. *acus*, a needle,] 53.

Adamantine, 56.

Adit. [Lat. *aditus*, an entrance.] The horizontal entrance to a mine.

Alkali. An oxyd having an acrid taste, and caustic; as potash, soda.

Alkaline. Like an alkali.

Alliaceous, [Lat. *allium*, garlic,] 66.

Alloy. A mixture of different metals (excluding mercury) by fusion together. Also, the metal used to deteriorate another metal by mixture with it.

Alluvial. [Lat. *alluo*, to wash over.] Of river or fresh-water origin.

Amalgam. [Gr. *malagma*, a softened substance.] A compound of mercury and another metal.

Amalgamation, 326.

Amorphous, [Gr. *a*, not, and *morphe*, shape,] 54.

Amygdaloidal, 339.

Anhydrous. [Lat. *a*, not, and *hudor*, water.] Containing no water.

Arborescent. [Lat. *arbor*, tree.] Branching like a tree.

Arenaceous. [Lat. *arena*, sand.] Consisting of, or having the gritty nature of, sand.

Argentiferous. [Lat. *argentum*, silver.] Containing silver.

Argillaceous. [Lat. *argilla*, clay.] Like clay; containing clay.

Arsenical odor, 66.

Asparagus green. Pale green, with much yellow.

Assay. [Same etymology as *essay*.] To test ores by chemical or blowpipe examination; said to be in *the dry way*, when done by means of heat, (as in a crucible,) and in *the wet way*, when by means of acids and liquid tests.

Assay. The material under chemical or blowpipe examination.

Astringent, 66.

Asteriated. [Gr. *aster*, star.]
ing the appearance of a
within.

Augitic. Containing augite.

Auriferous. [Lat. *aurum*, g
Containing gold.

Axes, 24; of double refraction

Basaltic, 339.

Bath stone. A species of limestone; called also Bath oolite; named from the locality, in England.

Bevelment, beveled, 35.

Bitter, 66.

Bittern, 106.

Bituminous. Containing bitumen; like bitumen.

Bladed. Thin blade-like.

Blast furnace, 233.

Blowpipe, 67; tests, 69, 70: implements, 68, 69.

Blue-john. Name for fluor spar, used in Derbyshire, where it often has a bluish-purple color.

Botryoidal, [Gr. *botrus*, a bunch of grapes,] 53.

Boulder, bowlder. Loose rounded mass of stone.

Breccia.

Brittle, 53, 65.

Calcine. [Lat. *calx*, burnt limestone.] To heat, in order to drive off volatile ingredients, and make easy to be broken or pounded.

Calcination. The process of calcining.

Carat, 82.

Carbon. Pure charcoal.

Carbonate. A salt containing carbonic acid. Carbonated; containing carbonic acid, as carbonated springs.

---

* The number after a word signifies the page where it is explained. The etymology is given in brackets, wherever it was deemed important.

Carbonize. To convert into charcoal.

Carburet. A compound of an element with carbon, not acid.

Catalan forge, 237.

Celandine green. Green with blue and gray; from the plant called *celandine*.

Cementation, 238.

Chalybeate. Impregnated with iron, 80.

Chert. A siliceous stone containing some lime; also, hornstone.

Chlorid. Combination of an element with chlorine.

Chloritic. Containing chlorite.

Chromate. A salt containing chromic acid.

Cinereous. [Lat. *cinis*, ashes.] Resembling ashes.

Cleavage 33.

Coke, 90.

Columnar, 52.

Compound crystals, 42.

Conchoidal, 65.

Coralloidal. Having a resemblance to coral.

Cretaceous. [Lat. *creta*, chalk.] Pertaining to chalk.

Cropping out. The rising of layers of rock to the surface.

Crucible. [Lat. *crux*, a cross.] A pot made of earth or clay for melting, or reduction.

Cruciform, [Lat. *crux*, a cross,] 43.

Crystal, [Greek *krustallos*, ice,] 19; systems of crystallization, 24, 32.

Cube, 25.

Cupel, cupellation, 317, 328.

Cupreous. [Lat. *cuprum*, copper.] Containing copper.

Curved crystals, 42.

Decrepitate. To crackle and fly apart when heated.

Deflagrate. To burn with vivid combustion.

Deliquesce. To change to a liquid, on exposure; arising from the attraction of moisture.

Dendrites. [Gr. *dendron*, tree.] Delicate delineations branching like a tree; due to infiltration of oxyd of iron or manganese.

Density. Specific gravity.

Desiccate. To dry, to exhaust of moisture.

Diaphaneity, 58.

Dichroism, 57.

Dimetric system, 32.

Dimorphism, 44.

Divergent, 53.

Disintegrate. To fall to pieces; a result of exposure and partial decomposition.

Disseminated. Scattered through a rock or gangue.

Dodecahedron, rhombic, 25; isosceles, 39, fig. 65; pentagonal 37; scalene, 40.

Dolomitic. Pertaining to dolomite.

Dressing of ores. The picking and sorting of ores, and washing preparatory to reduction.

Drusy, 54.

Dull, 56.

Earthy. Soft like earth, and without luster.

Ebullition. The state of boiling.

Effervescence, 67.

Effloresce. To change to a state of powder, by exposure; arises from the escape of water.

Elastic, 53, 65. Electricity of minerals, etc., 62.

Elements, 72.

Ellipsoid, 42.

Elutriation. [Lat. *elutrio*, to pour from one vessel to another.] Mixing a powdered substance (as powdered flint) with water, and then after the coarser particles have subsided, carefully decanting the liquid and putting it away to settle, in order to obtain the impalpable powder which is finally deposited.

Elvan. In Cornwall, the granite masses forming broad veins in the killas, and containing the stockwerks.

Enamel. A glass having an appearance like porcelain, or like the surface of a tooth.

Evaporate. To become a vapor; to cause to become a vapor.

Even fracture, 65.

Exfoliate. To separate into thin leaves, or to scale off.

Fault. Dislocation along a fissure, as often in coal beds, 87.

Feldspathic. Containing feldspar as a principal ingredient; consisting of feldspar.

Ferruginous. [Lat. *ferrum*, iron.] Containing iron.

Fetid, 66.

Fibrous, 52.

Filament. A thread-like fiber.

Finery furnace. A furnace used in the conversion of cast iron into bar iron.

Filiform, [Lat. *filum*, a thread,] 53.

Flexible, 53, 65.

Fluate. Containing fluoric acid.

Flux, [Lat. *fluo*, to flow,] 69.

Foliaceous, 53.

Forceps, Platinum, 69.

Fracture of minerals, 65.

Friable. Easily crumbling in the fingers.

Fundamental forms, 23.

Furnace, blast, 233; reverberatory, 327; Catalan, 237.

Gallery. A horizontal passage in mining.

Gangue, 204.

Gelatinize, 67.

Geniculate. [Lat. *genu*, knee.] Bent at an angle, 43.

Geode. [Gr. *gæodes*, earth-like.] A cavity studded around with crystals or mineral matter, or a rounded stone containing such a cavity.

Glance. [Germ. *glanz*, luster.] Certain lustrous metallic sulphurets of dark shades of color.

Glimmering. Glistening, 56.

Globular, 53.

Goniometer, common, 47; reflecting, 50.

Granular. Consisting of grains.

Granulate; to reduce to grains.

Hackly, 65.

Hardness, scale of, 64.

Hemihedral forms, 37.

Hepatic. [Gr. *hepar*, liver.] Having an external resemblance to liver.

Hexagonal prism, 27.

Hexagonal system, 33.

Homogeneous. Of the same texture and nature throughout.

Hyacinth red. Red with yellow and some brown.

Hyaline. [Gr. *hualos*, glass.] Resembling glass in transparency and luster.

Hydrated. [Gr. *hudor*, water.] Containing water.

Ignition. [Lat. *ignis*, fire.] The state of being so heated as to give out light; at a red or white heat.

Impalpable, 53.

Implanted crystals. Attached by one extremity.

Incandescence. White heat.

Incrustation. A coating of mineral matter.

Indurated. Hardened or solidified.

Infiltrate. To enter gradually, as water, through pores.

Infusible. In mineralogy, not fusible by means of the simple blowpipe.

Inspissate. To thicken.

Intumesce. To froth.

Investing. Coating or covering, as when one mineral forms a coating on another.

Irised. [Lat. *iris*, rainbow.] Having the colors of the spectrum.

Iridescence, 57.

Isomorphism, isomorphous, 74.

Juxtapose. To place contiguous.

Killas. In Cornwall, the schistose rock in which the lodes occur.

Lamellar, 53.

Lapidification. [Lat. *lapis*, a stone.] The process of changing to stone.
Lapilla. Small volcanic cinders.
Lavender-blue. Blue with some red and much gray.
Leek-green. The color of the leaves of garlic.
Lenticular. Thin, with acute edges something like a lens, except that the surface is not curved.
Leucitic. Containing leucite.
Levigation. [Lat. *levis*, light.] The process of reducing to a fine powder.
Liquation. [Lat. *liquo*, to melt.] The slow fusion of an alloy, by which the more fusible flows out and leaves the rest behind, 328.
Lithographic stone. A compact grayish or yellowish-gray limestone of very even texture and conchoidal fracture; used in lithography. That of Solenhofen, near Munich, is most noted.
Lithology. [Gr. *lithos*, stone, and *logos*, a discourse.] Mineralogy.
Lixiviate. [Lat. *lixivium*, lye.] To form a lye, by allowing water to stand upon earthy or alkaline material, and draining it off below, after it has dissolved the soluble ingredients present.
Lode. [Sax. *lædan*, to lead.] In mining, a vein of mineral substance; usually a vein of metallic ore. The lode is said to be *dead* when the material affords no metal.
Lodestone, 217.

Macle. A compound crystal, or one having a tesselated structure.
Magnesian. Containing magnesia.
Magnetism of minerals, 63.
Malleable, [Lat. *malleus*, a hammer,] 65.
Mammillary, [Lat. *mammilla*, a little teat,] 53.
Manganesian. Containing manganese.
Marly. Having the nature of marl; containing marl.
Massive. Compact, and having no regular form.
Matrix. [Lat. *matrix*, from *mater*, mother.] The rock or earthy material, containing a mineral or metallic ore.
Metallic, 55, 56. Metallic-pearly, 55. Metallic-adamantine, 56.
Metalliferous. Yielding metal.
Metallurgy. [Gr. *metallon*, and *ergon*, work.] The science of the reduction of ores.
Micaceous, 53.
Mineralized. Changed to mineral by impregnation with mineral matter. Also being disguised in character by combination with other substances; thus used with regard to metals when in combination with sulphur, arsenic, carbonic acid, or anything that affects their malleability and other qualities.
Molecules, 42.
Molybdate. A salt containing molybdic acid.
Monoclinate, 33.
Monometric, 32.
Mountain limestone. A limestone of the lower part of the coal series; called also carboniferous limestone.
Muffle, 317.

Nacreous. Like pearl.
Native metal, 202.
Nitrate. A salt containing nitric acid.
Nitriary, 102.
Nucleus. The center particle or mass around which matter is aggregated.

Ochreous. Like ocher.
Octahedron, pp. 23, 25, 26.
Octahedral. Having the form of an octahedron.
Odor of minerals, p. 66.
Oolite. [Gr. *oon*, egg,] p. 349.
Opalescence, p. 57.
Opaline. Like opal.
Opalized. Changed to opal.

Opaque, p. 58.
Ore, 202. Also, by miners, a disseminated ore and the including stone together; the term *metal* is often used for the pure ore.
Oxyd, 73.
Oxydizable. Capable of combining with oxygen.
Oxydating flame, 68.

Pearly 55.
Percolate. To pass gradually through pores.
Phosphorescence, 61.
Pisolitic, [Lat. *pisum*, a pea,] composed of large round grains or kernels, of the size of peas.
Pistachio-green. Green with yellow, and some brown.
Plastic. Adhesive, and capable of being moulded in the hands.
Plumose. Having the shape of a plume, or feather.
Polarisation, 60.
Polarity, 62.
Polychroism, 57.
Play of colors, 57.
Plutonic rocks. Granite and allied crystalline rocks.
Polyhedral. [Gr. *polus*, many, and *hedra* face.) Having many sides.
Polymorphism, 44.
Porous. Having minute vacuities, visible or invisible to the naked eye; a loose texture, allowing water to filtrate through.
Porphyritic. Like porphyry, 340.
Prisms, 23.
Pseudomorphous, 54.
Puddling Furnace. A reverberatory furnace, used in converting cast into bar iron, after the finery furnace.
Pulverize. [Lat. *pulvis*, dust,] to reduce to powder.
Pulverulent. Like a fine powder slightly compacted.
Pyritous. Having the nature of pyrites, 212.
Pyro-electric, 62.

Quartation, 318.
Quartzose. Containing quartz as a principal ingredient.

Radiated, 53.
Rake-vein. A perpendicular mineral fissure.
Rectangle, 24.
Reduction of ores, 204.
Reduction flame, 68.
Refraction, 58.
Refractory. Resisting the action of heat; infusible.
Refrigerate. To cool.
Regulus. The pure state of a metal, as regulus of antimony.
Reniform. [Lat. *ren*, kiduey,] 53.
Replacement, 35. Resinous, 55.
Resplendent. Having a brilliant luster.
Reticulated. [Lat. *rete*, a net,] 52, 54.
Reverberatory furnace, 327.
Rhombohedron, 27.
Riddling or sifting of ores. Putting the broken or pulverized ore in a seive, and plunging the seive into water, by which, the whole powdered material is raised by the water and the metallic part sinking first, may be separated to a great extent from the rest.
Roasting. Exposing to heat in piles, or in a furnace, and thus driving off any volatile ingredient.

Saccharoid. [Gr. *sakchar*, sugar.] Having a texture like loaf sugar.
Saline, (Lat. *sal*, salt.) Salt like; containing common salt.
Salt. In chemistry, any combination of an acid with a base, 74.
Scale of hardness, 64.
Schlich. The finely pulverized ore and gangue.
Schistose. Having a slaty structure.
Scopiform, (Lat. *scopa*, a broom.) Like a broom in form.
Scoria, (L. *scoria*, dross,) 205, 341.
Secondary forms, 34. Sectile, 65.
Semitransparent, 58.
Shaft. A vertical or much inclined pit, cylindrical in form.

Shale, 341. Shining, 56.
Silicate, 74.
Siliceous. Consisting of, or containing silex, or quartz.
Silky, 56.
Silurian. A term applied to the fossiliferous rocks, older than the coal series.
Slag, 205.
Smelting of iron ores, 233.
Spathic, (Germ. *spath.*) Like spar.
Spar. Any earthy mineral having a distinct cleavable structure and some luster, as calcareous spar.
Stalactitic, (Gr. *stalazo*, to drop or distil,) 54, 116.
Stalagmite, 116.
Specific gravity, 63.
Splendent, 56.
Splintery. Having splinters on a surface of fracture.
Stamping. Reducing to coarse fragments in a stamping mill.
Stellated, (Lat. *stella*, star,) 52.
Strata. A series of beds of rock.
Streak, streak-powder, 56.
Striated. Lined or marked with parallel grooves, more or less regular.
Stockwerks. In Cornwall, works in beds and veins of ore. The works in alluvial deposits are distinguished as stream-works.
Sub. In composition, signifies beneath; also, somewhat, or imperfectly, as submetallic, means imperfectly metallic.
Sublimation, (Lat. *sublimis*, high.) Rising in vapor, by heat, to be again condensed.
Submetallic, 55.
Subtranslucent, 58.
Subtransparent, 58.
Subterbrand. A name given to Bovey coal, or brown coal.
Subvitreous, 55.
Sulphate. A salt containing sulphuric acid.
Sulphurcous, 66.
Sulphuret. Combination of a metal with sulphur.

Tarnish, 57.
Tertiary strata. Strata more recent in age than the chalk, and antecedent to the recent epoch.
Tesselated, (Lat. *tesselatus*, chequered.) Chequered.
Tesseral system, (Lat. *tessera*, a four square tile, or dice,) 32.
Tetrahedron, (Gr. *tetra*, four, *hedra*, face,) 37.
Titaniferous. Containing titanium.
Transition rocks. The older silurian, which were formerly supposed to contain no trace of fossils.
Translucent, 58.
Transparent, 58.
Triclinate, 33.
Trimetric, 33.
Trimorphism, 44.
Truncation, truncated, 35.
Tufaceous. Like tufa, 347.
Tuyeres, or twiers, 234.
Twin crystals, 42.

Unctuous. Adhesive, like grease.
Ustulation. [L. *ustulatus*, scorched, or partly burnt.] Roasting of ores.

Veins. In miner's use, small lodes. In geology, any seams of rock material, intersecting strata crosswise.
Vein-stone. The gangue of a metal or mineral.
Verdigris-green. Green inclining to blue; the color of verdigris.
Vesicular. Containing small vacuities.
Viscous, 65.
Vitreous, (Lat. *vitrum*, glass,) 55.
Vitrification. Conversion to glass.
Volatile. Capable of passing easily to a state of vapor.
Washing of ores. Exposing them after stamping, (or before if in fragments,) to running water, which carries off the earthy material, it being lighter than the ore.

Zeolitic. Having the nature of a zeolite, 163.

MINERALOGY.

CHAPTER I.

GENERAL CHARACTERISTICS OF MINERALS.

Relations of the three Departments of Nature. Viewing the world around us, we observe that it consists of rocks, earth or soil, and water; that it is covered with a large variety of plants, and tenanted by myriads of animals. These three familiar facts lie at the basis of three primary branches of knowledge. The animals, of whatever kind, from the animalcule to man, give origin to that branch of science which is called *Zoology;* the various plants, to the science of *Botany;* and the rocks or minerals, to *Mineralogy.* The first two of these departments embrace all natural objects that have life, and treat of their kinds, their varities of structure, their habits, and relations.

The third branch of knowledge, Mineralogy, relates to inanimate nature. It describes the kinds of mineral material forming the surface of our planet, points out the various methods of distinguishing minerals, makes known their uses, and explains their modes of occurrence in the earth.

Importance of the Science of Mineralogy. To the unpracticed eye, the costly gem, as it is found in the rocks, often seems but a rude bit of stone; and the most valuable ores may appear worthless, for the metals are generally so disguised that nothing of their real nature is seen. There is an ore of lead which has nearly the color and luster of Glauber salt; an ore of iron that looks like sparry limestone; an ore of silver that might be taken for lead ore, and another that resembles wax. These are common cases, and

---

What classes of natural objects exist? Of what does Zoology treat? What Botany? Of what does Mineralogy treat? What advantages result from the study of minerals?

consequently much careful attention is required of the student to make progress in the science. Moreover, a great proportion of the mineral species are of no special value, and they occur under so many forms and colors that close study is absolutely necessary in order to be able to distinguish the useless, and avoid being deceived by them; for such deceptions are common and often lead to disastrous consequences in mining.

The science of Mineralogy is, therefore, eminently practical. Moreover, the very existence of many of the arts of civilized life, depends upon the materials which the rocks afford. Besides the metals and metallic ores, we here find the ingredients for many common pigments, and for various preparations used in medicine; also the enduring material so valuable for buildings and numberless other purposes: moreover, from the rocks comes the soil upon which we are dependent for food. At the same time, the student of Mineralogy who is interested in observing the impress of Infinite wisdom in nature around him, finds abundant pleasure in examining the forms and varieties of structure which minerals assume, and in tracing out the principles or laws which Creative power has established even throughout lifeless matter, giving it an organization, though simple, no less perfect than that characterizing animate beings.

What is a Mineral? It has been remarked that Mineralogy, the third branch of Natural History, embraces every thing in nature that has not life. Is, then, every different thing not resulting from life, a mineral? Are earth, clay, and all stones, minerals? Is water a mineral?

All the materials here alluded to properly belong to the mineral series. The minute grains which make up a bank of clay or earth, are all minerals, and if their characters could be accurately ascertained, each might be referred to some mineral species. It is evident, however, that the clay itself, unless the grains are all of one kind, is not a distinct species, though mineral in composition: it is a compound mass or an aggregate of different mineral grains; and this is true of all ordinary soil and earth. In the same manner very many rocks are aggregates of two or more minerals in intimate union. Mineralogy distinguishes the species, and enables us to point out the ingredients which are mixed in the constitution of such rocks. It searches for specimens that

---

Is clay a mineral? What is the nature of many rocks?

are pure and undisguised, ascertains their qualities and their varieties, and thus prepares the mind to recognize them under whatever circumstances they may occur.

Water has no qualities which should separate it from the mineral kingdom. All bodies have their temperature of fusion; lead melts at 612° F.; sulphur at 226° F.; water at 32°; mercury at —39°. No difference therefore of this kind can limit the mineral departments. Ice is as properly a rock as limestone; and were the temperature of our globe but a little lower than it is, we should rarely see water except in solid crystal-like masses or layers. Our atmosphere, and all gases occurring in nature, belong for the same reason to the mineral kingdom. Several of the gases have been solidified, and we can not doubt that at some specific temperature each might be made solid. We can not, therefore, exclude any substance from the class of minerals because at the ordinary temperature it is a gas or liquid. Quicksilver with such a rule would be excluded as well as water.

A mineral, then, *is any substance in nature not organized by vitality, which has a homogeneous structure.* The *first* limitation here stated—not organized by vitality—excludes all living structures, or such as have resulted from vital powers; and the *second*—a homogeneous structure—excludes all mixtures or aggregates. The different spars, gems, and ores are minerals, while granite rock, slate, clay and the like, are mineral aggregates. This compound character is apparent to the eye in granite, for there is no difficulty in picking out from the mass a shining scaly mineral, (mica,) and with more attention, semi-opaque whitish or reddish particles (feldspar) will be easily distinguished from others (quartz) that have a glassy appearance.

It is a popular belief, that stones grow. Yet the absence of any proper growth is the main point distinguishing minerals from objects that have life. Plants and animals are nourished by the circulation of a fluid through their interior; in plants, we call the fluid sap; in animals, blood; and increase or growth takes place by means of material secreted from this circulating fluid. The living being commences with the mere germ, and grows through youth to maturity;

Why should water and gases rank with minerals. What is a mineral? What limitations are here implied? What is the nature of granite?

and when this fluid finally ceases to circulate, it dies and soon decays.

Minerals, on the contrary, have no such nourishing fluid. The smallest particle is as perfect as the mountain mass. They increase in size only by additions to the surface from some external source. The deposit of salt forming in an evaporating brine, has layer after layer of particles added to it, and by this mode of accumulation, its thickness is attained.

Beds of an ore of iron, called *bog iron-ore*, are sometimes said to grow. They do in fact increase in extent. Rills of water running from the hills wash out the iron in the rocks they pass over, decomposing and altering the condition of the ore, and carry it to low marshy grounds. Here the water becomes stagnant, and gradually the iron is deposited. This bog ore, as the name implies, is found mostly in low marshy places, and often contains nuts, leaves, and sticks, changed to iron ore. The increase here is obviously by external additions.

In limestone caverns, and about certain lakes and streams, the water contains much carbonate of lime. As it evaporates, layer after layer of the lime is deposited, till thick beds are sometimes formed. In caverns, the water comes dripping through the roof, drop by drop, and each drop as it dries, deposits a little carbonate of lime. At first it forms but a mere wart on the surface; but it gradually lengthens, till it becomes a long tapering cylinder, and sometimes the pendant cylinder, or *stalactite*, as it is called, reaches the floor of the cave, and forms a column several feet in diameter.

It thus appears that minerals increase, or enlarge, by accretion, or additions to the surface only. They decrease, or the surface is worn away, by the action of running water and other agents. When they decay, as sometimes happens from contact with air and moisture, or some other cause, the change begins with the surface, and results in producing one or more different minerals. The line of demarkation, therefore, between living beings, and minerals or inorganic matter, is strongly drawn.

Characters of Minerals. In pursuing the subject of min-

What are the different modes of increase in the animate and mineral kingdoms? Mention examples of increase in mineral substances, and explain the mode.

erals, there are various qualities presented for our study. We observe that stones or minerals have *color;* they have *hardness* in different degrees, from being soft and impressible by the nail, to the extreme hardness of the diamond; they have *weight;* they have *luster,* from almost a total absence of the power of reflecting light to the brilliancy of a mirror. Some are as *transparent* as glass and others are *opaque.* A few have *taste.* These are the most obvious characters, and characters to which the mind would at once appeal in distinguishing species.

Other characters of equal importance are found in the internal and external structure of minerals. On examining a piece of coarse granite, we find that each scale of mica may be split by the point of a knife into thinner leaves. Here is evidence of a peculiar structure, called *cleavage;* and wherever mica is found, this peculiarity is constant. The feldspar in the same rock, if examined with care, will be found to break in certain directions with a smooth, or nearly smooth plain surface, showing a luster approaching that of glass, though somewhat pearly. It is true of feldspar also, that this cleavage is a constant character for the species, as regards direction and facility. In nearly all minerals, this kind of structure, more or less perfect in quality, may be distinguished. In a broken bar of iron the irregularity of the grains proceeds from this cause. In granular marble, although the mass as a whole has no such structure, the several grains if attentively examined will be seen to present a distinct cleavage structure and consequent angular forms. In finer varieties, the grains may be so small that the characters cannot be observed; or again the texture of the mass may be so compact that not even grains can be distinguished.

This cleavage, then, is a peculiarity of *internal structure.* It is intimately connected with another fact,—that these same minerals often occur under the form of some regular solid with neat plane surfaces; and are finished with a symmetry and perfection which art would fail to imitate. These forms are their *natural* forms, and every mineral has its own distinct system of forms. The beauty of a cabinet of minerals arises to a great extent from the variety of forms and

---

What physical characters are to be observed in the study of minerals? What character depends on internal structure? Mention examples and explain. What other character depends on structure?

high finish of these gems of nature's workmanship. The mineral quartz sometimes occurs in crystals consisting of two pyramids united by a short six-sided prism, and they have generally the transparency and almost the brilliancy of the diamond, whose name they bear in common language. The "diamonds" of central New York, and many other localities, are of this kind. In other cases a large surface of rock sparkles with a splendid grouping of the pyramidal glassy crystals. We might draw other illustrations from almost all the mineral species. But this will suffice to show that in addition to the physical characters above mentioned, there are others dependent on structure, which afford distinctions of species, apparent both in external form and internal cleavage.

Still other characters are derived from subjecting species to the action of heat, and to acids or other re-agents. One mineral, when heated, melts; another is infusible, or fuses only on the edges; another evaporates. By such trials, and others hereafter to be described, we study minerals in a different way, and ascertain their *chemical characters.* This mode of investigation more minutely pursued, leads to a knowledge of the constitution of minerals, a branch of study which belongs properly to Analytical Chemistry: the results are of the highest importance to the mineralogist.

It is perceived, therefore, that the learner may (1) examine into the *peculiarities of structure* among minerals; (2) he may attend to the *physical characters* depending on *light, hardness,* and *gravity;* (3) he may acquaint himself with the *effects of heat* and *chemical re-agents*—the *chemical characters.* These are three sources of distinctions giving mutual aid, and a knowledge of all is necessary to the mineralogist. To learn to distinguish minerals by their color, weight, and luster, is so far very well; but the accomplishment is of a low degree of merit, and when most perfect, makes but a poor mineralogist. But when the science is viewed in the light of Chemistry and Crystallography, it becomes a branch of knowledge, perfect in itself, and surprisingly beautiful in its exhibitions of truth. We are no longer dealing with pebbles of pretty shapes and tints, but with objects modeled by a Divine hand; and every additional fact becomes to the mind a new revelation of His wisdom.

Mention examples. What other characters are there? Enumerate the kinds of characters presented by minerals.

In the study of this science, the learner will be introduced first to the *structure* of minerals. The subject is treated of under its usual name, *crystallography*.

CHAPTER II.

CRYSTALLOGRAPHY: OR THE STRUCTURE OF MINERALS.

Crystals: Crystallization. The regular forms which minerals assume are called *crystals*, and the process by which their formation takes place, is termed *crystallization*.

Crystallization is the same as solidification. Whenever a liquid becomes solid there is actual crystallization. Under favorable circumstances regular crystals may form; but very commonly the solid is a mass of crystalline grains, as is the case in statuary marble, or a loaf of white sugar. In the case of the marble, crystallization commenced at myriads of points at the same instant, and there was no room for any to expand to a large size and regular outline. When on the contrary, the process is slow, simple crystals often increase to a large size.

We may understand this subject of crystallization by watching a solution of salt, as it evaporates over a fire. After a while, if the process is not too rapid, minute points of salt appear at the surface, and these continue enlarging. They are minute cubes when they begin, and they increase regularly by additions to their sides, till finally they become so heavy as to sink. In other cases, if the brine is boiled away too rapidly, a mass of salt may be formed at the bottom of the vessel, in which no regular crystals (cubes) can be seen. Yet it is obvious that the same power of crystallization was at work, and failed of yielding symmetrical solids, because of the rapidity of the evaporation. Crystals of salt have been found in the beds of this mineral a foot or more in breadth, which had been formed by natural evaporation; and the whole bed is in all cases crystalline in the structure of the salt. However finely the salt may be ground

---

Explain the terms crystal and crystallization. Are solidification and crystallization the same process? Explain the different results of crystallization by the example of salt. Is every grain, however minute, crystalline?

up, as that for our tables, still the grains were crystalline in their origin and are crystalline in structure.

This subject may be further illustrated by many other substances. A hot solution of sugar set away to cool, will form crystals upon the bottom, or upon any thread or stick in the vessel; and these crystals will continue increasing till a large part of the sugar has become crystals. It is a common and instructive experiment to place a delicate framework of a basket or some other object, in a solution of sugar or alum; after a while it becomes a basket of finished gems, the crystals glistening with their many polished facets. Again, if a quantity of sulphur be melted, it will crystallize on cooling. To obtain distinct crystals, the surface crust should be broken as soon as formed, and the liquid part within be poured out; the cavity, when cold, will be found to be studded with delicate needles. The crust in this case is as truly crystallized as the needles, although but faint traces of a crystalline texture are apparent on breaking it. This was owing to too rapid cooling. Melted lead and bismuth will crystallize in the same manner. There is a substance, iodine, which when heated passes into the state of a vapor; on cooling again, the glass vessel containing the vapor is covered with complex crystals, as brilliant as polished steel. During the cold of winter, the vapors constituting clouds, often become changed to snow; this is a similar process of crystallization, for every flake of snow is a congeries of crystals, and often they present the forms of regular six-sided stars. So also, our streams become covered with ice; and this is another form of the crystallization of water.

The power which solidifies, and the power which crystallizes, are thus one and the same. Crystallography, therefore, is not merely a science treating of certain regular solids in Mineralogy; it is the science of solidification in general.

Modes of Crystallization. In the above examples we have presented three different modes of crystallization. In *one case*, the substance is in solution in water, (or some solvent;) the particles are thus free to move, and as the solvent passes off by evaporation, they unite and form the crystal-

---

Explain the case of sulphur. Give instances of crystals forming from vapor. What does the science of crystallography embrace? What are the modes of crystallization alluded to in the examples given?

lizing solid. In a *second case*, the substance is fused by heat; here again the particles are free to move as long as the heat remains; and when it passes off solidification commences, under the power of crystallization. In a *third case*, the substance is reduced to a vapor by heat; and from this state—also one of freedom of motion among the particles—it crystallizes as the heated condition is removed.

In the hardening of steel, it is well known that the coarseness of the grains varies with the temperature used, and the manner in which the process is conducted. An increased coarseness of structure, implies that certain of the crystalline grains were enlarged at the expense of others. It teaches us that in some cases the powers of crystallization may act at certain temperatures, even without fusion or solution. The long continued vibration of iron, especially when under pressure, produces a similar change from a fine to a coarse texture; and this fact has been the cause of accidents in machinery, by rendering the iron brittle: it has led to the fracture of the axles of rail cars and of grindstones, and even the iron rails of a road may thus become weak and useless.

By these several processes, the various minerals and very many of the widely extended rocks of our globe, have been brought to their present state.

Perfect crystals are usually of moderate size, and gems of the finest water are quite small. As they enlarge they become less clear, or even opaque, and the faces lose their smoothness and much of their luster. The emerald, sufficiently pure for jewelry, seldom exceeds an inch in length, and is rarely as large as this; but a crystal of this species (of the variety *beryl*) was obtained a few years since at Acworth, New Hampshire, which measured 4 feet in length and 2½ feet in circumference; it was regular in its form, yet, except at the edges, opaque. The clear garnets, fit for setting, are seldom half an inch through; but coarse crystals have been found 6 inches in diameter. Transparent sapphires also, over an inch in length, are of extreme rarity; but opaque crystals occur a foot or more long.

Quartz crystals attain at times extraordinary dimensions. There is one at Milan which is 3¼ feet long and 5½ in circumference, and it weighs 870 pounds. From a single cav-

---

Is fluidity essential to the process of crystallization? What is said of steel and iron? What is said of the size and perfection of crystals?

ity at Zinken, in Germany, 1000 cwt. of crystals of quartz were taken above a century since. These facts indicate imperfectly the scale of operations in the laboratory of nature. The same process by which a single group, like that just alluded to, has been formed, has filled numberless similar cavities over various regions, and distributed the quartz material through vast deposits in the earth's structure. The same power presides alike over the solidification of liquid lavas, and the formation of a cube of salt, producing the crystalline grains constituting the former, and the structure and symmetrical faces of the latter.

Constancy of Crystalline Forms. Each mineral may be properly said to have as much a distinct shape of its own, as each plant or each animal, and may be as readily distinguished by the characters presented to the eye. Crystals are, therefore, the perfect individuals of the mineral kingdom. The mineral quartz has a specific form and structure, as much as a dog, or an elm, and is as distinct and unvarying as regards essential characters, although, owing to counteracting causes during formation, these forms are not always assumed. In whatever part of the world crystals of quartz may be collected, they are fundamentally identical. Not an angle will be found to differ from those of crystals obtained in any part of this country. The sizes of the faces vary, and also the number of faces, according to certain simple laws hereafter to be explained; but the corresponding angles of inclination are essentially the same, whatever the variations or distortions.

Other minerals have a like constancy in their crystals, and each has some peculiarity, some difference of angle, or some difference of cleavage structure, which distinguishes it from every other mineral. In many cases, therefore, we have only to measure an angle to determine the species. Both quartz and carbonate of lime crystallize at times in similar six-sided prisms with terminal pyramids; but the likeness here ceases; for the angles of the pyramids are quite different, and also the internal structure. Idocrase and tin ore crystallize in similar square prisms, with terminal pyramidal planes; but though similar in general form, each has its own characteristic angles of inclination between its planes, which angles

---

What is said of the generality of the power of crystallization? What is said of the constancy of the crystalline forms and structure of minerals? Explain by the mineral quartz, as an example.

admit of no essential variation. Upon this character, the constancy of crystalline forms, depends the importance of crystallography to the mineralogist.

FUNDAMENTAL FORMS OF CRYSTALS.

The forms of crystallized minerals are very various. To the eye there often seems to be no relation between different crystals of the same mineral. Yet it is true that all the various shapes are modifications according to simple laws of a few *fundamental forms.* There is perhaps no mineral which presents a greater variety of form than calc spar. Dog-tooth spar is one of its forms ; nail-head spar, as it is sometimes called, is another ; the *one*, a tapering pyrimadal crystal, well described in its name, the other broad and thin, and shaped much like the head of a wrought-nail. Yet both of these crystals and many others are derived from the same fundamental form. After a few trials with a knife, the student will find that slices may be readily chipped off from the crystals of this mineral in three directions ; and the process will obtain a solid from each, the one identical with the other in its angles. They consequently have the same nucleus or fundamental form.

The *fundamental forms* are those from which all the other forms of crystals are derived. The derivative forms, are called *secondary* forms, and their planes, *secondary planes.*

The number of fundamental forms indicated by cleavage, is *thirteen.* They are either *prisms,* octahedrons* or *dodecahedrons.*

The *prisms* are either *four-sided* or *six-sided.* The prisms are denominated *right* prisms, when they stand erect, and *oblique* prisms, when they are inclined. Figures 4, 5, 7, 8, are right prisms, and figures 12, 14, are oblique prisms. The sides in each case are called *lateral planes*, and the extremities *bases.*

An *octahedron*† has *eight* sides, and consists of two equal

How do the crystals of different minerals differ? Mention examples. What is said of the forms of crystals of the same mineral? What is understood by *fundamental* forms? What by secondary forms or planes? How many fundamental forms are there? What kinds of prisms are there? Explain the terms *lateral planes* and *bases.*

* Any column, however many sides it may have, is called a prism.

† From the Greek *okto*, eight, and *hedra*, face.

four-sided pyramids placed base to base. (Figs. 2, 6, 9) The plane in which the pyramids meet is called the *base* of the octahedron; (*bb*, fig. 6;) the edges of the base are called the *basal edges*, and the other edges the *pyramidal*.

The *dodecahedron** has twelve sides (fig. 3.)

The *axes* of these solids are imaginary lines connecting the centers of opposite faces, of opposite edges, or of opposite angles. The inclination of two planes upon one another is called an *interfacial* angle.†

The figures here added represent the forms of the bases and faces referred to in the following paragraphs.

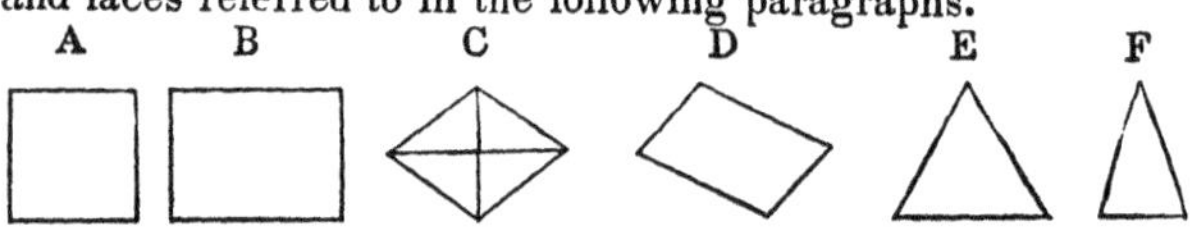

A, a *square*, having the 4 sides equal; B, a *rectangle*, differing from A, in having only the opposite sides equal; C, a *rhomb*, having the angles oblique and the sides equal; D, a *rhomboid*, differing from the rhomb in the opposite sides only being equal; E, an *equilateral triangle*, having all the sides equal; F, an *isosceles triangle*, having two sides equal. The lines crossing from one angle to an opposite are called diagonals.

The fundamental forms of crystals, though thirteen in number, constitute but six *systems of crystallization*, as follows:—

What is an octahedron? What is its base? How are the *basal* and *pyramidal* edges distinguished? What is a dodecahedron? What are axes? What are interfacial angles? Explain the terms square; rectangle; rhomb; rhomboid; equilateral triangle; isosceles triangle; diagonal. How many systems of crystallization are there?

* From the Greek *dodeka*, twelve, and *hedra*, face.

† An angle is the amount of divergence of two straight lines from a given point, or of two planes from a given edge. In the annexed figure, ACB is an angle formed by the divergence of two lines from C. If a circle be described with the angular point C as the center, and the circumference DABFE be divided into 360 equal parts, the number of these parts included between A and B will be the number of degrees in the angle ACB; that is, if 40 of these parts are included between A and B, the angle ACB equals 40 degrees (40°). DF being perpendicular to EB, these two lines divide the whole into 4 equal parts, and consequently the angle DCB equals 360°÷4 equals 90°. This is termed a *right* angle. An angle more or less than 90° is called an *oblique* angle; if less, as ACB, an *acute* angle; if more, as ACE, an *obtuse* angle.

I. The *first system* includes the cube (fig. 1 or 1*a*, the latter in outline ;) regular octahedron (fig. 2 ;) and the rhombic

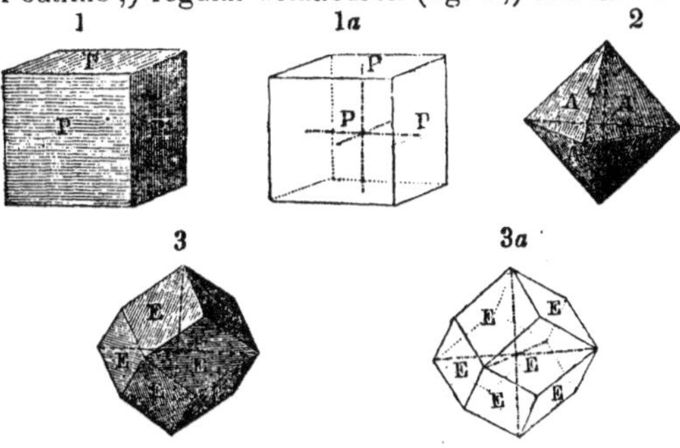
1 1*a* 2 3 3*a*

dodecahedron (fig. 3 or 3*a*.) They are symmetrical solids throughout, in all positions, being alike in having the height, breadth and thickness equal ; their three axes, represented by the dotted lines in the figures, are at right angles with one another and equal. In the cube, the axes connect the centers of opposite faces ; in the octahedron and dodecahedron, they connect the apices of solid angles. This is more fully explained on a following page.

The cube has its faces equal squares, and its angles all right angles.

The octahedron has its 8 faces equal *equilateral* triangles : its edges are equal ; its plane angles are 60° ; its *interfacial* angles (angles between adjacent faces) 109° 28′.

The dodecahedron has its 12 faces equal rhombs ; the edges are equal ; the plane angles of the faces are 109° 28′ and 70° 32′ ; its interfacial angles are 120°.

II. The *second system* includes the right square prism

4

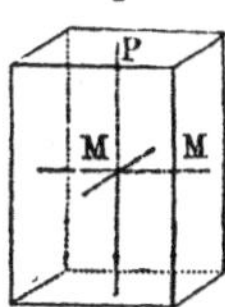

5

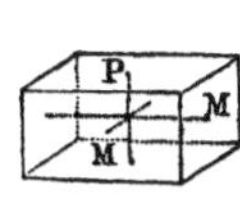

6

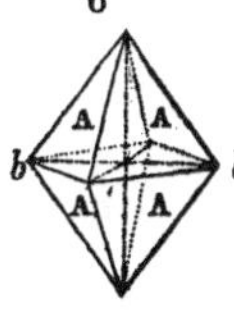

(figs. 4 and 5,) and square octahedron (fig. 6.) They have two equal lateral axes, and a vertical axis unequal to the

---

What forms does the first system include ? How are these forms related ? Describe the forms. What forms does the second system include, and how are they related ? Describe the forms.

lateral: that is, the width and breadth are equal, but the height is varying. All the axes are at right angles with one another. Fig. 4 is a square prism higher than its breadth, and fig. 5 is one shorter than its breadth.

The right square prism and square octahedron may be of any height, either greater or less than the breadth; but the dimensions are fundamentally constant for the same mineral species. The square prism has its base a square. The square octahedron has its base (*bb*) a square, and its 8 faces equal *isosceles* triangles. The lateral edges of the prism differ in length from the basal; and the terminal or pyramidal edges of the octahedron differ in length from the basal.

III. The *third system* includes the *rectangular prism* (fig. 7,) the *rhombic prism* (fig. 8,) and the *rhombic octahe-*

7

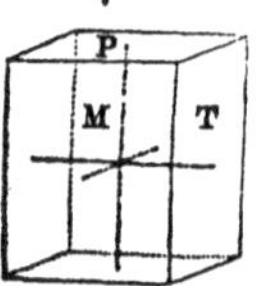

8

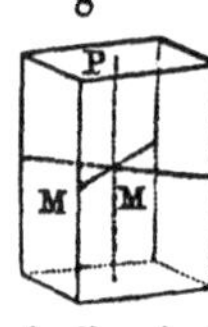

9

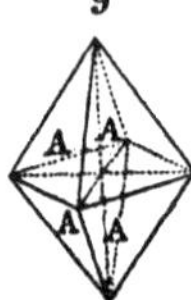

dron (fig. 9.) They are similar in having the three dimensions, or the three axes, unequal; and the axes at right angles with one another.

The rectangular prism has a rectangular base, and the axes connect the centers of opposite faces. The rhombic prism and rhombic octahedron have each a rhombic base, the angle of which differs for different species. The lateral axes of the prism connect the centers of opposite edges, and in the octahedron they connect the apices of opposite angles.

IV. The *fourth* system includes the *right rhomboidal* prism

10

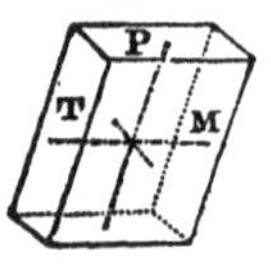

11

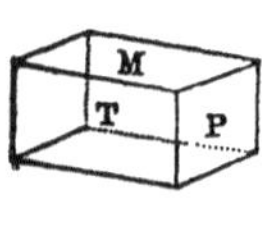

12

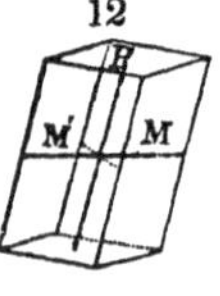

13

(figs. 10, 11,) and the *oblique rhombic* prism (figs. 12, 13.) The lateral axes are unequal, and at right angles as in the

What forms are included in the third system and how are they related? Describe the forms. What forms does the fourth system include and how are they related?

last system; but they are oblique to the vertical axes. Their positions are shown in the figures.

The right rhomboidal prism stands *erect* when on its rhomboidal base, as in fig. 11; but is *oblique* when placed on either of the other sides, as in fig. 10. The oblique rhombic prism is shown in a lateral view in fig. 12, and a front view in fig. 13.

V. The *fifth system* includes the *oblique rhomboidal* prism which has the three axes unequal, and all are oblique in their intersections. Fig. 14 represents a side view of this form, and fig. 15 a front view.

14 15

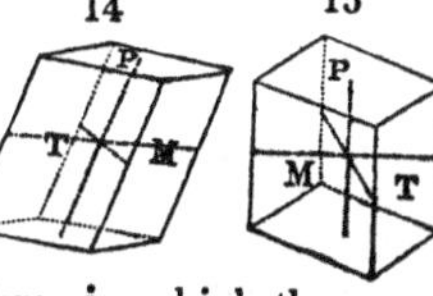

VI. The *sixth system* includes the rhombohedron and hexagonal prism, in which there are

16

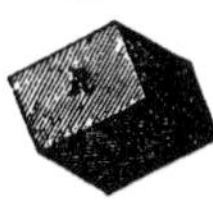

16*a*

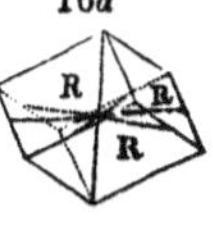

17

17*a*

18

three equal lateral axes and a vertical axis at right angles with the three. Fig. 16 is an obtuse rhombohedron, and 16*a* is the same in outline, showing the axes. Figs. 17, 17*a*, represent an acute rhombohedron. Fig. 18 is a hexagonal prism; it is bounded by six equal lateral planes; the lateral axes either connect the centers of opposite faces, as in the figure, or of opposite lateral edges.

To understand the rhombohedron, the student should have a model before him. On examining it he will find one solid angle made up of three equal plane angles, and another opposite one of the same kind; all the other solid angles are different from these. These two solid angles are called the *vertical solid angles*, and a line drawn from one to the other is the *vertical axis* of the rhombohedron. The rhombohedron should be held with this line vertical; it is then said to be *in position*. Thus placed, it will be seen to have *six lateral angles*, *six equal lateral edges*, and also *six equal terminal edges*, three of the terminal above and three below.

What forms does the fifth system include, and how does this system differ from the preceding? What does the sixth system include? What is said of the rhombohedron? of its position? its solid angles?

The lateral edges in figure 17*a*, are distinguished from the terminal by being made heavier. Figure 19 represents a vertical view of fig. 16; the three edges meeting at center are the terminal edges of one extremity: the exterior six are the lateral edges; and the six lateral angles are seen at their intersections. In fig. 19*a*, the same is seen in outline, and the dotted lines represent the three lateral or transverse axes, connecting the centers of opposite lateral edges. The lateral and terminal edges differ in one set being acute and the other obtuse; in the obtuse rhombohedron (fig. 16) the terminals are *obtuse*, and in the acute rhombohedron (fig. 17) they are *acute*.

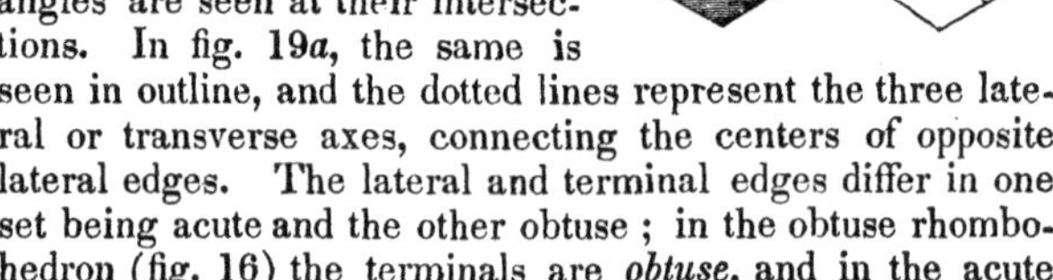

Several of the primary forms are easily cut from wood or chalk. Cut out a square stick, and then saw off a piece from one end as long as the breadth of the stick: this is the *cube*. Saw off other pieces longer or shorter than this, and they are different *right square prisms*. Shave off a piece of more or less thickness from one side of the square stick, and it then becomes a rectangular stick. From it, pieces may be sawn off, of different lengths, and they will be right *rectangular prisms*. Next cut a stick of a rhombic shape, (a section having the shape in figure C, page 26,) from it *right rhombic prisms* may be cut, of any length. Shave off more or less from one side of the rhombic stick, and it is changed to a rhomboidal form, (section as in fig. D, page 26,) and *rhomboidal prisms* may be sawn from it of any length. Take a *rhombic* stick again; and instead of sawing it off straight across, as before, saw off the end obliquely from one side-edge to the opposite; the base thus formed is oblique to the sides: then saw the stick again in parallel oblique directions, (accurately parallel,) and an *oblique rhombic prism* will be obtained. If the oblique direction is such that the basal plane equals the lateral, the solid is a *rhombohedron*. Proceeding in the same way with a *rhomboidal* stick, *oblique rhomboidal prisms* may be made. The student is advised to make these solids, either from wood, raw potatoes, or chalk,* in order to become familiar with them.

What is said of the lateral edges and angles of the rhombohedron?

* Models made of chalk become quite hard if washed over with a strong solution of gum Arabic, or varnish.

By means of such models, the student may trace out important relations between the fundamental forms.

Take a cube, and cut off each angle evenly, inclining the knife alike to the adjacent faces; this produces figure 20. Continue taking slice after slice equally from each angle, and the solid takes the form in fig. 20*a*, (called a cubo-octahedron;) still continue taking off regular slices from each angle alike, and it finally comes out a regular octahedron, the form represented in fig. 20*b*. The last diminishing point in each

20

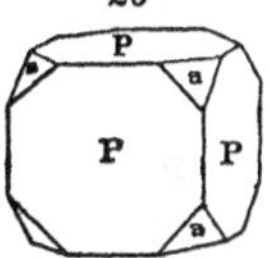

20*a*

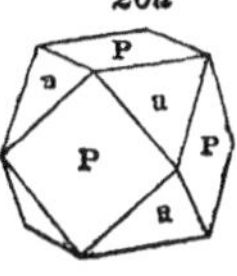

20*b*

face of the cube is the apex of each solid angle of the octahedron. It is hence apparent why the axes of the cube connect the opposite solid angles of the octahedron.

Take another cube (one of large size is preferable) and pursue the same process with each of the edges, keeping the knife, in cutting, equally inclined to the faces of the cube, and we obtain, in succession, the forms represented in figs.

21

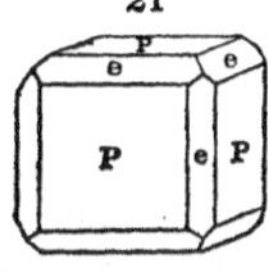

21*a*

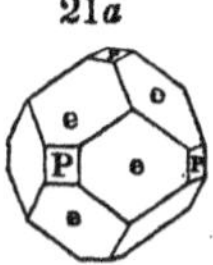

21*b*

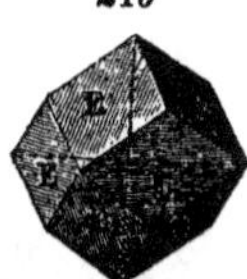

21 and 21*a*; and finally as the plane P disappears, it comes out the rhombic dodecahedron, (fig. 21*b*.) Hence the same axes which connect the centers of opposite faces in the cube, connect opposite acute solid angles in the dodecahedron.

So the cube, by reversing the process, may be made from an octahedron by cutting off its solid angles, passing in succession through the forms represented in figures 20*b*, 20*a*, 20, to figure 1. The dodecahedron also yields a cube in a similar manner, giving as the process goes on, the forms represented in figures 21*b*, 21*a*, 21, 1.

Moreover, the octahedron and dodecahedron are easily de-

How can you make an octahedron from a cube? How make a dodecahedron from a cube? How the cube from an octahedron? the cube from a dodecahedron? What relation hence exists between the solids of the *first* system?

rived from one another. Figure 22 represents an octahedron with the edges truncated. On continuing this truncation, the planes A are reduced in size, and the form in figure 22*a* is obtained; and another step beyond, we have the dodecahedron, (fig. 21*b*.) Figure 22*a* represents a dodecahedron with the obtuse solid angles replaced; and this replacement continued, produces finally an octahedron, the reverse of the preceding.

22 22*a*

These solids are, then, so related that they are all derivable from one another; and the three actually are often presented by the same mineral. All the figures above referred to, occur as forms of *galena*, *fluor-spar*, and several other species. Instead, therefore, of considering the three solids, the cube, regular octahedron, and dodecahedron, as independent forms, we properly speak of them as constituting together *one system*, or as belonging to the same series of forms.

Again: pursue the same mode of dissection on the angles of a *square prism*, taking care to move the knife parallel to a diagonal of the prism; the form in figure 23 is first obtained, and finally a square octahedron, figure 23*a*. The square prism and square octahedron (like the cube and regular octahedron) belong to one and the same *system*. The two often occur in the same mineral.

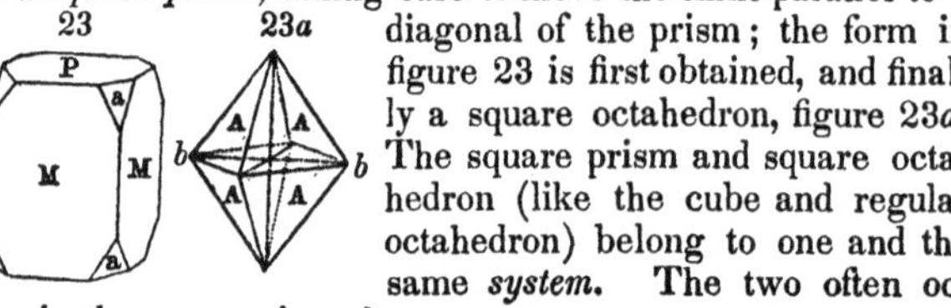

23 23*a*

Again: remove with a knife the basal edges of a *rhombic prism*, moving the knife parallel to a diagonal plane of the prism, figure 24 is at first obtained, and then a *rhombic* octahedron, (fig. 24*a*.) Remove the four lateral edges of a rhombic prism, (see fig. 26*a*,) keeping the knife parallel to a vertical diagonal plane: the form in figure 25 will first be obtained, and then a right rectangular prism, (fig. 25*a*); and conversely cut off the lateral edges

24 24*a*

How can you make a square octahedron from a square prism? How a rhombic octahedron from a rhombic prism? How a rectangular prism from a rhombic?

of a right rectangular prism, with the knife parallel to the ver-

25 25*a*

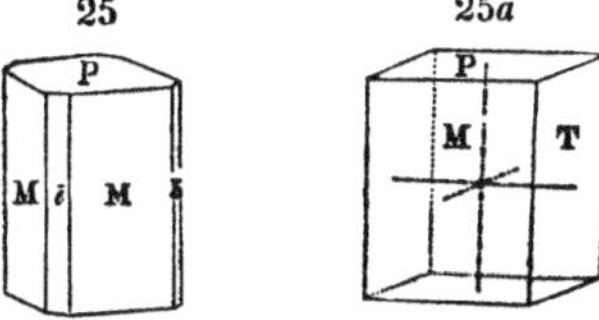

26

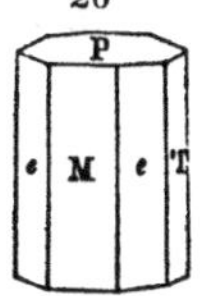

26*a*

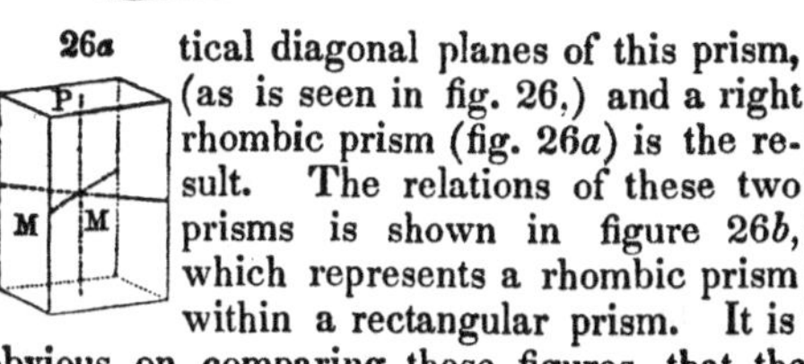

tical diagonal planes of this prism, (as is seen in fig. 26,) and a right rhombic prism (fig. 26*a*) is the result. The relations of these two prisms is shown in figure 26*b*, which represents a rhombic prism within a rectangular prism. It is obvious on comparing these figures, that the lateral axes which connect the centers of opposite faces in the rectangular prism, connect the centers of opposite lateral edges in the rhombic prism.

26*b*

These three forms, the right rhombic prism, rhombic octahedron, and rectangular prism, are so closely related, that one may give origin to the other, and all may occur in the same mineral. This is often the case, as in the minerals *celestine* and *heavy spar.*

Again: set the *right rhomboidal* prism on one of its lateral faces, and then slice off each lateral edge, (lateral, as so situated,) keeping the knife parallel with the diagonal plane, and an *oblique rhombic* prism is obtained. Figure 27 represents the process begun, and figure 13, as well as the interior of figure 27, the completed oblique rhombic prism.

27

Lastly: take a rhombohedron, and after placing it *in position*, fig. 16,) look down upon it from above, (fig. 19;) the six lateral edges are seen to form a regular six-sided figure around the axis. If these edges be cut off parallel to the axis, a six-sided prism (having a three-sided pyramid at each extremity) must, therefore, result. This process is shown begun in figure 28, and completed in figure

---

How is a rhombic prism derived from a rectangular? What relation hence between these prisms? How can you make an oblique rhombic prism from a right rhomboidal? How a right rhomboidal from an oblique rhombic? Explain the relation betw en the rhombohedron and hexagonal prism, and how one is reduced to he other.

28*a*. Looking down again on the model as before, the *lateral angles* are seen to form six equi-distant points around the axis; and if these angles are removed in the same manner, another six-sided prism is obtained, differing, however, from the former in having the faces of the pyramid at each end, *five-sided*, instead of *rhombic*. Figures 29, 30, illustrate the process. Conversely, we may make a rhombohedron out of

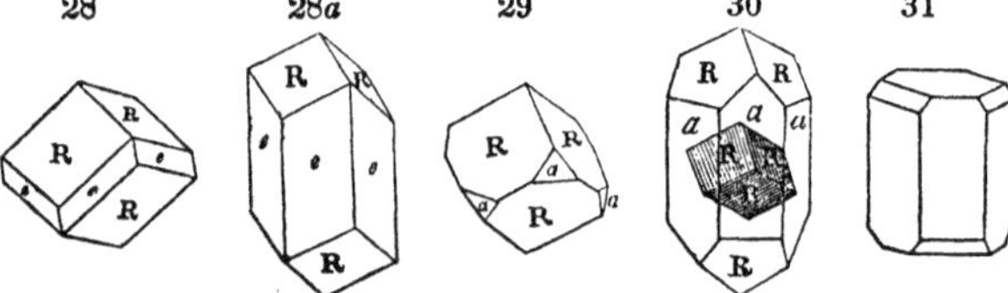

a hexagonal prism, by cutting off three alternate basal edges at one extremity of the prism, and similarly, three at the other extremity alternate with these, as in figure 31. In figure 30, the process is farther continued, and the rombohedron is shown as a nucleus to the prism. By cutting off slices parallel with R, the rhombohedron is at last obtained. The close relation of the rhombohedron and hexagonal prism is hence obvious. Calcareous spar has the rhombohedron as its primary, and very often occurs in hexagonal forms. The same is true of quartz and many other species.

From the above transformations, the study of which, with the aid of a knife and a few raw potatoes or lumps of chalk, may afford some amusement as well as instruction, the student will understand more fully the *six systems* of crystallization.* These six systems have received the following names:

1. *Monometric* or *tesseral system*, (from the Greek *monos*, one, and *metron*, measure, alluding to the three axes being equal in length.) Includes the cube, octahedron and dodecahedron, (figs. 1, 2, 3.)

2. *Dimetric system*, (from *dis*, two times, and *metron*, alluding to the vertical axis being unequal to the other two.)

Give the names of the systems of crystallization, and mention the forms each includes.

---

* In some text books, the student may read about certain *integral forms*, the *cube*, the *three-sided pyramid* and *three-sided prism*, from which it is stated all the other forms may be made. The idea of such forms has nothing to do with crystallography, or the actual constitution of crystals.

Includes the square prism and square octahedron, (figs. 4, 5, 6.)

3. *Trimetric system*, (from *tris*, three times, and *metron*, alluding to the *three* axes being unequal.) Includes the right rhombic prism, right rectangular prism and rhombic octahedron, (figs. 7, 8, 9.)

4. *Monoclinic* system, (from *monos*, one, and *klino*, to incline, one axis being inclined to the other two which are at right angles.) Includes the right rhomboidal prism and oblique rhombic prism, (figs. 10, 11, 12, 13.)

5. *Triclinic* system, (from *tris* and *klino*, the three axes being oblique to one another.) Includes the oblique rhomboidal prism, (figs. 14, 15.)

6. *Hexagonal system.* Includes the rhombohedron and hexagonal prism, (figs. 16, 17, 18.)

CLEAVAGE.

It has already been stated that crystals of calcareous spar may be chipped off easily in three directions, and by this means, the fundamental form, a rhombohedron, may be obtained. In all other directions only an irregular fracture takes place. This property of separating into natural layers, is called *cleavage*, and the planes along which it takes place, *cleavage joints.*

Cubes of fluor spar may be cleaved on the angles, with a slight pressure of the knife, and the process continued affords successively the forms represented in figures 20, 20*a*, and finally the completed octahedron, as already explained. A lead ore, called galena, yields cubes by cleavage. Mica—often improperly called isinglass—may be torn by the fingers into elastic leaves more delicate than the thinnest paper.

In many species cleavage is obtained with difficulty, and in others none can be detected. Quartz is an instance of the latter; yet it may sometimes be effected with this mineral by heating it and plunging it while hot into cold water.

The following are the more important laws with respect to this property:

Cleavage is uniform in all varieties of the same mineral.

It occurs parallel to the faces of a fundamental form or along the diagonals.

It is always the *same* in character parallel to *similar* faces

---

What is cleavage? How does it differ in different minerals? What are the laws relating to cleavage.

of a crystal, being obtained with equal ease, and affording planes of like luster: and conversely, it is *dissimilar* parallel to *dissimilar* planes. It is accordingly the same, parallel to all the faces of a cube; but in the square prism, the basal cleavage differs from the lateral, because the base is unequal to the lateral planes. Often there is an easy cleavage parallel to the base, and none distinct parallel to the sides, as in topaz; and so the reverse may be true.

The thirteen fundamental forms enumerated, are the solids obtained from the various minerals by cleavage.

Some minerals present peculiar cleavages of a subordinate character, independent of the principal cleavage. Calc spar, for example, has sometimes a cleavage parallel to the longer diagonal of its faces. The facts on this subject are of considerable interest, yet not of sufficient importance to be dwelt on in this place.

SECONDARY FORMS.

If crystals always assumed the shape of the primary form, there would be comparatively little of that variety and beauty which we actually find in the mineral kingdom. Nature first taught to heighten the brilliancy of the gem by covering its surface with facets. To the uninstructed eye, these cubes and prisms with their numberless brilliant surfaces, often appear as if they had been cut and polished by the lapidary: yet the skill and finish of the work, most perfect in the microscopic crystal, has but feeble imitation in art. Not unfrequently, crystals are found with one or two hundred distinct planes, and occasionally even a much larger number; and every edge and angle has the utmost perfection, and the surfaces an evenness of polish, that betrays no rude workmanship, even under the highest magnifying glass. Cavities are occasionally met with in the rocks, studded on every side with crystals—a crystal grotto in minature—sparkling when brought out to the sun like a casket of jewels. Even amid the apparent confusion, there is wonderful order of arrangement in the crystals: the corresponding planes generally face the same way, so that the sparkling effect appears in successive flashes over the surface, as every new set of facets comes in turn to the light. Add to this view, their delicate colors—the rich purple of the amethyst, the soft yellowish shades of the topaz, the deep green of the eme-

---

On what does the beauty of crystals to a great extent depend?

rald—and it will be admitted that the powers of crystallization scarcely yield to vitality in the forms of beauty they produce.

These results are not more wonderful than the simplicity of the laws that lead to them.

The various *secondary* forms proceed from the occurrence of planes on the angles or edges of the fundamental forms, which planes are called *secondary* planes. Figures 20, 21, are secondaries to the cube, and the planes a and e are secondary planes; figures 28, 29, 30, are secondaries to the rhombohedron, and the planes *e* and *a* are secondary planes.* These secondary planes however numerous, conform in their positions to a certain law called the *law of symmetry.* Previous to stating this law a few explanations are added.

The *cube*, it has been remarked, has six equal square faces. The *twelve edges* are therefore all equal, and so also the *eight angles.* In the *square prism* the vertical edges *differ* in length from the basal, and are therefore not similar. In the *rectangular prism*, not only the vertical differ from the basal, but two of the basal at each extremity differ from the other two basal. This will be seen at once in the models. In the right *rhombic* and *rhomboidal*, two of the lateral edges are *acute* and two *obtuse ;* these then are not similar to one another. In the oblique prisms some of the basal edges are acute and some obtuse. After tracing out the similar and dissimilar angles and edges in the primaries, with the models, the following laws may be easily applied: Either—

1. *All the similar parts of a crystal are similarly and simultaneously modified ;** or,—

Explain the relation of secondary planes to the fundamental form. What is said of the cube? of the square prism? the rectangular prism? the right rhombic and rhomboidal? the oblique prisms? What is the first law repecting secondary planes?

Note.—What is meant by replacement, bevelment, and truncation?

---

* To avoid circumlocutions, the following technical terms are employed in describing the modifications of crystals.

Replacement. An edge or angle is *replaced*, when cut off by one or more secondary planes, (figs. 20, 21, 32.)

Truncation. An edge or angle is *truncated*, when the replacing plane is equally inclined to the adjacent faces, (figs. 20, 21.)

Bevelment. An edge is *beveled*, when replaced by two planes, which are respectively inclined at equal angles to the adjacent faces, (fig. 32.) *Truncation and bevelment can occur only on edges formed by the meeting of equal planes.*

2. *Half the similar parts of a crystal, alternate in position, are modified independently of the other half.*

In the cube, octahedron, or dodecahedron, if *one* edge is replaced, *all the other* edges will be replaced, and by the same planes. If there are two planes on one edge, (fig. 32) there will be two on every other edge ; and the two on each will have the same inclinations. If there are three planes on one angle, (fig. 33) there will, in the same manner, be three on the other *seven* angles. Perfect symmetry is thus preserved, however numerous the added planes. The following figures illustrate this principle, that all the edges, and all the angles are modified alike.

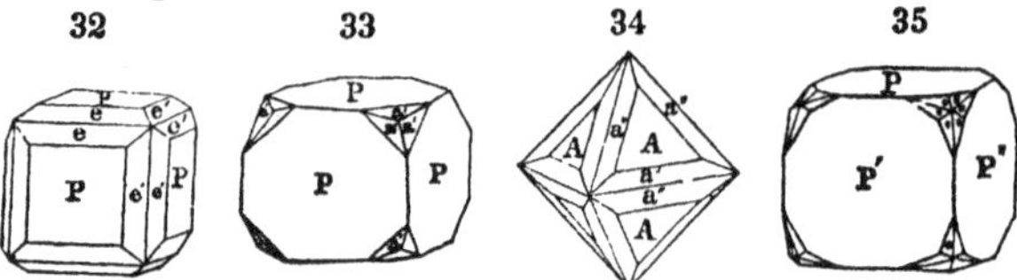

This symmetry is well seen in the solids which the secondary planes, in the above figures, produce, if enlarged till the primary planes are obliterated. Thus from figure 32, comes the form in figure 36, the planes e′ being enlarged till the planes P are obliterated ; from 33, comes the form in fig. 37 ; from 34, the form in 38 ; and from 35, the form in 39. The form in figure 37 has 24 faces, and is called a *trapezohedron.* It is common in garnet and leucite.

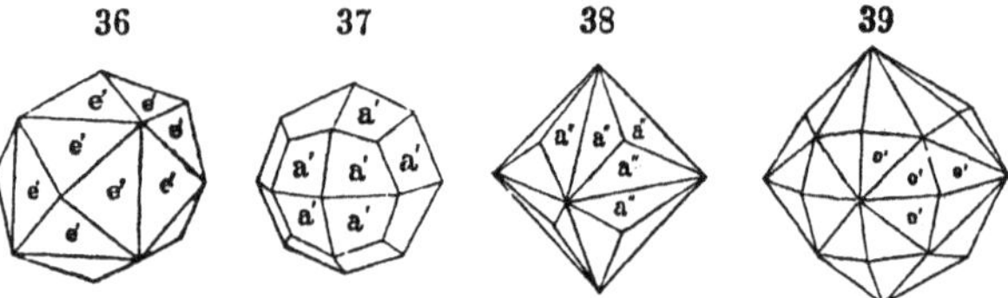

In figure 35, there are *six* planes on each angle, and as there are *eight* angles in the cube, the solid represented in figure 39 has *forty-eight* faces. Both 38 and 39 are forms of the diamond.

In connection with the law above given, it is stated that half the similar parts may be modified independently of the other half. The parts thus modified are alternate with one another and still produce symmetrical solids. Thus the

---

What second law is mentioned ? Explain the first law by examples.

cube may have only the *alternate angles* replaced; or only *one* of the *two beveling* planes shown in figure 32 may occur on each edge; or *three* of the *six* on each angle in figure 35. The following are examples; and each figure in the lower line, represents the completed form, produced by extending the secondary planes in the figure above, to the obliteration of the primaries, as explained on the preceding pages.

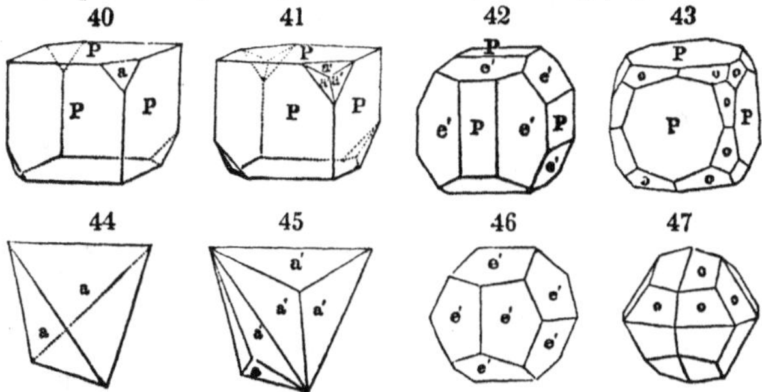

The replacement begun in figure 40, continued to the obliteration of the Ps, produces figure 44, which is a tetrahedron, or three-sided pyramid. So the planes á in figure 41, give rise to fig. 45; the planes é in 42, to figure 46, which is a pentagonal dodecahedron, so called because it has *twelve pentagonal* (or five-sided) faces. The forms represented in figures 40 and 41 are common in *boracite*, and those of figures 42, 43, in *iron-pyrites*. These forms with half the full number of planes are called *hemihedral* forms, from the Greek words for *half* and *face*.

The tetrahedron is sometimes placed among the primary forms; but it is properly a secondary form, derived from the cube, in the manner here explained, or from the octahedron by the extension of four faces to the obliteration of the other four. (Compare figs. 2 and 44.)

In the *right square prism*, the basal edges being unequal to the vertical, (because the prism, unlike the cube, is higher than broad,) these two kinds of edges are not replaced by similar planes, and the basal may be modified when the lateral are not modified, (figs. 48, 49.) The lateral edges may be *truncated*, because their including planes are equal;

Explain the second law. What are the resulting forms called? What is said of the tetrahedron?

the terminal cannot be truncated, but are replaced by planes *unequally inclined* to the including planes. The solid angles

48 49 50

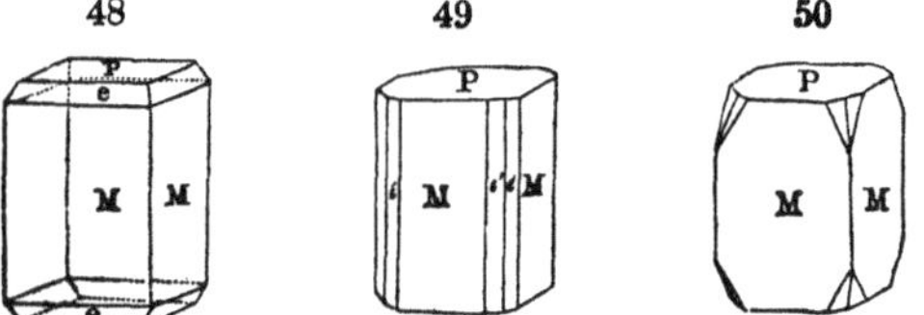

of the square prism are of one kind and are replaced alike, as in figures 23, 50; all the angles in these figures have the same number of planes, and the two adjacent planes in figure 50 are similar in their inclinations, because the lateral planes M, M, of a square prism, are equal.

In the rectangular and rhombic prisms the lateral axes are unequal. Consequently in the rectangular prism, two basal edges differ from the other two, and are therefore modified independently (figs. 51, 52.) The planes ĕ extended to the obliteration of T and P, would produce a rhombic prism (in a horizontal position,) as shown in figure 53, and another horizontal prism may be formed by the extension of the planes ē, fig. 52. In the rhombic prism the basal edges cor-

51 52 53 54

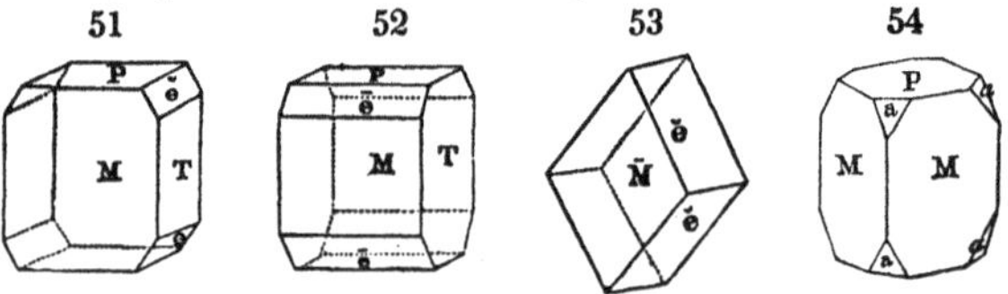

55

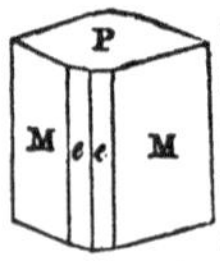

respond to the angles of the rectangular prism (see fig. 26*b*) and are similar and simultaneously replaced as in figure 24. The basal angles are unlike, one being obtuse and the other acute, and the planes of the two (fig. 54) differ in their inclinations. The lateral edges differ in the same manner, two being obtuse and two acute, and they are independently replaced, as in figure 55. The two planes *e* are similar planes, because, in a rhombic prism, M and M are equal; and the extension of these planes may produce another rhombic prism.

In an oblique rhombic prism the superior basal edges dif-

Explain these laws from the square prism; the rectangular and rhombic.

fer from the inferior in front, two being obtuse and two acute; consequently, they are independently replaced. Figure 56, shows the replacement of the obtuse basal. So also the front angles differ in the same manner, the upper (left side in fig. 57) being independent of the inferior in its modifications.

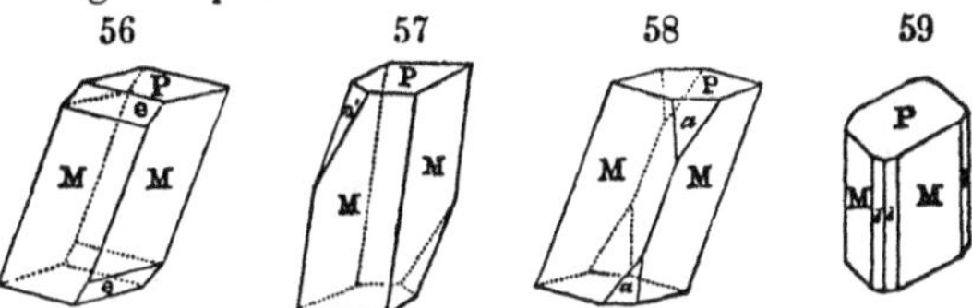

But the four lateral angles are similar (fig. 58.) Two of the lateral edges are obtuse and two acute, as in the right rhombic prism, and their secondary planes are therefore unlike (fig. 59.)

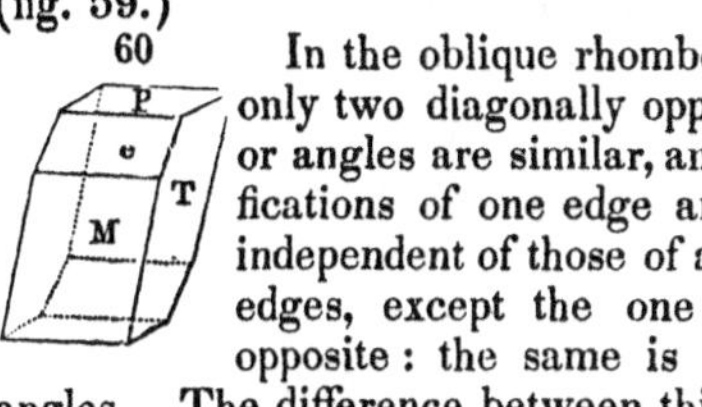

In the oblique rhomboidal prism, only two diagonally opposite edges or angles are similar, and the modifications of one edge are therefore independent of those of all the other edges, except the one diagonally opposite: the same is true of the angles. The difference between this prism and the oblique rhombic will thus be seen on comparing figures 56 and 60, and also figures 58 and 61:

In the rhombohedron, the distinction of *vertical* and *lateral solid angles* has already been explained, and also the difference between the *terminal* and *lateral edges*. The figures given will show how these distinctions are carried out in the

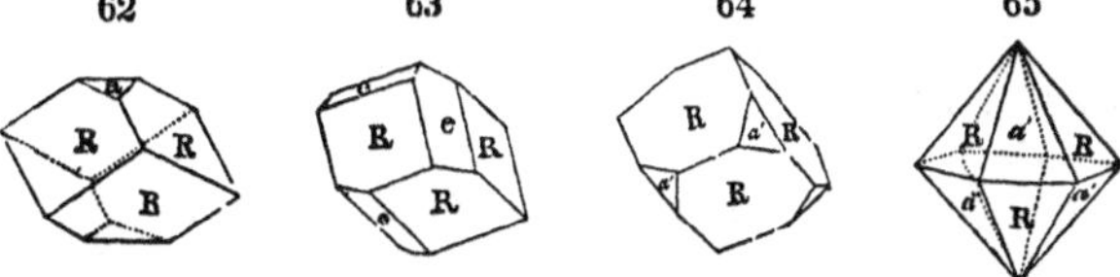

modifications. In figure 62, the *terminal* solid angles are replaced, but none of the lateral. In figures 64, 65 and 29, the *lateral* angles are replaced, but not the terminal. Figure 63, has the *terminal* edges replaced, and figures 68 and 28, the *lateral* edges.

Explain the laws with regard to secondary planes from the oblique rhombic prism; oblique rhomboidal; the rhombohedron.

When the planes a' in figure 64 are a little more extended, the form is changed to figure 65, or a double six-sided pyramid. It is in this way that the pyramidal form of crystals of quartz is produced from the primary rhombohedron. In figure 66,

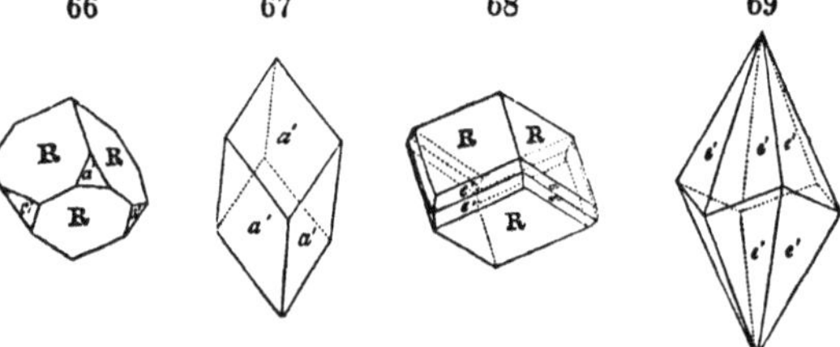

a', as is seen, is a different plane from a' in figure 64. By enlarging the planes a', till the planes R are obliterated, figure 67 is obtained, an acute rhombohedron. This may appear a singular result: but it will be understood on considering that there are *six* lateral angles; and *three* of the planes a' incline upward, and *three*, alternate, incline downward; they must therefore produce an oblong solid, bounded by six equal faces, which is nothing else than a rhombohedron. In figure 68, the lateral edges are beveled by the planes e'. The planes e' enlarged to the obliteration of the faces R, lead to the form in figure 69—a twelve-sided figure, or *dodecahedron*, and called from the shape of its faces, a *scalene dodecahedron*. It is the form of dog-tooth spar, a variety of calcareous spar. In figures 28, 29, the planes e and a are each parallel to the vertical axis, and they consequently produce prisms when extended, as explained on pages 31, 32.

In figure 3, under Tourmaline, we have an instance of a *hemihedral* modification in the hexagonal system. The extremities of the prism, as will be observed, have different secondary planes, there being in addition to the three faces R, three small triangular planes above, and three narrow linear planes below. Topaz crystals are also differently modified at the extremities, and are examples of hemihedral modifications in a right rhombic prism.

Another law gives still greater interest to the study of crystallography: but it can only be briefly alluded to in this place. When speaking of the right square prism it was

Mention some instances of hemihedral modifications, and explain.

stated that the basal edges were never truncated, but, when modified, were replaced by planes unequally inclined to the *basal* and *lateral* faces of the primary. These secondary planes do not however occur at random, at any possible inclination; but there is a direct relation, in all instances, to the comparative height and breadth of the fundamental form of the mineral. The same is true of planes on the angles, and in secondaries to all the fundamental forms.

Take a cube and cut off evenly one of the edges: this removes parts of two other edges, at each end of the plane. It is found that in cubic crystals these parts are either *equal* to one another, or one is *double* of the other, or *treble ;* or in some other simple ratio. The same is true in the other fundamental forms, except that, as stated, the relative height and breadth of the prism come into account, and influence the result.

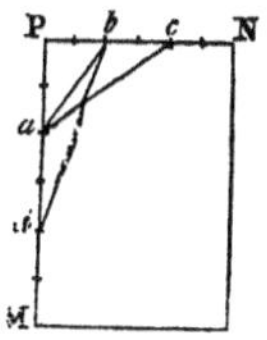

For example: in figure 70, (a section of a cube,) P M and P N are equal edges, divided into equal parts; now a plane on an edge of a *cube*, as *a b*, removes, as is seen, equal parts of P M and P N; another, as *a c*, removes twice as many parts of one edge as of the other; and so other planes have like simple ratios. In figure 71, a section of a *prism*, the lines P M and P N (height and breadth of the prism) are unequal: let them be divided into a like number of parts; then a plane on an edge, as *a b*, will cut off as many parts of P M as of P N; others, as *a c*, *b d*, twice as many parts of one as the other: and so on. *a b* truncates the edge in figure 70; but not so in figure 71. It is evident to the mathematical scholar that the inclination of a plane *a b* to P N or P M, is sufficient to determine the relative dimensions of P *a* and P *b*, or the relative height and breadth of the fundamental form.

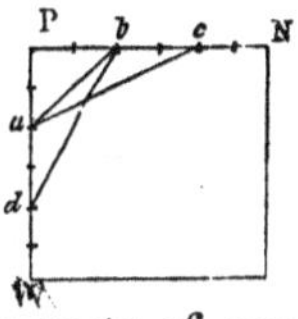

These principles give a mathematical basis to the science.

Thus we perceive that the attraction which guides each particle to its place in crystallization, produces forms of mathematical exactness. It covers the crystal with scores of facets of finished brilliancy and perfection; and these

---

What other law is there, respecting the occurrence of secondary planes? Explain by the figures.

facets are not only uniform in number on similar parts of a crystal, but are even fixed in every angle and every edge.*

COMPOUND CRYSTALS.

In the preceding pages, we have been considering simple crystals, and their secondary forms. The same forms are occasionally compounded so as to make what have been called *twin* or *compound crystals*. They will be understood

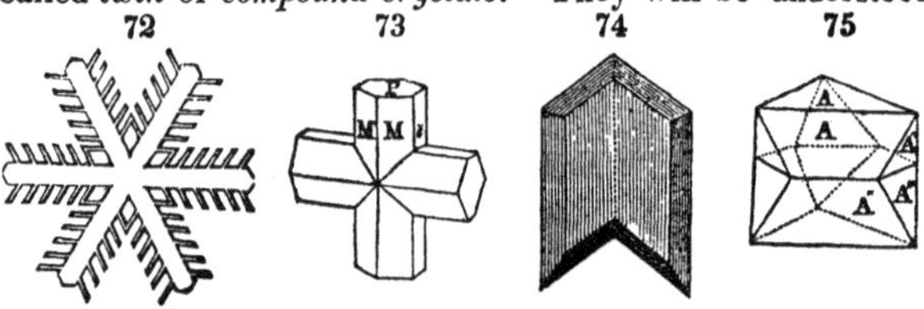

at once from the annexed figures. Figure 72 represents a crystal of snow of not unfrequent occurrence. It consists, as

What is a twin or compound crystal?

* On a preceding page, it has been explained that in *monometric* cystals the axes are equal; in *dimetric* and *hexagonal* crystals the lateral axes are equal, and the vertical is of a different length, shorter or longer. In the other systems, the *trimetric* and the two *oblique systems*, the three axes are all unequal. In the above paragraphs it has been shown that the relative lengths of the axes in a fundamental form of a crystal are fixed, and may be determined by simple calculations. These fixed relative dimensions are supposed to be the relative dimensions of the particles or molecules constituting crystals; that is if the fundamental form of a crystal is twice as long as broad, the same is true of its molecules. The molecules of a cube must therefore be equal in different directions; those of a square prism must be longer or shorter than broad, but equal in breadth and thicknesss; those of a rectangular prism must be unequal in three directions; and the relative inequality is determinable as just stated. The simplest and most probable view of the forms of molecules is that they are *spheres* for monometric solids; and *ellipsoids* of different axes for the other forms. Figure 1 represents a sphere.

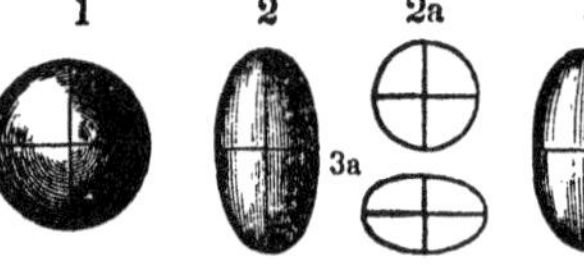

Figure 2 represents an ellipsoid with the lateral axes equal, as seen in the cross section 2a; it is the form in the *dimetric* and *hexagonal* systems.

Figure 3 represents an ellipsoid with the lateral axes unequal (fig. 3a), as in the *trimetric* and *oblique systems;* a variation in the length of the axes will vary the dimensions, according to any particular case.

is evident to the eye, either of six crystals meeting in a point, or of three crystals crossing one another. Besides, there are numerous minute crystals regularly arranged along the rays. Figure 73 represents a cross (cruciform) crystal of staurotide, which is similarly compound, but made up of fewer crystals. Figure 74, is a compound crystal of gypsum, and figure 75, one of spinel. These will be understood from the following figures.

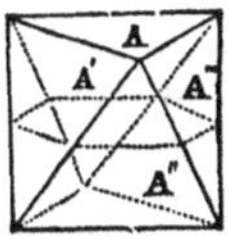

Figure 76 is a *simple* crystal of gypsum; if it be bisected along *a b*, and the right half be inverted and applied to the other, it will form figure 74, which is therefore a twin crystal, in which one half has a reverse position from the other. Figure 77, is a simple octahedron; if it be bisected through the dotted line, and the upper half, after being revolved half way around, be then united to the lower, it produces figure 75. Both of these therefore are similar twins, in which one of the two component parts is reversed in position.* Compound crystals are generally distinguished by their *reëntering angles*.

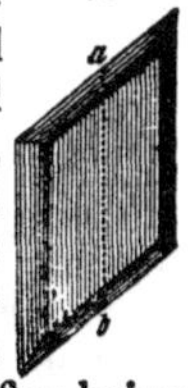

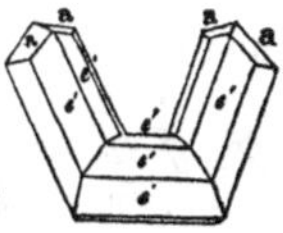

Besides the above, there are also geniculated crystals, as in the annexed figure. The bending has here taken place at equal distances from the center of the crystal; and it must therefore have been subsequent in time to the commencement of the crystal. The prism began from a simple molecule: but after attaining a certain length, an abrupt change of direction took place. The angle of geniculation is constant in the same mineral species; for the same reason that the angles of secondary planes are fixed; and it is such that a cross section directly through the geniculation is parallel to the position of a common secondary plane. In the figure given, the plane of geniculation is parallel to one of the terminal edges.

Mention illustrations. Explain their structure in the case of gypsum and spinel. What is said of geniculated crystals?

---

* Such crystals have proceeded from a compound nucleus in which one of the two particles was reversed. Compound crystals of the kind above described, thus differ from simple crystals in having been formed from a nucleus of two or more united molecules, instead of from a simple nucleus.

DIMORPHISM.—POLYMORPHISM.

It was formerly supposed that the same chemical compound could have but a single mode of crystallization. But later researches have discovered that there are many instances of substances crystallizing according to two distinct systems. Thus sulphur at different times crystallizes in oblique prisms and right rhombic octahedrons, or according to the two systems *monoclinic* and *trimetric*. Carbonate of lime at one time takes on the rhombohedral form, and is then called *calc spar;* at another, that of a rhombic prism, and it is then termed *aragonite*. Again, sulphuret of iron presents us both with cubical (monometric) crystals and rhombic prisms (trimetric.) As far as investigation has gone, it has appeared that one of these forms is assumed at a lower temperature than the other; and this takes place uniformly, so that the temperature attending solidification, in certain cases at least, determines the forms and system of crystallization. How far other causes operate is unknown.

This property is termed *dimorphism*, (from the Greek *dis*, two or twice, and *morphe*, form,) and a substance presenting two systems of crystallization is said to be *dimorphous*. In addition to the above, *garnet* and *idocrase*, the one dodecahedral, and the other square-prismatic, are different forms of the same substance. *Rutile*, which is dimetric, *anatase*, dimetric also, but of different dimensions, and *Brookite*, which is trimetric, are three distinct forms of the same substance, *oxyd of titanium*. In this last case, the property has been called *trimorphism*, (from the Greek *tris*, three times, and *morphe*, form.) As the number of forms may be still greater, the more general term *polymorphism* (*polus*, many, and *morphe*) has been introduced to include all cases, whatever the number of forms assumed.

A polymorphous substance in its different states presents not merely difference of form. There is also a difference in *hardness*, *specific gravity* and *luster*, in fact, in nearly all physical qualities. Aragonite has the specific gravity 2·93, and calc spar only 2·7; the hardness of aragonite is 3½, and that of calc spar but 3.

May the same substance crystallize under more than one fundamental form? Mention examples. What is this property called? What is said of oxyd of Titanium? What is trimorphism? polymorphism? What other differences beside that of form are connected with polymorphism?

The forms of a dimorphous substance differ in stability. Aragonite when heated gently falls to powder, arising from a change in the condition of its particles. Aragonite has been obtained by evaporating a solution of lime over a water bath, and calc spar when the same was evaporated at the ordinary temperature. When a *right rhombic prism* of sulphate of zinc (which is dimorphous) is heated to 126° F. certain points in its surface become opaque, and from these points, bunches of crystals shoot forth in the interior of the specimen; and in a short time the whole is converted into an aggregate of these crystals, diverging from several centers on the surface of the original crystal. These small crystals are *oblique rhombic prisms;* and the same form may be obtained by evaporating a solution at this temperature or above it. Many other similar cases might be cited, but these serve to explain the principle in view.

IRREGULARITIES OF CRYSTALS.

Before concluding this subject, a few remarks may be added on the irregularities of crystals.

Crystals of the same form vary much in length, and in the size of corresponding faces. The same mineral may occur in very short prisms, or in long and slender prisms: and some planes may be so enlarged as to obliterate others; a few figures of quartz crystals will illustrate these peculiarities.

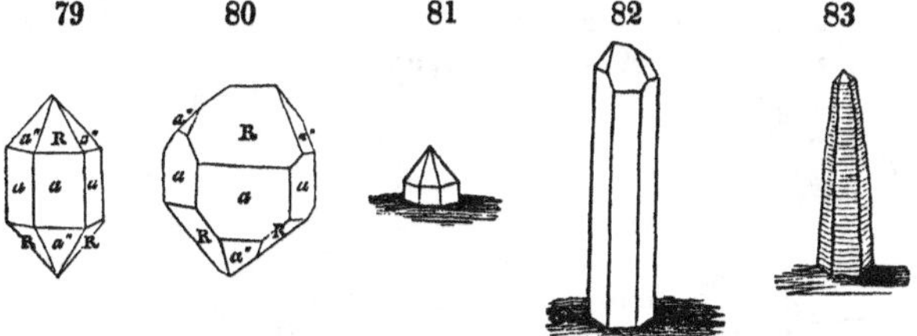

Figure 79 is the regular form of the crystal. Figure 80 is the same form with some faces very much enlarged, and others very small. Figure 81 is a very short prism and pyramid of quartz, such as is often seen attached to the surface of rocks; and figure 82 is a similar form very much elongated. Notwithstanding all these variations, every angle

What are some of the irregularities of crystals?

of inclination remains the same: and this is a general fact in all crystals, that whatever distortions take place, the angles are constant. Greater diversity is given to the shapes of crystals by these simple variations, without multiplying the number of distinct forms. Figure 83 is a tapering prism of the same mineral, with a minute pyramid at the apex. The faces of this pyramid have exactly the same inclinations as those of figure 79.

The constancy of the angles shows that the fundamental form of the crystal, or, in other words, the form of its molecules, is constant, amid all these variations of size and shape.

Crystals have sometimes curved faces. The faces of diamonds are usually convex, and some crystals are almost spheres. Figure 84 is one of these diamond crystals. It is the same form as is represented in figure 45. For cutting glass, they always select those crystals that have a natural curved edge, as others are much inferior for the purpose and sooner wear out. In figure 85 a different kind of curvature is represented. It is a curved rhombohedron, in which the opposite faces are parallel in their curving: it is a common form of *spathic* iron and *pearl spar*. The latter mineral from Lockport, New York, is always curved in this way.

84

85

Still more singular curvatures are sometimes met with. In the mammoth cave of Kentucky, leaves, vines and flowers are beautifully imitated in alabaster. Some of the "rosettes" are a foot in diameter, and consist of curving leaves, clustered in graceful shapes. The frostings on our windows in winter are often miniature pictures of forests and vines with rolled tendrils. It is one among the many singular results of crystallization. On the cool mornings of spring or autumn, in this climate, twigs of plants are occasionally found encircled by fibrous icy curls, (fig. 86,) which are attached vertically to the stem. They are formed during the night, and disappear soon after the appearance of the sun.

86

What is said of curved crystals? What of curved crystallizations of gypsum? of ice?

ON MEASURING ANGLES OF CRYSTALS.

As the angles of crystals are constant, minerals, as has been stated, may often be distinguished by measuring these angles. This is done by means of instruments called *goniometers*, a term meaning, literally, *angle-measurers*.* These are of two kinds; one is called the common goniometer, the other the reflecting goniometer.

87

The common goniometer depends on the very simple principle that when two straight lines cross one another, as A E, C D in the annexed figure 87, the parts will diverge equally on opposite sides of the point of intersection (O); that is, in mathematical language, the angle A O D is equal to the angle C O E, and A O C is equal to D O E.

The instrument in common use is here represented.

88

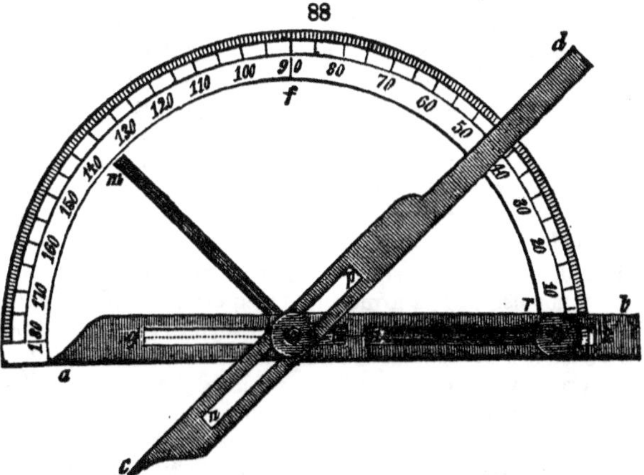

It consists of two arms, *a b*, *c d*, moving on a pivot at *o*: the arms open and shut, and their divergence, or the angle they make with one another, is read off on the graduated arc attached. In using it, press up between them, the edge of the crystal whose angle is to be measured, and continue opening the arms thus till the inner edges lie evenly against the faces that include

How are the angles of crystals measured? Explain the principle of the common goniometer from the figure. Explain the common goniometer and its use.

* From the Greek *gonu*, angle, and *metron*, measure.

the required angle. To insure accuracy in this respect, hold the instrument and crystal between the eye and the light, and observe that no light passes between the arm and the applied faces of the crystal. The arms may then be secured in position by tightening the screw at *o*; the angle will then be measured by the distance on the arc from *k* to the *left* or *outer* edge of the arm *c d*, this edge being in the line of *o*, the center of motion. As the instrument stands in the figure, it reads 45°. The arms have slits at *g h*, *n p*, by which they may be shortened so as to make them more convenient for measuring small crystals.

In some instruments of this kind the arc is detached from the arms. When this is the case, after the measurement is made and the screw at *o* tightened, the arc (which has the shape of *a f b* in the annexed figure, except that from *a* to *b* is a solid bar) is adjusted to the upper edge of one of the arms, bringing the mark at *o*, the center, exactly to the center of divergence of the arms. The angle is then read off as before.

With a little ingenuity the student may construct a goniometer for himself that will answer a good purpose. A semi-circle may be described on mica or a glazed card, of the shape in figure 88: it should then be divided into halves at *f*, and again each half subdivided into nine equal parts. Each of these parts measures 10 degrees; and if they are next divided into ten equal parts, each of these small divisions will be degrees. The semi-circle may then be cut out, and is ready for use. The arms might also be made of stiff card for temporary use; but mica, bone or metal is better. The arms should have the edges straight and accurately parallel, and be pivoted together. The instrument may be used like that last described, and will give approximate results, sufficiently near for distinguishing most minerals. The ivory rule accompanying boxes of mathematical instruments, having upon it a scale of sines for measuring angles, will answer an excellent purpose, and is as convenient as the arc. The annexed figure will illustrate the mode of using it. The scale is graduated along the margin, the middle point marking 90°, and the divisions either side 10 degrees (as in the figure) and also single de-

89

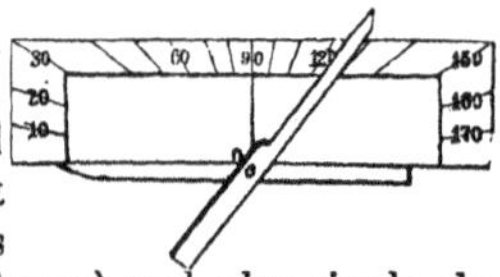

How is it used when the arms are detached? How may a temporary goniometer be made? How may a scale of sines be used?

grees. The arms are so applied to the scale, that the center of motion is exactly at the extremity of the middle line, marked 90; and the leg crossing the scale (or that edge of it in the line of the center of motion) will then indicate by its position over the graduated margin, the angle desired.*

In making such measurements it is important to remember that—

1. An angle A O D (figure 87) and A O C, together, equal 180°; so that if A O C be measured, A O D is ascertained by subtracting A O C from 180°.

2. In a rhomb or rhomboid, *b a b* and *a b a*, together, equal 180°; and one may be ascertained by subtracting the other from 180°. If an obtuse angle of a rhombic prism has been measured and found to be 110°, and the acute angle on measurement is ascertained to be 60°, the student should add the two together to find whether the sum is 180°; for if not, there is some error in the measurement, and it should be repeated. 110° added to 60° makes 170°, showing in this case an error of 10°.

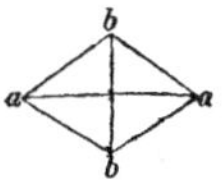

3. *In any polygon, the sum of the angles is equal to twice as many right angles as there are sides less two.* Let the number of sides, for example, be 6 : 6 less two is 4; and the angles together equal *twice* 4, (or 8,) right angles, which is equivalent to $8 \times 90° = 720°$. If we have a prism of six sides, and wish to ascertain the angles between these sides, the angles should be measured successively, and the whole added together to ascertain whether the measurements are correct. If the sum is 720°, there is good reason to confide in them. Crystals are at times a little irregular; and this should be looked to, as part of the apparent error may at times be thus accounted for. This general principle and the

What three points must be observed in making measurements?

* Another mode for approximate results consists in holding the crystal with the two faces (whose inclination is to be measured) in an exactly vertical position over a piece of paper: then place a small rule parallel, as near as the eye can judge, to one face, and draw a line; next do the same for the other face. The angle between the two lines, measured either by an arc or the ivory rule just mentioned, is the desired inclination. With practice, much skill may be acquired in such trials. They may be made with microscopic crystals under a microscope.

preceding, which is only a simpler case of the same, are of great importance in the measurements of crystals.

Reflecting Goniometer. The reflecting goniometer affords a more accurate method of measuring crystals that have luster, and may be used with those of minute size. The principle on which this instrument is constructed will be understood from the annexed figure (fig. 90) representing a crystal, whose angle *a b c* is required. The eye, looking at the face of the crystal *b c*, observes a reflected image of *m*, in the direction P *n*. On revolving, the crystal till *a b* has the position of *b c*, the same image will be seen again in the same direction P *n*. As the crystal is turned, in this revolution, till *a b d* has the present position of *b c*, the angle *d b c* measures the number of degrees through which it is revolved. But *d b c*, subtracted from 180°, equals the angle of the crystal *a b c*. The crystal is therefore passed in its revolution through a number of degrees, which, subtracted from 180°, give the required angle. This angle, in the reflecting goniometer of Wollaston, is measured by attaching the crystal to a graduated circle which revolves with it, as here represented (fig. 91.)

90

91

A B is the graduated circle. The wheel, *m*, is attached to the main axis, and moves the graduated circle together with the adjusted crystal. The wheel, *n*, is connected with an axis which passes through the main axis, (which is hollow for the purpose,) and moves merely the parts to which the crystal is attached, in order to assist in its adjustment. The contrivances for the adjustment of the crystal are at *p*, *q*, *r*, *s*. To use the instrument, it must be placed on a small stand or a table, and so elevated as to allow the observer to rest his elbows on the table. The whole, thus

---

Explain the principle of the reflecting goniometer. Explain the mode of using the instrument.

firmly arranged, is to be placed in front of a window, distant from the same from six to twelve feet, and with the axis of the instrument parallel to it. Preparatory to operation, a dark line must be drawn below the window near the floor, parallel to the bars of the window; or, what is better, on a slate or board placed before the observer on the table.

The crystal is attached to the movable plate, *q*, by a piece of wax, and so arranged that the edge of intersection of the two planes forming the required angle, shall be in a line with the axis of the instrument. This is done by varying its situation on the plate, *q*, or the situation of the plate itself, or by means of the adjacent joints and wheel, *r*, *s*, *p*, as will be readily understood from the instrument.

When apparently adjusted, the eye must be brought close to the crystal, nearly in contact with it, and on looking into a face, part of the window will be seen reflected, one bar of which must be selected for the trial. If the crystal is correctly adjusted, the selected bar will appear horizontal, and on turning the wheel, *n*, till this bar, as *reflected*, is observed to approach the dark line below, seen in a *direct* view, it will be found to be parallel to this dark line, and ultimately to coincide with it. If there is not a perfect coincidence, the adjustment must be altered until this coincidence is obtained. Continue then the revolution of the wheel, *n*, till the same bar is seen by reflection in the next face, and if here there is also a coincidence of the reflected bar with the dark line seen direct, the adjustment is complete; if not, alterations must be made, and the first face again tried. A few successive trials of the faces, will enable one to obtain a perfect adjustment.

The circle A B is usually graduated to half degrees, and by means of the vernier, *v*, minutes are measured. After adjustment, 180° on the arc must be brought opposite 0, on the vernier. The coincidence of the bar and dark line is then to be obtained, by turning the wheel, *n*. When obtained, the wheel, *m*, should be turned until the same coincidence is observed, by means of the next face of the crystal. If a line on the graduated circle now corresponds with 0 on the vernier, the angle is immediately determined by the number of degrees opposite this line. If no line corresponds with 0, we must observe which line on the vernier coincides with one on the circle. If it is the 18th on the vernier, and the line on the circle next below 0 on the vernier marks 125°, the required angle is 121° 18′; if this line marks 125° 30′, the required angle is 125° 48′.

Some goniometers are furnished with a small polished reflector, attached to the foot of the instrument below the part *s*, *q*, which is placed at an oblique angle so as to reflect a bar of the window. The reflected bar then answers the purpose of the line drawn below the window, (or on a slate,) and is more conveniently used.

Other modes of adjustment for the crystal, are also used; but they will explain themselves to the student acquainted with the above explanations, and need not here be dwelt upon.

MASSIVE MINERALS, OR IMPERFECT CRYSTALLIZATIONS.

Massive or imperfectly crystallized minerals either consist of fibers or minute columns, of leaves or laminæ, or of grains: in the *first*, the structure is said to be columnar; in the *second*, lamellar; in the *third*, granular. We have a familiar example of the lamellar structure in slate rocks and many minerals that occur in masses made up of separable laminæ. The *fibrous* or *columnar* structure is common in seams of rocks, and sometimes in incrustations covering exposed surfaces; the material of the seam or crust is made up of minute fibers or prisms closely compacted together, produced by a rapid crystallization on the supporting surface. The *granular* structure is well seen in loaf sugar and statuary marble.

1. Columnar Structure. The following are explanations of the terms used in describing the different kinds of columnar structure.

Fibrous; when the columns are minute and lie in the same direction; as *gypsum* and *asbestus*. Fibrous minerals very commonly have a silky luster: a fibrous variety of gypsum, and one of calc spar, have this luster very strongly, and each is often called *satin spar*.

Reticulated; when the fibers, or columns, cross in various directions, and produce an appearance having some resemblance to a net.

Stellated; when they radiate from a center in all directions, and produce a star-like appearance. Ex. *stilbite*, *gypsum*.

What kinds of structure exist in massive minerals? Explain the different varieties of columnar structure, fibrous; reticulated, &c.

Radiated, divergent; when the crystals radiate from a center, without producing stellar forms. Ex. *quartz, gray antimony.*

2. Lamellar Structure. In the lamellar structure, the laminæ or leaves may be thick, or very thin; they sometimes separate easily, and sometimes with great difficulty.

When the laminæ are thin and separate easily, the structure is said to be *foliaceous.* Mica is a striking example, and the term *micaceous* is often used to describe this structure.

When the laminæ are thick, the term *tabular* is often applied; quartz and heavy spar afford examples.

The laminæ may be *elastic,* as in mica, *flexible,* as in talc or graphite, or *brittle,* as in diallage.

Small laminæ are sometimes arranged in *stellar* shapes; this occurs in mica.

3. Granular Structure. When the grains in the texture of a mineral are coarse, it is said to be *coarsely granular,* as in granular marble; when fine, *finely granular,* as in granular quartz; and if no grains can be detected with the eye, the structure is described as *impalpable,* as in chalcedony.

Granular minerals, when easily crumbled by the fingers, are said to be *friable.*

Imitative Shapes.—Massive minerals also take certain *imitative shapes,* not peculiar to either of these varieties of structure. The following terms are used in describing imitative forms:

Globular; when the shape is spherical or nearly so: the structure may be columnar and radiating, or it may be concentric, consisting of coats like an onion. When they are attached, they are called *implanted* globules.

Reniform; kidney-shaped. In structure, they are like globular shapes.

Botryoidal; when a surface consists of a group of rounded prominences. The prominences or globules usually consist of fibers radiating from the center.

Mammillary; resembling the botryoidal, but consisting of larger prominences.

Filiform; like a thread.

Acicular; slender like a needle.

Explain the varieties of lamellar structure; of granular structure; the several imitative shapes, globular; reniform, &c.

Stalactitic; having the form of a cylinder, or cone, hanging from the roofs of cavities or caves. The term stalactite is usually restricted to the cylinders of carbonate of lime hanging from the roofs of caverns: but other minerals are said to have a stalactitic form when resembling these in their general shape and origin. Chalcedony and brown iron ore are often stalactitic.

Reticulated; net-like.

Drusy; a surface is said to be *drusy* when covered with minute crystals.

Amorphous; having no regular structure or form, either crystalline or imitative. The word is from the Greek, and means *without shape.*

PSEUDOMORPHOUS CRYSTALS.

A *pseudomorphous* crystal is one that has a form which is foreign to the species to which the substance belongs.*

Crystals sometimes undergo a change of composition from aqueous or some other agency, without losing their form; for example, octahedrons of spinel change to steatite, still retaining the octahedral form. Cubes of pyrites are changed to red or brown iron ore.

Again: crystals are sometimes removed entirely, and at the same time and with equal progress, another mineral is substituted; for example, when cubes of fluor spar are transformed to quartz. The petrifaction of wood is of the same kind.

Again: cavities left empty by a decomposed crystal, are refilled by another species by *infiltration*, and the new mineral takes on the external form of the original mineral, as a fused metal the form of the mould into which it is cast.

Again: crystals are sometimes *incrusted* over by other minerals, as cubes of fluor by quartz; and when the fluor is afterwards dissolved away, as sometimes happens, hollow cubes of quartz are left.

The first kind of pseudomorphs, are *pseudomorphs by alteration;* the second, *pseudomorphs by replacement;* the

---

What is a pseudomorphous crystal? What is the first, the second, the third and the fourth mode of pseudomorphism? What are they called?

---

* From the Greek *pseudes*, false, and *morphe*, form.

third, ***pseudomorphs by infiltration;*** the fourth, ***pseudomorphs by incrustation.****

Pseudomorphous crystals are distinguished by having a different structure and cleavage from that of the mineral imitated in form, and a different hardness, and usually little luster.

A large number of minerals have been met with as pseudomorphs. The causes of such changes have operated very widely and produced important geological results.

CHAPTER III.—PHYSICAL PROPERTIES OF MINERALS.

CHARACTERS DEPENDING ON LIGHT.

The characters depending on light are of *five* kinds, and arise from the power of minerals to *reflect*, *transmit*, or *emit* light. They are as follows:

1. *Luster;* 2. *Color;* 3. *Diaphaneity;* 4. *Refraction;* 5. *Phosphorescence.*

LUSTER.

90. The luster of minerals depends on the nature of their surfaces, which causes more or less light to be reflected. There are different degrees of *intensity of luster*, and also different *kinds of luster*.

a. The *kinds of luster* are six, and are named from some familiar object or class of objects.

1. *Metallic:* the usual luster of metals. Imperfect metallic luster is expressed by the term *sub-metallic.*

2. *Vitreous:* the luster of broken glass. An imperfect vitreous luster is termed *sub-vitreous.* Both the vitreous and sub-vitreous lusters are common. Quartz possesses the former in an eminent degree; calcareous spar often the latter. This luster may be exhibited by minerals of any color.

3. *Resinous:* luster of the yellow resins. Ex. opal, zinc blende.

4. *Pearly:* like pearl. Ex. talc, native magnesia, stilbite, &c. When united with sub-metallic luster, the term *metallic-pearly* is applied.

How are pseudomorphous crystals distinguished? What characters depend on light? Explain the varieties of luster, *metallic, vitreous,* &c.

---

* This subject is farther treated of by the author in the Amer. Jour. of Science, vol. xlviii, pp. 66, 81, 397.

5. *Silky:* like silk; it is the result of a fibrous structure. Ex. fibrous carbonate of lime, fibrous gypsum, and many fibrous minerals, more especially those which in other forms have a pearly luster.

6. *Adamantine:* the luster of the diamond. When sub-metallic, it is termed *metallic-adamantine.* Ex. some varieties of white lead ore.

b. The *degrees of intensity* are denominated as follows:

1. *Splendent:* when the surface reflects light with great brilliancy, and gives well defined images. Ex. Elba iron ore, tin ore, some specimens of quartz and pyrites.

2. *Shining:* when an image is produced, but not a well defined image. Ex. calcareous spar, celestine.

3. *Glistening:* when there is a general reflection from the surface, but no image. Ex. talc, copper pyrites.

4. *Glimmering:* when the reflection is very imperfect, and apparently from points scattered over the surface. Ex. flint, chalcedony.

A mineral is said to be *dull* when there is a total absence of luster. Ex. chalk.

COLOR.

In distinguishing minerals, both the external color and the color of a surface that has been rubbed or scratched, are observed. The latter is called the *streak,* and the powder abraded, the *streak-powder.*

The colors are either *metallic* or *non-metallic.*

The metallic are named after some familiar metal, as copper-red, bronze-yellow, brass-yellow, gold-yellow, steel-gray, lead-gray, iron-gray.

The non-metallic colors used in characterizing minerals, are various shades of *white, gray, black, blue, green, yellow, red* and *brown.*

There are thus snow-white, reddish-white, greenish-white, milk-white, yellowish-white;

Bluish-gray, smoke-gray, greenish-gray, pearl-gray, ash-gray;

Velvet-black, greenish-black, bluish-black;

Azure-blue, violet-blue, sky-blue, Indigo-blue;

Emerald-green, olive-green, oil-green, grass-green, apple-green, blackish-green, pistachio-green (yellowish);

What is observed respecting color?

Sulphur-yellow, straw-yellow, wax-yellow, ochre-yellow, honey-yellow, orange-yellow;

Scarlet-red, blood-red, flesh-red, brick-red, hyacinth-red, rose-red, cherry-red;

Hair-brown, reddish-brown, chesnut-brown, yellowish-brown, pinchbeck-brown, wood-brown.

A *play of colors:* this expression is used when several prismatic colors appear in rapid succession on turning the mineral. The diamond is a striking example; also precious opal.

Change of colors: when the colors change slowly on turning in different positions, as in labradorite.

Opalescence: when there is a milky or pearly reflection from the interior of a specimen, as in some opals, and in cat's eye.

Iridescence: when prismatic colors are seen within a crystal; it is the effect of fracture, and is common in quartz.

Tarnish: when the surface colors differ from the interior; it is the result of exposure. The tarnish is described as *irised*, when it has the hues of the rainbow.

*Pleochroism:** the property, belonging to some prismatic crystals, of presenting a different color in different directions The term *dichroism*† has been generally used, and implies different colors in *two* directions, as in the mineral *iolite*, which has been named *dichroite* because of the different colors presented by the bases and sides of the prism. *Mica* is another example of the same. The more general term has been introduced, because a different shade of color has been observed in more than two directions.

These different colors are observed only in crystals with *unequal* axes. *The colors are the same in the direction of equal axes, and often unlike in the direction of unequal axes.* This is the general principle at the basis of pleochroism.

What is a play of colors? change of colors? opalescence? iridescence? tarnish? dichroism and pleochroism? Mention examples of this last property; also the law relating to it.

---

* From the Greek *pleos*, full, and *chroa*, color.

† From the Greek *dis*, twice, and *chroa*.

DIAPHANEITY.

Diaphaneity is the property which many objects possess of *transmitting* light; or in other words, of permitting more or less light to pass through them. This property is often called *transparency*, but transparency is properly one of the degrees of diaphaneity. The following terms are used to express the different degrees of this property:

Transparent: a mineral is said to be transparent when the outlines of objects, viewed through it, are distinct. Ex. glass, crystals of quartz.

Subtransparent, or *semitransparent:* when objects are seen, but their outlines are indistinct.

Translucent: when light is transmitted, but objects are not seen. Loaf sugar is a good example; also Carrara marble.

Subtranslucent: when merely the edges transmit light faintly. When no light is transmitted, the mineral is described as *opaque*.

REFRACTION AND POLARIZATION.

Light is always bent out of its course on passing from one medium into another of different density: as from air into water, or from water into air. This bending of the rays of light is called *refraction*. Thus if a ray of light, as R S, pass into water at S, it becomes changed in direction to S U, instead of going straight in its course, R S T. The line *a* S *c* is a perpendicular to the surface of the water, and the greater refraction of the water is seen by the bending of the ray toward this perpendicular. If a circle be described about S as a center, and the lines R *a* and U *b* be drawn perpendicular to *a c*, or parallel to the surface of the water, we see by these lines the exact relation between the amount of refraction in these two cases; for the refraction in water is as much greater than in air as U *b* is less than R *a*.* This relation is called the

92

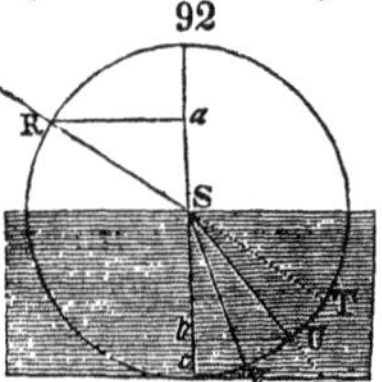

What is diaphaneity? Explain the terms *transparent*, &c. What is meant by refraction? Explain from the figure.

* In mathematical language, U *b* is the sine of the angle of refraction, and *a* R the sine of the angle *a* S R, the angle of incidence; the ratio between the two sines is constant, it being alike for every angle of incidence.

index of refraction. It is about 1⅓ for water, or more accurately, 1·335. With diamond, the ray would be bent in the direct S V, which indicates a much greater amount of refraction; its index is nearly 2½, or correctly, 2.439. The eye at R, looking into a diamond in the direction R S, would see an object in the direction of S V, and not in that of S T. The index of refraction has been obtained for many substances, of which the following are a few:

Air,	1·000	Calc spar,	1·654
Tabasheer,	1·211	Spinel,	1·764
Ice,	1·308	Sapphire,	1·794
Cryolite,	1·349	Garnet,	1·815
Water,	1·335	Zircon,	1·961
Fluor spar,	1·434	Blende,	2·260
Rock salt,	1·557	Diamond,	2·439
Quartz,	1·548	Chromate of lead,	2·974

Double Refraction.—Many crystals possess the property of refracting light in two directions, instead of one, and objects seen through them consequently appear double. This is called *double refraction.* It is most conveniently exhibited with a crystal of calc spar, and was first noticed in a pellucid variety of this mineral from Iceland, called from the locality Iceland spar. On drawing a line on paper and placing the crystal over it, two lines are seen instead of one—one by *ordinary* refraction, the other by an *extraordinary* refraction. If the crystal, as it lies over the line, be turned around, when it is in one position the two lines will come together. Instead of a line, make a dot on the paper, and place the crystal over the dot: the two dots seen will not come together on revolving the crystal, but will seem to revolve one around the other. The dot will, in fact, appear double through the crystal in every direction except that of the vertical axis, and this direction is called the *axis of double refraction.* To view it in this direction, the ends must be ground and polished. The divergence increases on passing from a view in the direction of the axis to one at right angles with it, where it is greatest. In some substances, the refraction of the extraordinary ray is greater in the latter direction than that of the ordinary ray, and in others it is less.

What is double refraction? What takes place on revolving a transparent rhomb of calc spar over a line or dot? In what direction is there no double refraction, and in which is it greatest?

In calc spar it is less, it diminishing from 1·654 to 1·483. In quartz it is greater, it increasing from 1·5484 to 1·5582. The former is said to have a *negative* axis, the latter a *positive.*

This property of double refraction belongs to such of the fundamental forms as have unequal axes; that is, to all except those of the monometric system. Those forms in which the lateral axes are equal, (the dimetric and hexagonal systems,) have one axis of double refraction; and those in which they are unequal, (the trimetric, monoclinic and triclinic systems,) have two axes of double refraction.*

Both rays in the latter are rays of extraordinary refraction. In niter, the two axes are inclined about 5° to each other; in arragonite, 18° 18′; in topaz, 65°. The positions of the axes thus vary widely in different minerals.

Polarization.—The extraordinary ray exhibits a peculiar property of light, termed polarization. Viewed by means of another doubly-refracting crystal, or crystalline plate, (called from this use of it an analyzing plate,) the ray of light becomes alternately visible and invisible as the latter plate is revolved. If the polarized light be made to pass through a crystal possessed of double refraction, and then be viewed in the manner stated, rings of prismatic colors are developed,

93 94 95 96

and on revolving the analyzing plate, the colored rings and

What is meant by positive and negative double refraction? What crystalline forms exhibit double refraction? which have one and which two axes of double refraction? What are the effects due to polarization?

* The figures in the note to page 42, represent the form of the molecules corresponding to these three conditions: 1, *a sphere;* 2, *an ellipsoid with equal transverse axes;* 3, *an ellipsoid with unequal lateral axes.*

intervening dark rings successively change places. If crystalline plates, having one axis of double refraction, be viewed in the direction of the axis, the rings are circles, and they are crossed by a dark or light cross. Figure 93 shows the position of the colored rings and cross in calc spar, and figure 94, the same at intervals of 90° in the revolution of the plate. With a crystal having two axes of double refraction, there are two series of elliptical rings, as in figures 95, 96; these figures show the character of the rings in niter, the latter alternating with the former in the revolution of the plate.

The same results are produced when the light is polarized by other means. For example, if a ray of light be reflected from a plate of glass at a certain angle, (56° 45′,) it is polarized; and on causing this ray to pass through crystals, as above, similar rings are shown with the same succession of changes on revolving the analyzing plate.

97

There are some monometric crystals which have the property of polarization. The accompanying figure of a crystal of analcime, by Sir David Brewster, exhibits a singular symmetrical arrangement of lines of prismatic colors and dark alternating lines with cross bands, producing a very brilliant effect. An irregular polarization has also been detected in some diamonds.

PHOSPHORESCENCE.

Several minerals give out light either by friction or when gently heated. This property of emitting light is called *phosphorescence.*

Two pieces of white sugar struck against one another give a feeble light, which may be seen in a dark place The same effect is obtained on striking together fragments of quartz, and even the passing of a feather rapidly over some specimens of zinc blende, is sufficient to elicit light.

Fluor spar is the most convenient mineral for showing phosphorescence by heat. On powdering it, and throwing

What is said of the appearance of certain crystals in polarized light? What is phosphorescence? Mention examples explaining the different modes of exhibiting it.

it on a shovel heated nearly to redness, the whole takes on a bright glow. In some varieties, the light is emerald green; in others, purple, rose, or orange. A massive fluor, from Huntington, Connecticut, shows beautifully the emerald green phosphorescence.

Some kinds of white marble, treated in the same way, give out a bright yellow light.

After being heated for a while, the mineral loses its phosphorescence; but a few electric shocks will, in many cases, to some degree, restore it again.

ELECTRICITY AND MAGNETISM.

Electricity.—Many minerals become electrified on being rubbed, so that they will attract cotton and other light substances; and when electrified, some exhibit positive, and others negative electricity, when brought near a delicately suspended magnetic needle. The diamond, whether polished or not, always exhibits positive electricity, while other gems become negatively electric in the rough state, and positive only in the polished state. Friction with a feather is sufficient to excite electricity in some varieties of blende. Some minerals, thus electrified, retain the power of electric attraction for many hours, as topaz, while others lose it in a few minutes.

Many minerals become electric when heated, and such species are said to be *pyro-electric*, from the Greek *pur*, fire, and electric.

If a prism of tourmaline, after being heated, be placed on a delicate frame, which turns on a pivot like a magnetic needle, on bringing a magnet near it, one extremity will be attracted, the other repelled, thus indicating the polarity alluded to. The same is better shown if the ends of the crystal be brought near the poles of a delicately suspended magnetic needle. The prisms of tourmaline have different secondary planes at the two extremities, or, as it is expressed, are hemihedrally modified (page 37.)

Several other minerals have this peculiar electric property, especially boracite and topaz, which, like tourmaline, are *hemihedral* in their modifications. Boracite crystallizes in

Will electricity restore the phosphorescent property when it is lost by heating a mineral? What two modes are there of exciting electricity in minerals? What is said of the diamond as compared with other gems? What is a pyro-electric? What is said of tourmaline? what of topaz and boracite?

cubes, with only the alternate solid angles similarly replaced (figs. 40, 41, page 37.) Each solid angle, on heating the crystals, becomes an electric pole; the angles diagonally opposite, are differently modified and have opposite polarity.

MAGNETISM.—*Lodestone* includes certain specimens of an ore of iron, called magnetic oxyd of iron, having the power of attraction like a magnet; it is common in many ore beds where this ore of iron occurs. When mounted like a horseshoe magnet, a good lodestone will lift a weight of many pounds. This is the only mineral that has decided magnetic attraction. But several ores containing iron are *attracted* by the magnet, or, when brought near a magnetic needle, will cause it to vibrate; and moreover, the metals nickel, cobalt, manganese, palladium, platinum and osmium, have been found to be slightly magnetic.

Many minerals become attractable by the magnet after being heated, that are not so before heating. This arises from a partial reduction, developing the protoxyd of iron.

SPECIFIC GRAVITY.

The specific gravity of a mineral is its weight compared with that of some substance, taken as a standard. For solids and liquids, *distilled water* at 60° F. is the standard ordinarily used; and if a mineral weighs twice as much as water, its specific gravity is 2; if three times, it is 3. It is then necessary to compare the weight of the mineral with the weight of an equal bulk of water. The process is as follows:

First weigh a fragment of the mineral in the ordinary way, with a delicate pair of scales: next suspend the mineral by a hair or fiber of silk to one of the scales, immerse it thus suspended in a tumbler of water, (keeping the scales clear of the water,) and weigh it again: subtract the *second* weight from the *first*, to ascertain the loss by immersion, and divide the *first* by the difference obtained: the result is the specific gravity. The loss by immersion is

98

What ore is at times possessed of magnetic attraction? What is said of other minerals as regards magnetism? What is specific gravity? Explain. Mention the mode of ascertaining specific gravity.

equal to the weight of the same bulk of water as the mineral.*

A better and more simple process than the above, and one available for *porous* as well as compact minerals, is performed with a light glass bottle, capable of holding exactly a thousand grains (or any known weight) of distilled water. The specimen should be reduced to a coarse powder. Pour out a few drops of water from the bottle, and weigh it; then add the powdered mineral till the water is again to the brim, and reweigh it: the difference in the two weights, divided by the loss of water poured out, is the specific gravity sought. The weight of the glass bottle itself is here supposed to be balanced by an equivalent weight in the other scale.

HARDNESS.

The comparative hardness of minerals is easily ascertained, and should be the first character attended to by the student in examining a specimen. It is only necessary to draw the file across the specimen, or to make trials of scratching one with another. As standards of comparison, the following minerals have been selected, increasing gradually in hardness from *talc*, which is very soft and easily cut with a knife, to the *diamond*, which nothing will cut. This table is called the scale of hardness.

1, *talc*, common foliated variety; 2, *rock salt;* 3, *calc spar*, transparent variety; 4, *fluor spar*, crystallized variety; 5, *apatite*, transparent crystal; 6, *feldspar*, cleavable variety; 7, *quartz*, transparent variety; 8, *topaz*, transparent crystal; 9, *sapphire*, cleavable variety; 10, *diamond.*

If on drawing a file across a mineral, it is impressed as easily as *fluor spar*, the hardness is said to be 4; if as easily as *feldspar*, the hardness is said to be 6; if more easily than

What other mode is fitted for porous as well as compact minerals? How is the hardness of minerals ascertained? What is the scale of hardness? Explain its use. What directions are given for trials of hardness?

* For perfectly accurate results, the most delicate scales and weights should be used, and great care be observed in the trial. The purity and temperature of the water should also be attended to, and the height of the barometer. For the latter, an allowance is made for any variation from a height of 30 inches. The temperature of water at its maximum density, or at 39° 1 F., is recommended as preferable to 60° F.

feldspar, but with more difficulty than apatite, its hardness is described as 5½ or 5·5.

The file should be run across the mineral three or four times, and care should be taken to make the trial on angles equally blunt, and on parts of the specimen not altered by exposure. Trials should also be made by scratching the specimen under examination with the minerals in the above scale, as sometimes, owing to a loose aggregation of particles, the file wears down the specimen rapidly, although the particles are very hard.

STATE OF AGGREGATION.

Solid minerals may be either *brittle*, *sectile*, *malleable*, *flexible* or *elastic*. Fluids are either *gaseous* or *liquid*.

1. *Brittle:* when parts of the mineral separate in powder on attempting to cut it.
2. *Sectile:* when thin pieces may be cut off with a knife but the mineral pulverises under a hammer.
3. *Malleable:* when slices may be cut off, and these slices will flatten out under the hammer. Example, native gold and silver.
4. *Flexible:* when the mineral will bend, and remain bent after the bending force is removed. Example, talc.
5. *Elastic:* when after being bent, it will spring back to its original position. Example, mica.

A liquid is said to be *viscous*, when on pouring it the drops lengthen and appear ropy. Example, petroleum.

FRACTURE.

The following are the several kinds of fracture in minerals:

1. *Conchoidal:* when the mineral breaks with a curved, or concave and convex surface of fracture. The word conchoidal is from the Latin *concha*, a shell. Flint is a good example.
2. *Even:* when the surface of fracture is nearly or quite flat.
3. *Uneven:* when the surface of fracture is rough with numerous small elevations and depressions.
4. *Hackly:* when the elevations are sharp or jagged, as in broken iron.

Explain the use of the term brittle; sectile; malleable, &c. Explain the use of the term conchoidal; even; uneven.

TASTE.

Taste belongs only to the soluble minerals; the kinds are—

1. *Astringent:* the taste of vitriol.
2. *Sweetish-astringent:* the taste of alum.
3. *Saline:* taste of common salt.
4. *Alkaline:* taste of soda.
5. *Cooling:* taste of saltpeter.
6. *Bitter:* taste of epsom salts.
7. *Sour:* taste of sulphuric acid.

ODOR.

Excepting a few gases and soluble minerals, minerals in the dry, unchanged state, do not give off odor. By friction, moistening with the breath, the action of acids and the blowpipe, odors are sometimes obtained, which are thus designated:

1. *Alliaceous:* the odor of garlic. It is the odor of burning arsenic, and is obtained by friction and more distinctly by means of the blowpipe from several *arsenical* ores.

2. *Horse-radish odor:* the odor of decaying horse-radish. It is the odor of burning selenium, and is strongly perceived when ores of this metal are heated before the blowpipe.

3. *Sulphureous:* odor of burning sulphur. Friction will elicit this odor from pyrites, and heat from many sulphurets.

4. *Fetid:* the odor of rotten eggs or sulphuretted hydrogen. It is elicited by friction from some varieties of quartz and limestone.

5. *Argillaceous:* the odor of moistened clay. It is given off by serpentine and some allied minerals when breathed upon. Others, as pyrargillite, afford it when heated.

CHAPTER IV.—CHEMICAL PROPERTIES OF MINERALS.

ACTION OF ACIDS.

Acids are used in distinguishing certain minerals that are decomposed by them. The acids employed are either the sulphuric, muriatic, or nitric. Carbonate of lime, (calca-

---

What taste is astringent? sweetish astringent? saline? What will develop odor in some minerals? What is understood by an alliaceous odor? What mineral when heated produces this odor? What is the odor of fumes of selenium? How is a sulphureous odor obtained from certain minerals? What gas has a fetid odor? What is an argillaceous odor?

reous spar,) when dropped into either of these acids gives off bubbles of gas, which effect is called *effervescence.* The same result takes place with some other minerals. The acid used in these tests, should be half water; and to avoid error, it is best to put a little of it in a test tube, and drop in small fragments of the coarsely powdered mineral. Sometimes heat will cause an effervescence, which does not take place with cold acid. Often effervescence arises from some impurity present, which is discontinued before the solution of the mineral in the acid is complete.

Other minerals, that do not effervesce in the acids, become changed to a jelly-like mass. For trials of this kind, the strong acids should generally be used. The powdered mineral is allowed to remain for a while in the acid, and gradually a jelly-like mass is formed. Often heat is required, and in that case, the jelly appears, as the solution cools. The minerals belonging to the zeolite family more especially undergo this change from the action of acids, and it arises from the separation of their silica in a gelatinous state.

BLOWPIPE.

To ascertain the effect of heat on minerals, a small instrument is used called a blowpipe. In its simplest form, (fig. 100,) it is merely a bent tube of small size, 8 to 10 inches long, terminating at one end in a minute orifice, not larger than a pin hole. It is used to concentrate the flame of a candle or lamp on a mineral, and this is done by blowing through it while the smaller end is just within the flame.

100 101 102

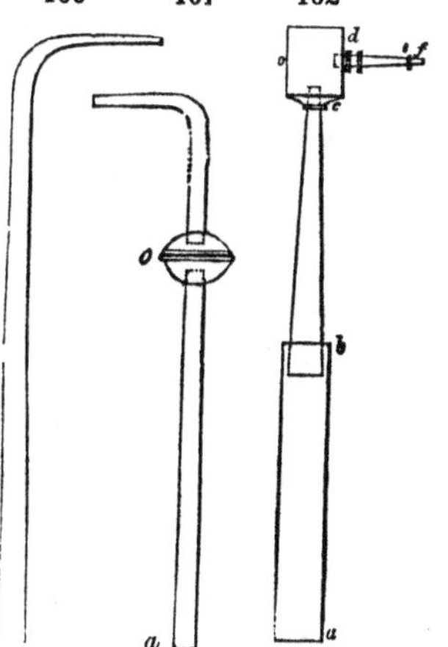

Figures 101 and 102 are other forms of the blowpipe, containing air chambers (*o*) to receive the moisture which is condensed in the tube

What is effervescence, and how produced? How should the acid be used? How are some minerals made to gelatinize? On what does this property depend? What is the object of a blowpipe?

during the blowing; the moisture, unless thus removed, is often blown through the small aperture and interferes with the experiment. The air chamber in figure 102 is a cylinder, into which the tube *a b c* is screwed at *c*, and the smaller piece *d e f*, at *d*. For the convenience of packing it away, there is a screw at *b*. The part *b c*, after unscrewing it, may be run into the part *a b*, through the large end, (*a*,) and screwed up again, and thus it is half the length it has when arranged for use. The mouth piece *e f* screws off, and is made of platinum in order that it may be cleaned when necessary by immersion in an acid. The best material for the blowpipe is silver, or if a cheaper material is desired, tinned iron with the piece *e f* of brass. Brass gives a disagreeable smell to the moist fingers.

In using the blowpipe, it is necessary to breathe and blow at the same time, that the operator may not interrupt the flame in order to take breath. Though seemingly absurd, the necessary tact may easily be acquired. Let the student first breathe a few times through his nostrils, while his cheeks are inflated and his mouth closed. After this practice, let him put the blowpipe to his mouth, and he will find no difficulty in breathing as before; while the muscles of the inflated cheeks are throwing the air they contain through the blowpipe. When the air is nearly exhausted, the mouth may again be filled through the nose without interrupting the process of blowing.

A lamp with a large wick, so as to give a broad flame, and fed with olive oil, is best; but a candle is more conveniently carried about when travelling. The wick should be bent in the direction the flame is to be blown.

The flame has the form of a cone, yellow without and blue within. The heat is most intense just beyond the extremity of the blue flame. In some trials, it is necessary that the air should not be excluded from the mineral during the experiment, and when this is the case, the *outer* flame is used. The outer is called the *oxydating** flame, and the inner the *reducing* flame.

Explain the structure and mode of use. What is said of the flame of a candle before the blowpipe? Which is the oxydating, and which the reducing flame?

---

* It is so called because when thus heated, oxygen, one of the constituents of the atmosphere, combines in many cases with some parts of the assay (or substance under experiment.)

The mineral is supported in the flame, either on charcoal, or by means of steel forceps, (fig. 103,) with platinum extremities (*a b*); the forceps are opened by pressing on the pins *p p*. The charcoal should be firm and well burnt. Charcoal is especially necessary when the reduction of the assay needs the presence of carbon; and platinum when simple heat is required. Platinum foil for enveloping the mineral, and small platinum cups are also used. When nothing better is at hand, the mineral mica or kyanite may be employed. The fragment of mineral under trial should be less than half a pea in size, and often a thin splinter is required.

103

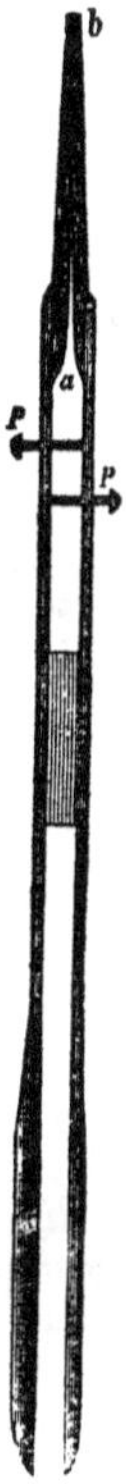

To test the presence of water or a volatile ingredient, the mineral is heated in a glass tube or test vial. The tube may be three or four inches long and as large as a quill. The flame is directed against the exterior of the tube beneath the assay, and the volatilized substance usually condenses in the upper part of the tube. By inserting into the upper end of the tube a strip of litmus or other test paper, it is ascertained whether the fumes are acid or not.

Some species require for fusion the aid of what are called *fluxes*. Those more commonly used are *borax*, *salt of phosphorus*, and *carbonate of soda*. They are fused to a clear globule, to which the mineral is added; or powdered and made up into a ball with the moistened mineral in powder. In this way some minerals are fused that cannot be attacked otherwise, and nearly all species, as they melt, undergo certain changes in color, arising from changes in composition, which are mentioned in describing minerals.

The above mentioned fluxes also are often required in order to obtain the metals from the metallic ores. On heating a fragment of copper pyrites with borax, a globule of copper is obtained; and tin ore heated with soda yields a globule of tin.

What instruments or appliances are used for holding minerals before the blowpipe? How is the presence of water ascertained? How may its acidity be tested? How are the common fluxes employed, and what is their use?

The following table contains the reactions of some of the metallic oxyds with the ordinary fluxes :*

	Borax.	*Salt of Phosphorus.*	*Soda.*
Titanic acid	O, colorless or milky	O, colorless, trp	Deep yw, hot; w or gyh, cold
Oxyd of iron	O, red, hot; ywh or colorless, cold R, green or bh gn	O, red, hot; paler or colorless, cold	
Oxyd of cerium	O, r; yw on cooling; w enamel on flaming R, colorless or w enamel	O, fine r, hot; colorless, cold	
Oxyd of manganese	O, amethystine	O. amethystine	*Pl.* trp gn, hot; bh-gn, cold
Oxyd of cobalt	O, trp blue	O, blue	*Pl.* pale r, hot; gray, cold
Oxyd of chrome	O, bn, hot; pale gn, cold R, emerald-gn, cold	O, green R, green	O. *Pl.* dull orange; op & yw on cooling
Oxyd of copper	O, green R, colorless, hot; but suddenly opaque and rdh on cooling	O, green R, colorless, hot; r on solidifying	*Pl.* gn, hot; col, op, cold

The following are other reactions :

Nitrate of cobalt in solution added to the assay after heating to redness, and then again heated, produces before fusion a blue color for alumina and a pale-red for magnesia.

Boracic acid fused with a phosphate produces a globule, into which if the extremity of a small iron wire be inserted, and the whole heated in the reduction flame, the globule attached to the wire will be brittle, as proved by striking it with a hammer on an anvil. Before this trial it should be ascertained that no sulphuric or arsenic acid is present, which also may form a brittle globule with the iron; nor any metallic oxyd reducible by the iron.

For what is nitrate of cobalt used? How and for what is boracic acid used?

* *O* stands for *oxydating* flame; *R* for *reducing* flame; *Ch* for charcoal; *trp* for transparent; *bh* bluish; *yw* yellow; *gn* green; *r* red; *gyh* grayish; *w* white; *Pl* in platinum forceps; *op* opaque.

Tin-foil is used to fuse with certain peroxyds of metals to reduce them to protoxyds. The assay, previously heated in the reducing flame, should be touched with the end of the tin foil; a very minute quantity of a metallic oxyd is thus detected.

Saltpeter added along with a flux to a compound containing manganese, gives the amethystine color, when the quantity is too small to be detected without it.

Potash salts, if there is no soda present, give a slightly violet tinge to the flame.

Soda salts give the flame a deep yellow color.

Lithia salts give the flame a reddish tinge; the silicates require the addition of some fluor spar and bisulphate of potash. By adding soda and heating on platinum, the lithia stains the platinum brown.

Sulphurets, Sulphates. A glass made of soda and silica becomes red or orange yellow when sulphur is present. Heated on charcoal with soda, and then adding a drop of water, they yield sulphuretted hydrogen, which blackens a test paper containing acetate of lead. Sulphurets heated in a glass tube closed below, with litmus paper above, redden the litmus paper, and yield usually a sulphureous odor.

Seleniets give off a horse-radish odor.

Arseniurets give off an odor like garlic, which is brought out by heating with soda in the reduction flame, if not otherwise perceptible; heated in a tube, orpiment is condensed.

Fluorids. Heated with salt of phosphorus, previously melted in a glass tube, the glass is corroded; and Brazil paper placed in the tube becomes yellow. The salt of phosphorus for this trial should be free from all chlorids.

Nitrates detonate on burning coals.

CHAP. V.—CLASSIFICATION OF MINERALS.

Under the term *mineral*, as explained, are included all inorganic substances occurring in nature. These substances have been found to consist of various elements, some few

---

How and for what is tin-foil used? saltpeter?—What is said of the constitution of minerals?

---

* For full information on the use of the blowpipe and its reactions there is no better work than Berzelius on "the Use of the Blowpipe," translated by J. O. Whitney. 238 pp. 8vo. Boston, 1845.

species being each a simple element alone, and others consisting of two or more elements in a state of combination. The various native metals, as native gold, silver, copper, mercury, are some of the elements. Iron ores are compounds of the element iron with some other element or elements, as oxygen, sulphur, or oxygen and carbon, &c. Marble is a compound of three elements, calcium, oxygen and carbon. Water consists of two elements, hydrogen and oxygen. Diamond is the simple element carbon, which is identical with pure charcoal. All the so-called elements of matter are found in the mineral kingdom, either in a pure or combined state; and it is the object of chemical analysis to ascertain the proportions of each in the constitution of the several minerals. Upon these results depends to a great degree our knowledge of those relations of the species upon which the classification of minerals is based.

The number of elemental substances in nature, according to the most recent results of chemistry, is sixty. Of these, forty-seven are metals, and five are gases; the remainder, as, for instance, *sulphur* and *carbon*, are solids without a metallic luster, excepting one (*bromine*) which is a liquid at the ordinary temperature. Of these sixty elements, very much the larger part are of rare occurrence in nature. The rocks of the globe, with their most common minerals, are made up of about *thirteen* of the elements. These are the gases *oxygen*, *hydrogen*, *nitrogen*, *chlorine*; the non-metallic elements *carbon*, *sulphur*, *silicon*; the metals *calcium*, (basis of lime,) *sodium*, (basis of soda,) *potassium*, (basis of potash,) *magnesium*, (basis of magnesia,) *aluminium*. (basis of alumina, the principle constituent of clay,) with *iron*. The element silicon combined with oxygen, forms *silica*. In this state, it is the mineral quartz, the most common in the constitution of the rocks of the globe: it is a constituent of granite, mica slate and the allied rocks, of the hard granular quartz rock; and it is the essential part of all sandstones and millstone grits, as well as the principal ingredient of the sands of the sea shore and of most soils. Combined with lime, potash or soda, magnesia or alumina, and often with iron, it forms nearly all the other mineral in-

---

What is the number of elements, and how many are metals? How many constituents are essential to the rocks of the globe, and what are they? What is said of quartz?

gredients of granite, mica slates, volcanic rocks, shales, sandstones and various soils. No element is therefore more important than this in the constitution of the earth's strata. and it is specially fitted for this preëminence by its superior hardness, a character it communicates to the rocks in which it prevails. Next to silica, rank *lime* and *carbon;* for carbon with oxygen constitutes *carbonic acid,* and this combined with *lime,* produces *carbonate of lime,* the ingredient which, when occurring in extended beds, we call *limestone* and *marble.* Again, *lime* combined with sulphur and oxygen, (sulphuric acid,) makes sulphate of lime, or common *gypsum. Iron* is very generally diffused; it is one of the constituents of many siliceous minerals, and forms vast beds of ore. *Oxygen,* as has been implied, is a constituent in all the rocks above mentioned, and besides, is an essential part of the atmosphere and water; it is the most universally diffused of the elements. It is united with *hydrogen* in the constitution of water, and with *nitrogen* in the constitution of the atmosphere. *Chlorine* combined with *sodium* constitutes *common salt,* which occurs in sea water and brine springs, and is also found in vast beds in some rock strata.

It is thus seen how few are the elements essential to the framework of our globe. The various metallic ores, of less general diffusion, are however of vast economical importance to man, and multiply considerably the number of mineral species. Those important to the general student, however, are comparatively few. The whole number of well established species in the mineral kingdom is about 600; of these, more than two-thirds are known only to the mineralogist.

It is the province of chemistry to discuss fully the nature of the elements, and their modes of combination. It is sufficient to add here, for the benefit of any who may not have the requisite elementary chemical knowledge, how the chemical names of minerals indicate their composition. Terms such as *oxyd of iron, chlorid of iron,* express a combination of iron with the element oxygen, or chlorine; so also *sulphuret of iron* is a compound of iron with sulphur. The force of the terminations *id* or *uret* is always as here explained. Protoxyd and peroxyd imply different proportions

---

Which are the next most common ingredients of rocks? Mention the other ingredients alluded to. What is an oxyd? a chlorid? a sulphuret? a carbonate?

of oxygen, the latter the highest. Terms such as *carbonate of lime*, *sulphate of lime*, indicate that the substance is composed of an acid—carbonic acid, or sulphuric acid in the instances cited, with lime. So *silicate of soda* is a compound of soda and silicic acid (or silica); and all such compounds are theoretically said to consist of an *acid* and a *base*—lime and soda, in the cases mentioned, being bases.

The true foundation of a species in mineralogy must be derived from crystallization, as the crystallizing force is fundamental in its nature and origin; and it is now generally admitted that *identity of crystalline form and structure* is evidence of *identity of species*. This principle unites certain distinct chemical compounds into the same species:—for example, a *silicate of magnesia* and a *silicate of iron* crystallizing alike, constitute but one species in mineralogy, though chemically so different. Oxyd of iron and magnesia are themselves nearly identical in molecular *form* and *size*, and on this fact depends their power of replacing one another even in complex compounds. They are therefore said to be *isomorphous* (from the Greek *isos*, similar, and *morphe*, form.)

There are many groups of these isomorphous substances, and some knowledge of them is necessary to enable the reader to understand why different varieties of a mineral species may differ so widely, as they often do, in composition. Some of these groups are as follows:

1. Alumina, peroxyd of iron, peroxyd of manganese.
2. Lime, magnesia, protoxyds of iron, manganese and zinc.
3. Baryta, strontia, oxyd of lead.
4. Sulphur, selenium, tellurium.
5. Tungsten, molydenum.
6. Phosphoric acid, arsenic acid.

In epidote the alumina may be replaced by peroxyd of iron or manganese, and the magnesia in part or wholly by lime, or the protoxyds of iron or manganese. The same is true of garnet and several other minerals. The rhombohedrons of carbonate of lime, carbonate of iron, and carbonate of magnesia, are very nearly identical in angle, because the bases are ismorphous. This subject is illustrated by the greater part of mineral species.

What is a sulphate? a silicate? What is the test of identity of species in mineralogy? What are isomorphous substances? What are the common groups of isomorphous substances in minerals? Explain by examples.

GENERAL VIEW OF THE CLASSIFICATION OF MINERALS.

The classification adopted in this work is based on the constitution of minerals. The following is a general view of it:

Class I. Gases: consisting of or containing nitrogen or hydrogen.

Class II. Water.

Class III. Carbon, and compounds of carbon.

Class IV. Sulphur.

Class V. Haloid minerals: compounds of the alkalies and earths, with the soluble acids (sulphuric, nitric, carbonic, &c. or water,) or of their metals with chlorine or fluorine. 1, Salts of ammonia; 2, of potash; 3, of soda; 4, of baryta; 5, of strontia; 6, of lime; 7, of magnesia; 8, of alumina.

Class VI. Earthy minerals: silica and siliceous or aluminous compounds of the alkalies and earths—1, silica; 2, lime; 3, magnesia; 4, alumina; 5, glucina; 6, zirconia; 7, thoria.

Class VII. Metals and metallic ores, (exclusive of the metals of the alkalies and earths): 1, *Metals easily oxydizable*—cerium, yttrium, titanium, tin, molybdenum, tungsten, vanadium, tellurium, bismuth, antimony, arsenic, uranium, iron, manganese, chromium, nickel, cobalt, zinc, cadmium, lead, mercury, copper; 2, *Noble metals:* platinum, iridium, palladium, gold, silver.

Explain the classification adopted.

CLASS I.—GASES.

The gases occurring native are as follows: 1. *containing or consisting of nitrogen:* atmospheric air, nitrogen. 2. *containing hydrogen:* carbureted hydrogen, phosphureted hydrogen, sulphureted hydrogen, muriatic acid. 3. *containing carbon or sulphur:* carbonic acid, sulphurous acid.

ATMOSPHERIC AIR.

1. Atmospheric air is the air we breathe. It consists of oxygen 21 per cent. by weight, and nitrogen 79 per cent., with a small proportion of carbonic acid. It has neither color, odor, nor taste. It supports life and combustion through the oxygen which it contains, this gas being used or absorbed in respiration as well as in the burning of wood or a candle. The oxygen thus consumed is restored to the air again by vegetation which gives out oxygen through the day, and in this way the quality of the atmosphere requisite for life is sustained. It is about 815 times lighter than water, and 11,065 times lighter than mercury. A hundred cubic inches weigh about 31 grains.

NITROGEN GAS.

Nitrogen destroys life, and has neither color, odor nor taste. It is one of the constituents of the atmosphere. It bubbles up through the waters of many springs, having been derived from air by some decompositions in progress within the earth, by which the oxygen of the air is absorbed.

Lebanon springs in Columbia county, New York, and a region in the town of Hoosic, Rensselaer county, afford large quantities of this gas. There is another locality at Canoga, Seneca county, where the water is in violent ebullition from the escape of the gas; its temperature is 40° F. There are other nitrogen springs in Virginia, west of the Blue Ridge at Warm and Hot Springs; in Buncombe county, N. C.; and on the Washita in Arkansas. At Bath, in England, nitrogen is escaping from the tepid springs at the

---

What gases occur in nature? What is the constitution of the atmosphere? its general characters? the weight? What is said of the characters of nitrogen? Where does nitrogen occur in nature?

rate of 267 cubic inches a minute, or 222 cubic feet a day. The gas from these nitrogen springs contains only 2 or 3 per cent. of oxygen, and often a very little carbonic acid.

CARBURETTED HYDROGEN.

Carburetted hydrogen consists of carbon 75, hydrogen 25; burns with a bright yellow flame. It is the same gas nearly that is used for lighting the streets in some of our cities. It issues abundantly from some coal beds and beds of bituminous slate. At Fredonia, in western New York, near Lake Erie, it is given out so freely from a slate rock, that it is used for lighting the village. A vessel containing 220 cubic feet is filled in about 15 hours. A light-house at Portland harbor, on Lake Erie, four miles from Fredonia, is also lighted with the same gas from other springs.

Another carburetted hydrogen, burning with a pale blue flame, rises in bubbles through pools of water, owing to vegetable decomposition in the soil beneath.

PHOSPHURETTED HYDROGEN.

Phosphuretted hydrogen consists of phosphorus 91·29, and hydrogen 8·71. It takes fire spontaneously. The phosphoric matter, called Jack-o'-lantern, sometimes seen floating over marshy places, is supposed to be phosphureted hydrogen.

SULPHURETTED HYDROGEN.

Sulphureted hydrogen consists of sulphur 94·2, hydrogen 5·8. It has the odor and taste of putrescent eggs and burns with a bluish flame. It is abundant about sulphur springs, issuing freely from the waters, as in western New York and in Virginia. It is sometimes found about volcanoes. It blackens silver and also a common cosmetic made of oxyd of bismuth.

MURIATIC ACID.—*Hydrochloric Acid.*

Muriatic acid gas consists of hydrogen 2·74, chlorine 97·26. It has a very pungent odor and is acrid to the skin.

What is the composition of carbureted hydrogen? its general characters? mode of occurrence in nature? What is said of Fredonia? Mention the characters of phosphureted hydrogen; the characters of sulphureted hydrogen; its mode of occurrence. What is said of muriatic acid?

It is rapidly dissolved by water. If passed into a solution of nitrate of silver, it produces a white precipitate which soon blackens on exposure. It is given out occasionally by volcanoes.*

CLASS II.—WATER.

Water (oxyd of hydrogen) is the well known liquid of our streams and wells. The purest natural water is obtained by melting snow, or receiving rain in a clean glass vessel; but it is absolutely pure only when procured by distillation. It consists of hydrogen 1 part by weight, and oxygen 8 parts. It becomes solid at 32° Fahrenheit, (or 0° Centigrade) and then crystallizes, and constitutes ice or snow. Flakes of snow consist of a congeries of minute crystals, and stars like the annexed figure may often be detected with a glass. Various other allied forms are also assumed. The rays meet at an angle of 60°, and the branchlets pass off at the same angle with perfect regularity. The density of water is greatest at 39° 1 F.; below this it expands as it approaches 32°, owing to incipient crystallization. It boils at 212 F. A cubic inch of pure water at 60° F. and 30 inches of the barometer, weighs 252·458 grains. A pint, United States standard measure, holds just 7342 troy grains of water, which is little above a pound avoirdupois (7000 grains troy.)

Water as it occurs on the earth, contains some atmospheric air, without which the best would be unpalatable. This air, with some free oxygen also present, is necessary to the life of water animals. In most spring water there is a minute proportion of salts of lime, (sulphate, chlorid or carbonate,) often with a trace of common salt, carbonate of magnesia and some alumina, iron, silica, phosphoric acid, carbonic acid, and certain vegetable acids. These impurities constitute usually from $\frac{1}{10}$ to 10 parts, in 10,000 parts by weight. The Long Pond water, used in Boston,

Of what does water consist? What is said of snow and ice? What of the density of water? its boiling temperature? the weight of a pint? What are the usual impurities of common spring or river water?

* *Carbonic acid* and *sulphurous* acid gases, are described, one under *carbon*, and the other under *sulphur*.

contains about ½ a part in 10,000; the Schuylkill of Philadelphia, about 1 part in 10,000; the Croton, used in New York city, 1 to 1½ parts in 10,000. In the Schuylkill water the constituents of the 1 part of solid ingredients were, chlorid of sodium 1·47, chlorid of magnesium 0·094, sulphate of magnesia 0·57, silica 0·8, carbonate of lime 18·72, carbonate of magnesia 3·51, carbonate of soda and loss 16·44.* The water towards the surface is always purer than that below.

Sea water contains 32 to 37 parts of solid substances in solution in 1000 parts of water. The largest amount in the Atlantic, 36·6 parts, is found under the equator, away from the land or the vicinity of fresh water streams; and the smallest in narrow straits, as Dover Straits where there are only 32·5 parts. In the Baltic and the Black Sea, the proportion is only one-third that in the open ocean. Of the whole, one-half to two-thirds is common salt (chlorid of sodium.) The other ingredients are magnesian salts, (chlorid and sulphate,) amounting to four-fifths of the remainder, with sulphate and carbonate of lime, and traces of bromids, iodids, phosphates and fluorids. The water of the British channel affords, water 964·7 parts in 1000, chlorid of sodium 27·1, chlorid of potassium 0·8, chlorid of magnesium 3·7, sulphate of magnesia 2·30, sulphate of lime 1·4, carbonate of lime 0·03, with some bromid of magnesium, and probably traces of iodids, fluorids and phosphates. The bitter taste of sea water is owing to the salts of magnesia present.

The waters of the Dead Sea contain 200 to 250 parts of solid matter in 1000 parts, (or 20 to 25 per cent.,) including 7 to 10 per cent. of common salt, the same proportion of magnesian salts principally the chlorid, 2½ to 3½ per cent. of carbonate and sulphate of lime, besides some bromids and alumina. The density of these waters is owing to this large proportion of saline ingredients. The brine springs of New York and other states south and west, are well known sources of salt, (see beyond under common salt.) Many of the springs afford bromine, and large quantities of it are manufactured for making daguerreotype plates and other purposes.

What proportion of solid substances in sea water, and of this what proportion is common salt? What proportion magnesian salts? What is the bitter taste of sea water owing to?

---

* Chem. Exam. by B. Silliman, Jr, Jour. Sci., ii ser., ii, 218.

Mineral waters vary much in constitution. They often contain carbonate of iron, like those of Saratoga and Ballstown, and are then called *chalybeate* waters, from the ancient name for iron or steel, *chalybs*, derived from the name of a country on the Baltic. The water of Congress Spring, according to Dr. Steel, contains in a pint, chlorid of sodium 48·1, bicarbonate of magnesia 12·0, carbonate of lime 12·3, carbonate of iron 0·6, silica 0·2, iodid of sodium nearly 0.5, with a trace of bromid of potash ; of carbonic acid 39·0 cubic inches and nearly 1 cubic inch of atmospheric air.

Minute traces of salts of zinc and arsenic, lead, copper, antimony and tin, have been found in some waters. Whatever is soluble in a region through which waters flow, will of course be taken up by them, and many ingredients are soluble in minute proportions, which are usually described as insoluble.

CLASS III.—CARBON AND COMPOUNDS OF CARBON.

Carbon occurs crystallized in the diamond. In a massive form, and more or less pure state, it constitutes the various kinds of mineral coal. Combined with hydrogen, or hydrogen and oxygen, it forms bitumen, amber, and a number of native mineral resins.

DIAMOND.

Monometric. In octahedrons, dodecahedrons and more complex forms. Faces often curved, as in the annexed figures. Cleavage octahedral ; highly perfect.

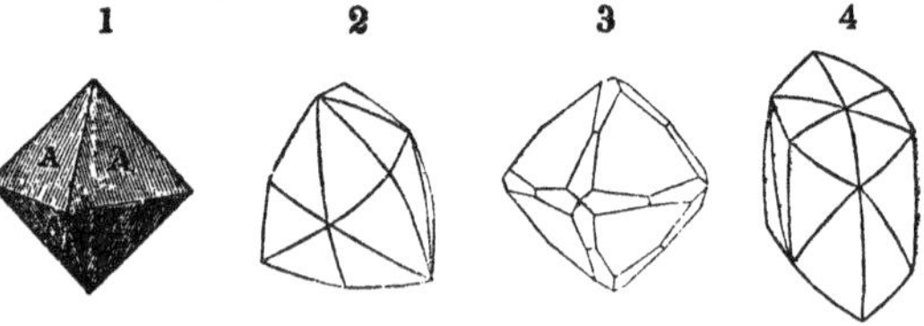

Color white or colorless ; also yellowish, red, orange,

What are chalybeate waters? What is the difference between the diamond and charcoal? What is the crystallization of the diamond? What other characters are mentioned?

green, brown or black. Luster adamantine. Transparent; translucent when dark colored. H = 10. Gr = 3·48—3·55.

Composition. Pure carbon. It burns and is consumed at a high temperature, producing carbonic acid gas. Exhibits vitreous electricity when rubbed. Some specimens exposed to the sun for a while, give out light when carried to a dark place. Strongly refracts and disperses light.

Dif. Diamonds are distinguished by their superior hardness; their brilliant reflection of light and adamantine luster; their vitreous electricity when rubbed, which is not afforded by other gems unless they are polished; and by the practiced ear, by means of the sound when rubbed together.

Obs. Diamonds occur in India, in the district between Golconda and Masulipatam, and near Parma, in Bundelcund, where some of the most magnificent specimens have been found; also on the Mahanuddy, in Ellore. In Borneo, they are obtained on the west side of the Ratoos mountain, with gold and platina. The Brazilian mines were first discovered in 1728, in the district of Serra do Frio, to the north of Rio de Janeiro; the most celebrated are on the river Jequitinhonha, which is called the Diamond river, and the Rio Pardo; twenty-five to thirty thousand carats are exported annually to Europe from these regions. In the Urals of Russia they had not been detected till July, 1829, when Humboldt and Rosè were on their journey to Siberia. The river Gunil, in the province of Constantine, in Africa, is reported to have afforded some diamonds. In the United States, the diamond has been met with, in Rutherford county, North Carolina, (fig. 4,) and Hall county, Georgia.

The original rock in Brazil appears to be either a kind of laminated granular quartz called *itacolumite;* or a ferruginous quartzose conglomerate. The itacolumite occurs in the Urals, and diamonds have been found in it; and it is also abundant in Georgia and North Carolina. In India, the rock is a quartzose conglomerate. The origin of the diamond has been a subject of speculation, and it is the prevalent opinion that the carbon, like that of coal, is of vegetable origin. Some crystals have been found with black uncrystallized particles or seams within, looking like coal; and this fact has been supposed to prove their vegetable origin.

How is the diamond distinguished? What are its principal localities?

Diamonds with few exceptions are obtained from alluvial washings. In Brazil, the sands and pebbles of the diamond rivers and brooks (the waters of which are drawn off in the dry season to allow of the work) are collected and washed under a shed, by a stream of water passing through a succession of boxes. A negro washer stands by each box, and inspectors are stationed at intervals. When a diamond is found weighing 17½ carats, the negro is entitled to his liberty.

The largest diamond of which we have any knowledge is mentioned by Travernier, as in the possession of the Great Mogul. It weighed originally 900 carats, or 2769·3 grains, but was reduced by cutting to 861 grains. It has the form and size of half of a hen's egg. It was found in 1550, in the mine of Colone. The diamond which formed the eye of a Braminican idol, and was purchased by the Empress Catharine II. of Russia from a French grenadier who had stolen it, weighs 193 carats, and is as large as a pigeon's egg. The Pitt or regent diamond is of less size, it weighing but 136·25 carats, or 419¼ grains; but on account of its unblemished transparency and color, it is considered the most splendid of Indian diamonds. It was sold to the Duke of Orleans by Mr. Pitt, an English gentleman, who was governor of Bencolen, in Sumatra, for £130,000. It is cut in the form of a brilliant, and is estimated at £125,000. The Rajah of Mattan has in his possession a diamond from Borneo, weighing 367 carats. The Koh-i-noor, on its arrival in England, weighed 186.016 carats. It has since been re-cut and reduced one-third in weight.

The diamonds of Brazil are seldom large. Maure mentions one of 120 carats, but they rarely exceed 18 or 20. One weighing 254½ carats, called the "*Star of the South*," was found in 1854. It will be reduced one-half in cutting.

Diamonds are valued according to their color, transparency and size. When limpid (of pure water) and no extraordinary magnitude, the value of a wrought diamond is estimated by first ascertaining the weight in carats.* The

How are diamonds obtained? How are diamonds valued?

* A carat is a conventional weight, and is divided into 4 grains, which are a little lighter than 4 grains troy; 74 1-16 carat grains are equal to 72 troy grains. The term *carat* is derived from the name of a bean in Africa, which, in a dried state, has long been used in that country for weighing gold. These beans were early carried to India and were employed there for weighing diamonds.

rule given is as follows: double the weight in carats, and multiply the square of the product by £2. Thus a wrought diamond weighing 1 carat, would be worth £8; one of 4 carats, £128; one of 10 carats, £800. Above 20 carats. the prices rise much more rapidly. A flaw, however minute, or the slightest smokiness, diminishes very much the value. The average price of rough diamonds, of first quality, of 1 carat, is £2; of 2 carats, £8, since it loses half its weight in cutting, and becomes then one of 1 carat wrought.

The rule just given is scarcely regarded in market, as so much depends upon the purity of water. In different countries, moreover, the standard of taste as regards diamonds is very different, the market in England demanding the very first quality, while in other countries a somewhat inferior kind satisfies the purchaser.

The rose diamond is more valuable than a snow-white diamond, owing to the great beauty of its color and its rarity. The green diamond is much esteemed on account of its color. The blue is prized only for its rarity, as the color is seldom pure. The black diamond, which is uncommonly rare and without beauty, is highly prized by collectors. The brown, gray and yellow varieties are of much less value than the pure white or limpid diamond.

The diamond is cut by taking advantage of its cleavage, and also by abrasion with its own powder and by friction with another diamond. The flaws are first removed by cleaving it; or else by sawing it with an iron wire, which is covered with diamond powder—a tedious process, as the wire is generally cut through after drawing it across five or six times. After the portion containing flaws has thus been cut off, the crystal is fixed to the end of a stick, in a strong cement, leaving the part projecting which is to be cut; and another being prepared in the same manner, the two are rubbed together till a facet is produced. By changing the position, other facets are added in succession till the required form is obtained. A circular plate of soft iron is then charged with the powder produced by the abrasion, and this, by its revolution, finally polishes the stone. To complete a single facet often requires several hours. Diamonds were first cut in Europe, in 1456, by Louis Berquen, a citizen of Bruges;

How are diamonds cut?

but in China and India, the art of cutting appears to have been known at a very early period.

By the above process, diamonds are cut into *brilliant, rose* and *table* diamonds. The brilliant has a *crown* or upper part, consisting of a large central eight-sided facet, and a series of facets around it; and a *collet*, or lower part, of pyramidal shape, consisting of a series of facets, with a smaller series near the base of the crown. The depth of a brilliant is nearly equal to its breadth, and it therefore requires a thick stone. Thinner stones, in proportion to the breadth, are cut into rose and table diamonds. The surface of the *rose* diamond consists of a central eight-sided facet of small size, eight triangles, one corresponding to each side of the table, eight trapeziums next, and then a series of sixteen triangles. The collet side consists of a minute central octagon, surrounded by eight trapeziums, corresponding to the angles of the octagon, each of which trapeziums is subdivided by a salient angle into one irregular pentagon and two triangles. The *table* is the least beautiful mode of cutting, and is used for such fragments as are quite thin in proportion to the breadth. It has a square central facet, surrounded by two or more series of four-sided facets, corresponding to the sides of the square.

Diamonds have also been cut with figures upon them. As early as 1500, Charadossa cut the figure of one of the Fathers of the church on a diamond, for Pope Julius II.

Diamonds are employed for cutting glass; and for this purpose only the natural edges of crystals can be used, and those with curved faces are much the best. Diamond dust is used to charge metal plates of various kinds for jewelers, lapidaries and others. Those diamonds that are unfit for working, are sold for various purposes, under the name of *bort.* Fine drills are made of small splinters of bort, which are used for drilling other gems, and also for piercing holes in artificial teeth and vitreous substances generally.

The diamond is also used for lenses for microscopes. When ground plano-convex, they have but slight chromatic aberration, and consequently a larger field, and but little loss of light, compared with similar lenses of other materials. They often have an irregularity of structure when perfectly

---

What are the three forms usually given the diamond? For what purposes are diamonds used?

pellucid, which unfits them for this purpose, and such lenses therefore are seldom made.

MINERAL COAL.

Massive. Color black or brown, opaque. Brittle or sectile. H=1—2·5. Gr=1·2—1·75.

Composition. Carbon, with usually a few per cent. of silica and alumina, and sometimes oxyd of iron; often contains a large proportion of bitumen. The bituminous varieties burn with a bright flame and bituminous odor; while those destitute of bitumen afford only a pale blue flame, arising from the decomposition of the water present and the formation of the gas called carbonic oxyd.

VARIETIES.—1. *Without bitumen.*

Anthracite. Anthracite (called also *glance coal* and *stone coal*) has a high luster, and is often iridescent. It is quite compact and hard, and has a specific gravity from 1·3 to 1·75. It usually contains 80 to 90 per cent. of carbon, with 4 to 7 of water, the rest consisting of earthy impurities. There is often some bitumen present, in which case it burns with considerable flame.

Besides the use of anthracite for fuel, it is often made into inkstands, small boxes, and other articles, which have a high polish, and fine specimens of this kind of ware may be obtained in Philadelphia.

2. *Bituminous varieties.*

Bituminous coal varies much and indefinitely in the amount of bitumen it contains, and there is a gradual passage in its varieties into varieties of anthracite. It is softer than anthracite and less lustrous. The specific gravity does not exceed 1·5.

Pitching or *caking coal*, as it is distinguished in England, at first breaks when heated, into small pieces, which, on raising the heat, again unite into a solid mass. Its color is velvet or grayish black. It burns readily with a lively yellow flame, but requires frequent stirring to prevent its caking, and so clogging the fire. The principal beds at Newcastle, England, afford this kind of coal. *Cherry coal* resembles pitch coal in appearance, but does not soften and cake. It

---

Of what does mineral coal consist? How does anthracite differ from other varieties?

is very brittle, and in mining there is consequently much waste. It burns with a clear yellow flame. It occurs at the Glasgow coal beds, and is named from its luster and beauty. The *splint coal* (or hard coal) of the same region is harder than the cherry coal.

Cannel coal is very compact and even in texture, with little luster, and breaks with a large conchoidal fracture. It takes fire readily, and burns without melting with a clear yellow flame, and has hence been used as candles—whence the name. It is often made into inkstands, snuff-boxes and other similar articles.

Brown coal, *wood coal*, *lignite*, are names of a less perfect variety of coal, usually having a brownish black color, and burning with an empyreumatic odor. It has often the structure of the original wood. The term *brown coal* is, however, applied generally to any coal more recent in origin than the era of the great coal beds of the world, although it may not have any distinct remains of a woody structure, or burn with an empyreumatic odor. The name *lignite* has sometimes the same general application, though without strict propriety.

Jet resembles cannel coal, but is harder, of a deeper black color, and has a much higher luster. It receives a brilliant polish, and is set in jewelry. It is the *Gagates* of Dioscorides and Pliny, a name derived from the river Gagas, in Syria, near the mouth of which it was found, and the origin of the term jet, now in use.

Obs. Mineral coal occurs in extensive beds or layers, interstratified with different rock strata. The associate rocks are usually clay shales (or slaty beds) and sandstones; and the sandstones are occasionally coarse grit rocks. There are sometimes also beds of limestone alternating with the other deposits. In a vertical section through the *coal measures*—as the series of rocks and coal seams are usually called—there may be below, sandstones and shales in alternating layers, or sandstones alone and then shales; there may next appear upon the shale a bed or layer of coal, one, two or even thirty feet thick; then above the coal, other layers of shale and sandstone; and then another layer of coal; again shale and sandstones in various alternations, or

---

What is cannel coal? brown coal or lignite? jet? How do beds of coal occur, and what are the associated rocks?

perhaps layers of limestone; and then a third bed of coal, and so on. By such alternations the series is completed. Immediately in the vicinity of the coal, the rock is generally rather a shale than a sandstone, and these shales are usually full of impressions of leaves and stems of plants. The clay shales are sometimes quite soft and earthy, and of a light clay color; but in most coal regions they are hard and firm, with a brownish or black color, in the vicinity of the coal layer. The sandstones are either of a grayish, bluish, or reddish color.

These various layers constituting coal beds, are sometimes nearly or quite horizontal in position, as in New Holland and west of the Appalachians. They are very often much tilted, dipping at various angles and sometimes vertical, as is generally the case throughout central Pennsylvania; and in some cases the beds are raised in immense folds, as the leaves of a book may be folded, by a sidewise pressure. They are very commonly intersected by fractures, along which the coal seam on one side is higher or lower than on the other, owing to a dislocation, (then said to be *faulted*); and miners working in a bed for a while, in such a case, find it to terminate abruptly, and have to explore above or below for its continuation. These are points of great importance in the mining of coal.

There is no infallible indication of the presence of coal distinguishable in the mineral nature of rocks; for just such rocks as are here described occur where no coal is to be found, and where none is to be expected. The presence of fossil leaves of ferns, and of plants having jointed stems or a scarred or embossed surface, in the shales or sandstone, is a useful hint; the discovery of the coal itself a much better one. The geologist ascertains the absence of coal from a region by examining the fossils in the rocks; these fossils being different in rocks of different ages, they indicate at once whether the beds under investigation belong to what is called the *coal series*. If they contain certain trilobites, and other species which are found only in more ancient rocks, there is no longer a doubt that coal is not to be obtained in any workable quantities; and he arrives at the same conclusion if the remains are those of more recent

---

What is said of the position of the beds? How do the rocks indicate whether coal is to be expected in a region or not?

rocks, such as fossil fish of certain genera, or the remains or traces of birds or quadrupeds, or of such species of shells as never occur as low in the rocks as true coal beds. But if the fossils are such as have been described as characterizing a coal series, there is then reason for exploration. It is impossible in this place to give such knowledge as will be practically useful. The inquirer must refer to treatises on geology, or better to the practical geologist, whose judgment in such questions might often have saved much useless mining and wasted expenditure.

Mineral coal is very widely distributed over the world. England, France, Spain, Portugal, Belgium, Germany, Austria, Sweden, Poland and Russia, have their beds of mineral coal. It is also abundant in India, China, Madagascar, Van Diemen's Land, Borneo and other East India Islands, New Holland, and at Conception in Chili. But no where is the coal formation more extensively displayed than in the United States, and in no part of the world are its beds of greater thickness, more convenient for working, or more valuable in quality. There are four extensive areas occupied by this formation. One of these areas commences on the north, in Pennsylvania and southeastern Ohio, and sweeping south over western Virginia and eastern Kentucky and Tennessee, to the west of the Apalachians, or partly involved in their ridges, it continues to Alabama near Tuscaloosa, where a bed of coal has been opened. It has been estimated to cover 63,000 square miles. It embraces several isolated patches in the eastern half of Pennsylvania. A second coal area (the Illinois) lies adjoining the Mississippi, and covers the larger part of Illinois, the western part of Indiana, and a small northwest part of Kentucky; it is but little smaller than the preceding. A third occupies a portion of Missouri west of the Mississippi. A fourth covers the central portion of Michigan. Besides these, there is a smaller coal region (a fifth) in Rhode Island, which appears near Portsmouth, not far from the railroad to Boston, and also in Mansfield, Massachusetts. Out of the borders of the United States, on the northeast, commences a sixth coal area, that of Nova Scotia and New Brunswick, which covers 10,000 square miles,

---

What is said of the distribution of coal over the globe? How many coal areas are there in the United States, and what their positions? What is said of the Nova Scotia and New Brunswick coal beds?

2500 square miles of which are in Nova Scotia. At Cape Breton is still another field of coal.

The coal of Rhode Island and eastern Pennsylvania is anthracite. Going west in Pennsylvania, the anthracite becomes more and more bituminous; and at Pittsburg, at its western extremity, as also throughout the western states, it is wholly of the *bituminous* kind. The Rhode Island variety is so hard and compact and free from all volatile ingredients, that for many years it had been deemed unfit for use. The anthracite of eastern Pennsylvania affords 3 to 6 per cent. of aqueous vapor, and 1 to 4 per cent. of volatile combustible matter. In the Bradford coal field, lying near the eastern limits of the bituminous coal deposits, Prof. Johnson obtained 1 to 8 per cent. of moisture, 9 to 15 per cent. of incondensable gas, 5 to 17 of earthy matter, and 62 to 75 of carbon. In the bituminous coal of the Portage railroad, Cambria county, Penn., he obtained 18·2 per cent. of volatile combustible matter; in that of Caseyville, Ky., and Cannelton, Indiana, 30 to 34 per cent.; and in a coal from Osage river, Missouri, 41·35 per cent. The general fact that the proportion of bitumen increases as we go westward, is here well exhibited.

Some of these results, derived from an extensive series of experiments, are thus averaged by Prof. Johnson:

	Moisture.	*Vol. Combustible Matter.*	*Ashes and Clinker.*	*Fixed Carbon.*
Pennsylvania anthracites,	1·34	3·84	7·37	87·45
Maryland free burning bituminous coal	1·25	15·80	9·94	73·01
Pennsylvania free burning bituminous coal,	0·82	17·01	13·35	68·82
Virginia bituminous,	1·64	36·63	10·74	50·99
Cannelton, Indiana, bituminous,	2·20	33·99	4·97	58·44

It has also been shown that this fact is connected with the geological condition of the country, the anthracite occurring in the east where the rocks are variously uplifted and thrown out of position by subterranean forces, evincing also other

What is the relative geographical position of the anthracite and bituminous coal in the United States? What has probably made the difference in these two kinds of coal?

effects of heat besides this debituminisation of the coal; while the bituminous coal occurs where such disturbances of the rocks have not taken place: and the amount of bitumen increases as we recede from the region of greatest disturbance. The heat and attendant siliceous solutions have therefore been the means of giving unusual hardness to the Rhode Island coal.

Owing to the various upliftings or foldings of the strata and subsequent denudations, the beds are often exposed to view in the sides of hills or ridges, and the coal in Pennsylvania is in most cases rather quarried out than mined. The layers are at times 20 to 35 feet thick, without any slaty seams, and the excavations appear like immense caverns, whose roofs are supported by enormous columns of coal, "into which a coach and six might be driven and turned again with ease."

Besides the great coal beds of the *coal era*, as it is significantly called, there are small beds, sometimes workable, of a more recent date. The bed near Richmond, Va., belongs to a subsequent period; there are also beds in Yorkshire, and at Brora in Sutherland. Tertiary coal occurs in Provence, and also in Oregon on the Cowlitz. These beds of more recent coals are seldom sufficiently extensive to pay for working, and are often much contaminated by pyrites.

The amount of anthracite worked in 1820, in Pennsylvania, was only 380 tons; in 1847, it amounted to more than 3,000,000 tons; and the whole amount of both anthracite and bituminous coal worked in that state, in 1847, was not less than 5,000,000 tons. In Great Britain, the annual amount of coal mined is about 35,000,000 of tons.

The uses of mineral coal are well known. The Pennsylvania anthracite was first introduced into blacksmithing in 1768 or 1769, by Judge Obadiah Gore, a blacksmith, who early left Connecticut for Wilkesbarre. It is now employed in smelting iron ores, and for nearly every purpose in the arts for which charcoal was before employed.

The formation of *coke* from pit coal, for smelting iron, is done in close furnaces or ovens. After heating up, the coal (about two tons) is thrown in at a circular opening at top, and remains for 48 hours; the doorway is gradually closed to shut off the air as the combustion increases, and finally the atmosphere is wholly shut off, and in this condition it

---

How is coke prepared?

remains for 12 hours. The volatile matter is thus expelled, and the cokes produced are ponderous, extremely hard, of a light gray color, and having a metallic luster. To make another kind of coke, like charcoal, the pit coal is placed in a receptacle more like a baker's oven, and the air has more free access. Both of these kinds of coke are used in smelting.

GRAPHITE.—*Plumbago.*

Occasionally in six-sided prisms, with a transversely foliated structure. Usually foliated, and massive; also granular and compact.

Luster metallic, and color iron black to dark steel gray. Thin laminæ flexible. H=1—2. Gr=2·09. Soils paper, and feels greasy.

Composition. 90 to 96 per cent. of carbon, with the rest iron. Some specimens from Brazil contain scarcely a trace of iron. It is often called *carburet of iron*, but is not a chemical compound. It is infusible before the blowpipe, both alone and with reagents; it is not acted upon by acids.

Dif. Resembles molybdenite, but differs in being unaffected by the blowpipe and acids. The same characters distinguish the granular varieties from any metallic ores they resemble.

Obs. Graphite (called also *black lead*) is found in crystalline rocks, especially in gneiss, mica slate and granular limestone; also in granite and argillite, and rarely in greenstone. Its principal English locality is at Borrowdale, in Cumberland. Ure observes that this mineral became so common a subject of robbery, a century ago, as to have enriched many living in the neighborhood; a body of miners would break into the mine and hold possession of it for a considerable time. The place is now protected by a strong building, and the workmen are required to put on a working dress in an apartment on going in and take it off on coming out. In an inner room two men are seated at a large table assorting and dressing the graphite, who are locked in while at work and watched by the steward from an adjoining room, who is armed with two loaded blunderbusses. This is deemed necessary to check the pilfering spirit of the Cum-

---

What is the appearance of graphite? What is its prominent characteristic? its composition? Where does it occur? Where is it worked in England?

berland mountaineers. In some years the net produce of the *six weeks'* annual working of the mine, has amounted to £40,000.

In the United States, graphite occurs in large masses in veins in gneiss at Sturbridge, Mass. It is also found in North Brookfield, Brimfield and Hinsdale, Mass.; at Roger's rock, near Ticonderoga; near Fishkill landing in Dutchess county; at Rossie, in St. Lawrence county, and near Amity, in Orange county, N. Y.; at Greenville, L. C.; in Cornwall, near the Housatonic, and in Ashford, Ct.; near Attleboro, in Buck's county, Penn.; in Brandon, Vermont; in Wake, North Carolina; on Tyger river, and at Spartanburg, near the Cowpens furnace, South Carolina.

For the manufacture of pencils the granular graphite has been preferred, and it is this character of the Borrowdale graphite which has rendered it so valuable. At Sturbridge, Mass., it is rather coarsely granular and foliated, and has been extensively worked; the mine yields annually about 30 tons of graphite. The mines of Ticonderoga and Fishkill landing, N. Y.; of Brandon, Vt.; and of Wake, North Carolina, are also worked; and that of Ashford, Ct., formerly afforded a large amount of graphite, though now the works are suspended.

The material for lead pencils, when of the finest quality, is first calcined and then sawn up into strips of the requisite size and commonly set in wood, (usually cedar,) as they appear in market. It is much used now in small cylinders without wood for ever-pointed pencil cases. Graphite of coarser quality, according to a French mode, is ground up fine and calcined, and then mixed with the finest levigated clay, and worked into a paste with great care. It is made darker or lighter and of different degrees of hardness, by varying the proportion of clay and the degree of calcination to which the mixture is subjected; and the hardness is also varied by the use of saline solutions. Lampblack is sometimes addded with the clay.

A superior method in use at Taunton, Mass., where the Sturbridge graphite is extensively employed, consists in finely pulverising it, and then by a very heavy pressure obtained by machinery, condensing it into thin sheets. These

---

How are the best lead pencils made? How are they manufactured from the Sturbridge bed?

sheets are then sawn up of the size required. The pencil is pure graphite, and the foliated variety is preferred on account of its being freer from impurities.

Graphite is extensively employed for diminishing the friction of machinery; also for the manufacture of crucibles and furnaces, and as a wash for giving a gloss to iron stoves and railings. For crucibles it is mixed with half its weight of clay.

CARBONIC ACID.

Carbonic acid is the gas that gives briskness to the Saratoga and many other mineral waters, and to artificial soda water. Its taste is slightly pungent. It extinguishes combustion and destroys life. *Composition:* carbon 27·65, oxygen 72·35.

Besides occurring in mineral waters, it is common about some volcanoes. The *Grotto del Cane* (Dog cave) near Naples, is a small cavern filled to the level of the entrance with this gas. It is a common amusement for the traveler to witness its effects upon a dog kept for the purpose. He is held in the gas a while and is then thrown out apparently lifeless; in a few minutes he recovers himself, picks up his reward, a bit of meat, and runs off as lively as ever. If continued in the carbonic acid gas a short time longer life would have been extinct.

Carbonic acid combined with lime forms carbonate of lime or common limestone; with oxyd of iron it constitutes spathic iron, one of the common ores of iron; with oxyd of zinc, it forms calamine, the most profitable ore of zinc. It is found in combination also in various other minerals.

AMBER.

In irregular masses. Color yellow, sometimes brownish or whitish; luster resinous. Transparent to translucent. H=2—2·5. Gr=1·18. Electric by friction.

Composition. Carbon 79·0, hydrogen 10·5. oxygen 10·5, Burns with a yellow flame and aromatic odor.

Obs. Occurs in alluvium and on coasts, in masses from a very small size to that of a man's head. In the Royal Museum at Berlin, there is a mass weighing 18 pounds. On

---

For what other purposes is it used? What is carbonic acid? Combined with lime, what does it form? What is the appearance of amber? Where does it occur?

the Baltic coast it is most abundant, especially between Königsberg and Memel. It is met with at one place in a bed of bituminous coal; it also occurs on the Adriatic, in Poland, on the Sicilian coast near Catania, in France near Paris in clay, in China. It has been found in the United States, at Gay Head, Martha's Vineyard, Camden, N. J., and at Cape Sable, near the Magothy river, in Maryland.

It is supposed with good reason to be a vegetable resin, which has undergone some change while inhumed, a part of which is due to acids of sulphur proceeding from decomposing pyrites or some other source. It often contains insects, and specimens of this kind are so highly prized as frequently to be imitated for the shops. Some of the insects appear evidently to have struggled after being entangled in the then viscous resin, and occasionally a leg or a wing is found some distance from the body, having been detached in the struggle for escape.

Amber is the *elektron* of the Greeks; from its becoming electric so readily when rubbed, it gave the name electricity to science. It was also called *succinum*, from the Greek *succum*, juice, because of its supposed vegetable origin.

Uses. Amber admits of a good polish and is used for ornamental purposes, though not very much esteemed, as it is wanting in hardness and brilliancy of luster, and moreover is easily imitated. It is much valued in Turkey for mouthpieces to their pipes.

Amber is the basis of an excellent transparent varnish. After burning, there is left a light carbonaceous residue, of which the finest black varnish is made. Amber affords by distillation an oil called *oil of amber*, and also *succinic acid;* and as the preparation of amber varnish requires that the amber be heated or fused, these products are usually obtained at the time.

MINERAL CAOUTCHOUC.—*Elastic Bitumen.*

In soft flexible masses, somewhat resembling caoutchouc or India rubber. Color brownish black; sometimes orange red by transmitted light. Gr=0·9—1·25.

Composition: carbon 85·5, hydrogen 13·3. It burns readily with a yellow flame and bituminous odor.

What is said of the origin of amber? What term has it given to science? For what is amber used? What is mineral caoutchouc?

Obs. From a lead mine in Derbyshire, England, and a coal mine at Montrelais. It has been found at Woodbury, Ct., in a bituminous limestone.

RETINITE.—*Retinasphaltum.*

In roundish masses. Color light yellowish brown, green, red; luster earthy or slightly resinous in the fracture. Subtransparent to opaque, Often flexible and elastic when first dug up, but loses these qualities on exposure. H=1—2·5. Gr=1·135.

Composition: vegetable resin 55, bitumen **41**, earthy matter 3. Takes fire in a candle and burns with a bright flame and fragrant odor. The whole is soluble in alcohol except an unctuous residue.

Obs. Accompanies Bovey coal at Devonshire; also found with brown coal at Wolchow in Moravia, and near Halle.

BITUMEN.

Both solid and fluid. Odor bituminous. Luster resinous; of surface of fracture often brilliant. Color black, brown or reddish when solid; fluid varieties nearly colorless and transparent. H=0—2. Gr=0·8—1·2.

VARIETIES:

Mineral pitch or *Asphaltum.* The massive variety, often breaking with a high luster like hardened tar. The *earthy* mineral pitch includes less pure specimens.

Petroleum. A fluid bitumen of a dark color, which oozes from certain rocks and becomes solid on exposure. A less fluid variety is called *maltha*, or *mineral tar.*

Naphtha, or *mineral oil.* A limpid or yellowish fluid, lighter than water; specific gravity 0·7—0·84. It hardens and changes to petroleum on exposure. It may be obtained from petroleum by heat, which causes it to pass off in vapor.

Composition of naphtha: carbon 82·2, hydrogen 14·8. The above varieties burn readily with flame and smoke.

Obs. Asphaltum is met with abundantly on the shores of the Dead Sea, and in the neighborhood of the Caspian. A very remarkable locality occurs on the island of Trinidad, where there is a lake of it about a mile and half in circumference. The bitumen is solid and cold near the shores; but gradually increases in temperature and softness towards

---

Describe bitumen. What is asphaltum? petroleum? naphtha? What is said of the asphaltum of Trinidad?

the center, where it is boiling. The appearance of the solidified bitumen is as if the whole surface had boiled up in large bubbles and then suddenly cooled. The ascent to the lake from the sea, a distance of three quarters of a mile, is covered with the hardened pitch, on which trees and vegetation flourish, and here and there about Point La Braye, the masses of pitch look like black rocks among the foliage.

Large deposits of asphaltum occur in sandstone in Albania. It is also found in Derbyshire, and with quartz and fluor in granite in Cornwall; in cavities of chalcedony and calc spar in Russia and other places.

Naphtha issues from the earth in large quantities in Persia and the Birman empire. At Rangoon, on one of the branches of the Irawady river, there are upwards of 500 naphtha and petroleum wells which afford annually 412,000 hogsheads. In the peninsula of Apcheron on the western shore of the Caspian, naphtha rises through a marly soil in vapor, and is collected by sinking pits several yards in depth, into which the naphtha flows. Near Amiano in the state of Parma, there is an abundant spring.

In the United States petroleum is common. The salines of Kenawha, Va.; Scotsville, Ky.; Oil creek, Venango county, Penn.; Duck creek, Monroe county; near Hinsdale in Allegany county, N. Y., and Liverpool, Ohio, are among its localities. It was formerly collected for sale by the Seneca and other Indians; the petroleum is therefore commonly called *Genesee* or *Seneca oil*, under which name it is sold in market. The *Rock oil* of commerce is Naphtha.

Uses. Bitumen in all its varieties was well known to the ancients. It is reported to have been employed as a cement in the construction of the walls of Babylon. At Agrigentum it was burnt in lamps and called *Sicilian oil.* The Egyptians made use of it in embalming.

The asphaltum of Trinidad mixed with grease or common pitch is used for pitching (technically, *paying*) the bottoms of ships; and it is supposed to protect them from the Teredo. Two ship loads of the pitch were sent to England by Admiral Cochrane; but it was found that the oil required to fit it for use exceeded in expense the cost of pitch in England;

Where is naphtha obtained? What is Seneca oil? For what is asphaltum used?

and consequently the project of employing it in the arts was abandoned.

Asphaltum is a constituent of the kind of black varnish called Japan. It is used in France in forming a cement for covering the roofs and lining water cisterns. A limestone, thoroughly dried, is ground up fine and stirred well in a vessel containing about one-fifth its weight of hot melted bitumen. It is then cast into rectangular moulds, which are first smeared with loam to prevent adhesion. When cold, the frame of the mould is taken apart and the block removed.

Petroleum is used in Birmah as lamp oil; and when mixed with earth or ashes, as fuel. Naphtha affords both fuel and light to the inhabitants of Batku on the Caspian. The vapor is made to pass through earthen tubes and is inflamed as it passes out and used in cooking. The spring near Amiano is used for illuminating the city of Genoa. Both petroleum and naphtha have been employed as a lotion in cutaneous eruptions, and as an embrocation in bruises and rheumatic affections. Naphtha is often substituted for oil in oil paint, on account of its drying quickly. It is also employed for preserving the metals of the alkalies, potassium and sodium, which, owing to their tendency to unite with oxygen, cannot be kept in any liquid that contains this gas.

The petroleum or Seneca oil of western New York, Pennsylvania and Ohio, as it appears in the market, is of a dark brown color, and a consistency between that of tar and molasses.

The following are the names of other kinds of fossil resin or wax:—*Fossil Copal, Middletonite, Piauzite,* which are resinous and nearly or quite insoluble in alcohol; *Guyaquillite* and *Berengelite,* from South America, resinous and soluble in alcohol like Retinite; *Scheererite, Hatchetine, Dysodile, Hartite, Ixolyte, Ozocerite, Fichtelite, Konlite, Branchite,* found with coal, especially brown coal, and resembling wax or tallow. *Idrialine* is grayish or brownish black with a grayish luster, and occurs at the Cinnabar mines of Idria.

CLASS IV.—SULPHUR.

Sulphur exists abundantly in the native state. It occurs combined with various metals, forming sulphurets and sulphates; and the sulphurets especially are very common ores. The sulphuret of iron is common iron pyrites; sulphuret of copper is the yellow copper ore of Cornwall and other regions; sulphuret of mercury is cinnabar, the ore from which mercury is mostly obtained; sulphuret of lead is galena, the usual ore of lead. It is also sparingly met with in the condition of sulphuric and sulphurous acids.

NATIVE SULPHUR.

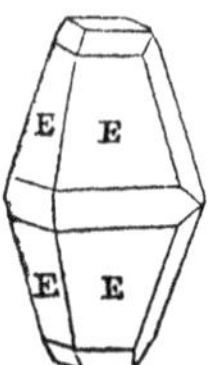

Trimetric. In acute octahedrons, and secondaries to this form, with imperfect octahedral cleavage. Also massive.

Color and streak sulphur yellow, sometimes orange yellow. Luster resinous. Transparent to translucent. Brittle. H=1·5—2·5. Gr=2·07.

Native sulphur is either pure or contaminated with clay or bitumen. It sometimes contains selenium, and has then an orange yellow color.

Dif. It is easily distinguished by burning with a blue flame and a sulphur odor.

Obs. The great repositories of sulphur are either beds of gypsum and the associate rocks, or the regions of active or extinct volcanoes. In the valley of Noto and Mazzaro in Sicily, at Conil near Cadiz in Spain, Bex in Switzerland, and Cracow in Poland, it occurs in the former situation. Sicily and the neighboring volcanic islands, Vesuvius and the Solfatara in its vicinity, Iceland, Teneriffe, Java, Hawaii, New Zealand, Deception island, and most active volcanic regions afford more or less sulphur. The native sulphur of commerce is brought mostly from Sicily, where it occurs in beds along the central part of the south coast and to some distance inland. It is often associated with fine crystals of sulphate of strontian. It undergoes rough purification by fusion before exportation, which separates the earth and clay with which it occurs. Sixteen or seventeen thousand tons are annually imported from Sicily into England alone. Sulphur is also exported from the crater of Vulcano, one of the Lipari islands, and from the Solfatara near Naples.

On the Potomac, 25 miles above Washington, fine specimens of sulphur are found associated with calc spar in a gray compact limestone. Sulphur is also found as a deposit about springs where sulphureted hydrogen is evolved, and in cavities where iron pyrites have decomposed. Localities of the

---

What is the crystallization of sulphur? Mention its other characters. Where is the sulphur of the arts obtained?

former kind are common in the state of New York, and of the latter in the coal mines of Pennsylvania, the gold rocks of Virginia and elsewhere.

The sulphur of commerce is also largely obtained from copper and iron pyrites, it being given off during the roasting of these ores, and collected in chambers of brick work connected with the reverberatory furnace. It is afterwards purified by fusion and cast into sticks.

Sulphur when cooled from fusion, or above 232° F., crystallizes in oblique rhombic prisms. When poured into water at a temperature above 300° F. it acquires the consistency of soft wax, and is used to take impressions of gems, medals, &c., which harden as the sulphur cools.

The uses of sulphur for gunpowder, bleaching, the manufacture of sulphuric acid, and also in medicines, are well known. Gunpowder contains 9 to 20 per cent.—9 or 10 per cent. for the best shooting powder, and 15 to 20 for mining powder.

SULPHURIC AND SULPHUROUS ACIDS.

Sulphuric acid is occasionally met with around volcanoes, and it is also formed from the decomposition of sulphureted hydrogen about sulphur springs. It is intensely acid. *Composition*, sulphur, 40·14, oxygen 59·86. It is said to occur in the waters of Rio Vinagro, South America; also in Java, and at Lake de Taal on Luzon in the East Indies; in Genesee Co., N. Y.; and at Tuscarora, St. Davids and elsewhere, Canada West.

Sulphurous acid is produced when sulphur burns, and causes the odor perceived during the combustion. It is common about active volcanoes. It destroys life and extinguishes combustion. *Composition*, sulphur 50·00, oxygen 49·00.

SELENIUM, ARSENIC. Selenium has close relations to sulphur. Its most striking characteristic is the horse-radish odor perceived when it is heated. It occurs in nature combined like sulphur with various metals, and these ores, called *seleniets* or *seleniurets*, are at once distinguished by the odor when subjected to the heat of the blowpipe flame.

Arsenic is also near sulphur in a chemical point of view, although metallic in luster. It forms similar compounds with the metals and metallic oxyds, which are called *arseniurets* and are often highly important ores. The *arseniurets* of nickel and cobalt are the main sources of these metals. Its ores are distinguished by giving off when heated an odor resembling garlic.

What is said of sulphuric acid? What is said of sulphurous acid?

Tellurium and *Osmium* are other metals having chemical relations to sulphur. They form similar compounds with the metals. They are of rare occurrence.

The minerals containing the elements arsenic, selenium, tellurium and osmium, are described under Class VII, including metals and metallic ores.

CLASS V.—HALOID MINERALS.

1. AMMONIA.

The salts of ammonia are more or less soluble, and are entirely and easily dissipated in vapor before the blowpipe. By this last character they are distinguished from other salts.

SAL AMMONIAC.—*Muriate of Ammonia.*

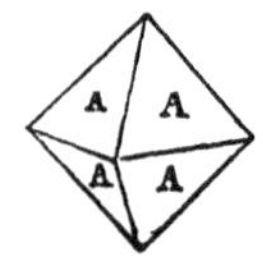

Occurs in white crusts or efflorescences, often yellowish or gray. Crystallizes in regular octahedrons. Translucent—opaque; taste saline and pungent. Soluble in three parts of water.

Composition: ammonium 33·7, chlorine 66·3. Gives off the odor of *hartshorn* when powdered and mixed with quicklime.

Dif. Distinguished by the odor given off when heated along with quicklime.

Obs. Occurs in many volcanic regions, as at Etna, Vesuvius, and the Sandwich Islands, where it is a product of volcanic action. Occasionally found about ignited coal seams.

But the sal ammoniac of commerce is manufactured from animal matter or coal soot. It is generally formed in chimneys of both wood and coal fires. In Egypt, whence the greater part of this salt was formerly obtained, the fires of the peasantry are made of the dung of camels; and the soot which contains a considerable portion of the ammoniacal salt is preserved and carried in bags to the works, where it is obtained by sublimation. Bones and other animal matters are used in France, and a liquor condensed from the gas works, in England.

What are general characters of the salts of ammonia? What is a distinctive character of sal ammoniac? What is its composition? From what is it manufactured? How is it manufactured in Egypt?

Uses. It is a valuable article in medicine, and is employed by tinmen in soldering; also, mixed with iron filings or turnings to pack the joints in steam apparatus.

Mascagnine—Sulphate of Ammonia. In mealy crusts, of a yellowish-gray or lemon-yellow color. Translucent. Taste pungent and bitter. *Composition,* sulphuric acid 53·3, ammonia 22·8, water 23·9. Easily soluble in water. Occurs at Etna, Vesuvius, and the Lipari Islands. It is one of the products from the combustion of anthracite coal

Phosphate of ammonia, bicarbonate of ammonia, and *phosphate of magnesia and ammonia* have been found native in guano. The last is identical with *struvite.*

Struvite. A phosphate of ammonia and magnesia, containing 13 per cent. of water. It occurs in yellowish subtransparent rhombic crystals. G=1·7. H=1. Slightly soluble in water. Found on the site of an old church in Hamburg, where there had been quantities of cattle dung.

2. POTASSA.

NITER.—*Nitrate of Potash.*

Trimetric. In modified right rhombic prisms. M : M 118° 50′. Usually in thin white subtransparent crusts, and in needleform crystals on old walls and in caverns. Taste saline and cooling.

Composition: potassa 46·56, nitric acid 53·44. Burns vividly on a live coal.

Dif. Distinguished readily by its taste and its vivid action on a live coal; and from nitrate of soda, which it most resembles, by its not becoming liquid on exposure to the air.

Uses. Niter, called also saltpeter, is employed in making gunpowder, forming 75 to 78 per cent. in shooting powder, and 65 in mining powder. The other materials are sulphur (12 to 15 per cent.) and charcoal, (9 to 12½ for shooting powder, and 20 for mining.) It is also extensively used in the manufacture of nitric and sulphuric acids; also for pyrotechnic purposes, fulminating powders, and sparingly in medicine.

Obs. Occurs in many of the caverns of Kentucky and other Western States, scattered through the earth that forms the floor of the cave. In procuring it, the earth is lixiviated, and the lye, when evaporated, yields the saltpeter. India is its most abundant locality, where it is obtained largely for

---

What does niter consist of? What effect is produced when it is put on a live coal? What are its uses? Where does it occur?

exportation. It is there used for making a cooling mixture, an ounce of powdered niter in five ounces of water reduces the temperature 15° F.

Spain and Egypt also afford large quantities of niter for commerce. This salt forms on the ground in the hot weather succeeding copious rains, and appears in silky tufts or efflorescences; these are brushed up by a kind of broom, lixiviated, and after settling, evaporated and crystallized. In France, Germany, Sweden, Hungary and other countries, there are artificial arrangements called *nitriaries* or niter-beds, from which niter is obtained by the decomposition mostly of the nitrates of lime aud magnesia which form in these beds. Refuse animal and vegetable matter putrified in contact with calcareous soils produces nitrate of lime, which affords the niter by reaction with carbonate of potash. Old plaster lixiviated affords about 5 per cent. This last method is much used in France.

Chlorid of potassium, or *sylvine,* has been observed with salt at Saltzburg.

3. SODA.

The following salts of soda are all more or less soluble: they are in general distinguished by giving a deep yellow light before the blowpipe. Hardness below 3; specific gravity below 2·9.

GLAUBER SALT.—*Sulphate of Soda.*

Monoclinic. In oblique rhombic prisms. Occurs in efflorescent crusts of a white or yellowish-white color; also in many mineral waters. Taste cool, then feebly saline and bitter. *Composition,* soda 19·3, sul. acid 24·8, water 55·9.

Dif. It is distinguished from Epsom salt, for which it is sometimes mistaken, by its coarse crystals, and the yellow color it gives to the blowpipe flame.

Uses. It is used in medicine, and is known by the familiar name of "salts."

Obs. On Hawaii, one of the Sandwich Islands, in a cave at Kailua, glauber salt is abundant, and is constantly forming. It is obtained by the natives and used as medicine. Glauber

---

What is a *nitriary?* What effect is produced on the blowpipe flame by soda? What is its composition? How is it distinguished from Epsom salt? Where does Glauber salt occur native?

salt occurs also in efflorescences on the limestone below Genesee Falls, near Rochester, N. Y. It is also obtained in Austria, Hungary and elsewhere in Europe.

The artificial salt was first discovered by a German chemist by the name of Glauber. It is usually prepared for the arts from sea water.

NITRATE OF SODA.

Rhombohedral; R: R=106° 33′. Also in crusts or efflorescences, of white, grayish and brownish colors; taste cooling. Soluble and very deliquescent.

Composition: nitric acid 63·5, soda 36·5. Burns vividly on coal, with a yellow light.

Dif. It resembles niter, (saltpeter,) but deliquesces, and gives a deep yellow light when burning.

Obs. In the district of Tarapaca, the dry Pampa for an extent of forty leagues is covered with beds of this salt, mixed with gypsum, common salt, Glauber salt and remains of recent shells. The country appears to have been under the sea at no very remote period.

Uses. It is used extensively in the manufacture of nitric acid or aqua fortis.

NATRON.—*Carbonate of Soda.*

Monoclinic. Generally in white efflorescent crusts, sometimes yellowish or grayish. Taste alkaline. Effloresces on exposure, and the surface becomes white and pulverulent.

Composition: a simple hydrous carbonate of soda. Effervesces strongly with nitric acid.

Dif. Distinguished from other soda salts by effervescing, and from Trona, by efflorescing on exposure.

Obs. Abundant in the soda lakes of Egypt, situated in a barren valley called Bahr-bela-ma, about 30 miles west of the Delta. Also in lakes at Debreczin in Hungary; in Mexico, north of Zacatecas, and elsewhere. Sparingly dissolved in the Seltzer and Carlsbad waters.

Trona is a *sesquicarbonate* of soda. In the province of Suckenna in Africa, between Tripoli and Fezzan, it forms a

---

How does nitrate of soda differ in composition from niter? What are other peculiarities distinguishing it? For what is it used? Where does it occur native? What are the distinctive characters of carbonate of soda?

fibrous layer an inch thick beneath the soil, and several hun dred tons are collected annually. At a lake in Maracaibo, 48 miles from Merido, it is very abundant.

Uses. Carbonate of soda is used extensively in the manufacture of soap. The powders put up for making soda water consist of this salt and tartaric acid. On mixing the two, the tartaric acid unites with the soda and the carbonic acid of the carbonate of soda escapes as a gas producing the effervescence. In Mexico, this salt (or the sesquicarbonate, trona) occurs in such abundance over extensive districts that it is employed as a flux in smelting ores of silver, especially the chlorid of silver which is a common ore.

COMMON SALT.

Monometric. In cubes (fig 1) and its secondaries, as the following. Sometimes crystals have the shape of a shallow

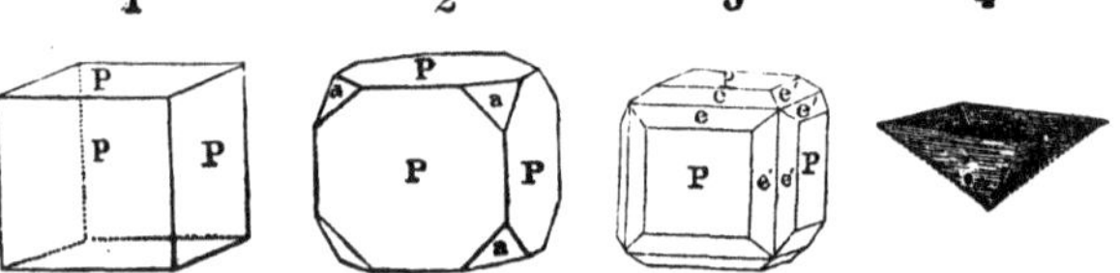

cup like figure 4, and are called hopper shaped crystals. They were formed floating; the cup receiving its enlargement at the margin, this being the part which lay at the surface of the brine where evaporation was going on. Common salt is usually white or grayish, but sometimes presents rose red, yellow and amethystine tints. H=2. Gr=2·257. Taste saline.

Composition: chlorine 60·7, sodium 39·3. Crackles or decrepitates when heated.

Dif. Distinguished by its taste, solubility, and blowpipe characters.

Obs. Salt is usually associated with gypsum, and clays or sandstone. It occurs in extensive beds in Spain, in the Pyrenees, in the valley of Cardona and elsewhere, forming hills 300 to 400 feet high; in Poland at Wieliczka; at Hall in the Tyrol, and along a range through Reichenthal in Bavaria,

---

For what is it used? What happens when tartaric acid and carbonate of soda are mixed? What are the forms of crystals of common salt? Of what does it consist? Where are some of the most remarkable deposits of rock salt?

Hallein in Saltzburg, Hallstadt, Ischel and Ebensee in Upper Austria, and Aussee in Stiria; in Hungary at Marmoros and elsewhere; in Transylvania; Wallachia, Gallicia and Upper Silesia; at Vic and Dieuze in France; at Bex in Switzerland; in Cheshire, England; in northern Africa in vast quantities, forming hills and extended plains; in northern Persia at Teflis; in India in the province of Lahore, and in the valley of Cashmere; in China and Asiatic Russia; in South America, in Peru and the Cordilleras of New Grenada.

The most remarkable deposits are those of Poland and Hungary. The former, near Cracow, has been worked since the year 1251, and it is calculated that there is still enough salt remaining to supply the whole world for many centuries. Its deep subterranean regions are excavated into houses, chapels and other ornamental forms, the roof being supported by pillars of salt; and when illuminated by lamps and torches, they are objects of great splendor.

The salt is often impure with clay, and is purified by dissolving it in large chambers, drawing it off after it has settled and evaporating it again. The salt of Norwich (in Cheshire) is in masses 5 to 8 feet in diameter, which are nearly pure, and it is prepared for use by crushing it between rollers.

Beds of salt have lately been opened in Virginia in Washington county, where as usual it is associated with gypsum. The Salmon mountains of Oregon also afford rock salt.

Salt beds occur in rocks of various ages: the brines of the United States come from a red sandstone below the coal; the beds of Norwich, England, occur in magnesian limestone; those of the Vosges in marly sandstone beds of the lower secondary; that of Bex in the lias or middle secondary; that of the Carpathian Alps in the upper oolite; that of Wieliczka, Poland and the Pyrenees, in the cretaceous formation or upper secondary; that of Catalonia in tertiary: and moreover there are vast deposits that are still more recent, besides lakes that are now evaporating and producing salt depositions.

Vast lakes of salt water exist in many parts of the world. Lake Timpanogos, or Youta, called also the Great Salt Lake, has an area of 2000 square miles, and is remarkable for its extent, considering that it is situated towards the sum-

---

What is said of the beds of Cracow? How is this salt purified? Where do beds occur in North America? What is said of salt lakes?

mit of the Rocky Mountains, at an elevation of 4200 feet above the sea. The dry regions of these mountains and of the semideserts of California abound in salt licks and lakes. There is a small spring on the Bay of San Francisco. In northern Africa large lakes as well as hills of salt abound, and the deserts of this region and Arabia abound in saline efflorescences. The Dead and Caspian seas, and the lakes of Khoordistan, are salt. Over the pampas of La Plata and Patagonia there are many ponds and lakes of salt water.

The greater part of the salt made in this country is obtained by evaporation from salt springs. Those of Salina and Syracuse are well known; and many nearly as valuable are worked in Ohio and other western states. At the best New York springs a bushel of salt is obtained from every 40 gallons.—(Beck.) The springs of Onondaga county, New York, afforded in 1841 upwards of three millions of bushels of salt, and it is estimated that three hundred and twenty-two millions of gallons of brine were raised and evaporated during that year.—(Beck.) To obtain the brine, wells from 50 to 150 feet deep are sunk by boring. It is then raised by machinery, carried by troughs to the boilers, which are large iron kettles set in brickwork, and there evaporated by heat. As soon as the water begins to boil, the water becomes turbid from the deposit of calcareous salts which are also contained in salt waters, and are less soluble than the salt. These are removed with ladles, called *bittern* ladles, with the exception of what adheres firmly to the sides of the boiler. The salt is next deposited; it is then collected and carried away to drain. The liquid which remains contains a large proportion of magnesian salts, and is called *bittern* from the bitter taste of these salts. Some of the brine is also evaporated by exposure to the sun in broad, shallow vats.

This last process is extensively employed in hot climates for making salt from sea water, which affords a bushel for every 300 or 350 gallons. For this purpose a number of large shallow basins are made adjoining the sea; they have a smooth bottom of clay, and all communicate with one another. The water is let in at high tide and then shut off for the evaporation to go on. This is the simplest mode, and is

---

What is the source of the salt manufactured in the United States? How much water is necessary to procure a bushel of salt? How is the salt obtained from the brine? How much salt is afforded by sea water, and how is it obtained?

used even in uncivilized countries, as among the Pacific Islands. It is better to have a large receiving basin for the salt water, which shall detain the mechanical impurities of the water.

Martinsite is a compound of 91 per cent. of chlorid of sodium and 9 of sulphate of magnesia. It is from the salines of Stassfurth.

BORAX.—*Borate of Soda.*

Monoclinic. In right rhomboidal prisms, (see fig. 11, page 26); M : T=106 35′. Cleavage parallel with M perfect. The crystals are white and transparent with a glassy luster. H=2—2·5. Gr=1·716. Taste sweetish-alkaline.

Composition : soda 16·25, boracic acid 36·58, water 47·17. Swells up to many times its bulk and becomes opaque white before the blowpipe, and finally fuses to a glassy globule.

Obs. Borax was originally brought from a salt lake in Thibet, where it is dug in considerable masses from the edges and shallow parts of the lakes. The holes thus made in a short time become filled again with borax. The crude borax was formerly sent to Europe under the name of *tincal*, and there purified for the arts. It has also been found in Peru and Ceylon. It has of late been extensively made from the boracic acid of the Tuscan lagoons by the reaction of this acid on carbonate of soda.

Uses. Borax is used as a flux not only by the mineralogist in blowpipe experiments, but extensively in metallurgical operations, in the process of soldering, and in the manufacture of gems.

Boracic acid. Occurs in small scales, white or yellowish. Feel smooth and unctuous. Taste acidulous and a little saline and bitter. G=1·48. *Composition*, boracic acid 56·38, water 43·62. Fuses easily in the flame of a candle, tinging the flame at first green.

Found at the crater of Vulcano, and also at Sasso in Italy, whence it was called *Sassolin*. The hot vapors of the lagoons of Tuscany afford it in large quantities. The vapors are made to pass through water, which condenses them ; and the water is then evaporated by the steam of the springs, and boracic acid obtained in large crystalline flakes. It

---

What are some of the characters of borax? What is its composition? What are its effects before the blowpipe? What is it used for? Where was it originally obtained? How is it procured in Tuscany? What is boracic acid? What is said of the boracic acid lagoons of Tuscany?

still requires purification, as the best thus procured contains but 50 per cent. of the pure acid.

It is employed in the manufacture of borax. Boron occurs in nature also, in datholite, tourmaline, and a few other species, but these are not a sufficient source to be employed in the arts.

Thenardite. Thenardite is an anhydrous sulphate of soda from Espartine in Spain.

Gay-Lussite. Occurs in oblong crystals, in a lake in Maracaibo S. A.; it is a hydrous compound of the carbonates of lime and soda.

Glauberite. In oblique cystals, (usually flattened, with sharp edges,) nearly transparent and yellowish-gray in color. Taste weak, slightly saline; consists of 49 per cent. of sulphate of lime and 51 of sulphate of soda. Occurs in rock salt at Villa Rubia, Spain, and also at Aussee in Upper Austria, and Vic in France.

4. BARYTA.

The salts of baryta are distinguished by their high *specific gravity,* which ranges from 3·5 to 4·8. They resemble the salts of strontia, and some of the metallic salts. From the latter they are distinguished by giving no odor nor metallic reaction before the blowpipe, when pure. Hardness below 4.

HEAVY SPAR.—*Sulphate of Baryta.*

Trimetric. In modified rhombic and rectangular prisms, (figs. 1, 2) M : M=101° 40′; P : a=141° 10; P : *a*=127° 18′. Crystals usually tabular. Massive varieties often coarse lamellar; also columnar, fibrous, granular and compact. Luster vitreous; color white and sometimes tinged yellow, red, blue or brown. Transparent or translucent. H=2·5—3·5. Gr=4·3—4·8. Some varieties are fetid when rubbed.

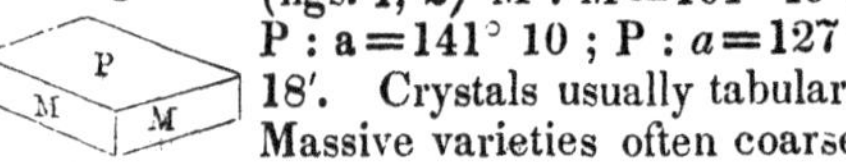

Composition: sulphuric acid 34, baryta 66. Decrepitates before the blowpipe and fuses with difficulty.

Dif. Distinguished by its specific gravity from celestine and arragonite, and also by not effervescing with acids from the various carbonates; from the metallic salts, by no metallic reaction before the blowpipe.

Obs. Heavy spar is often associated with the ores of

What is a striking character of the salts of baryta? How are they distinguished from salts of the metals? What are the forms of the crystals of heavy spar? What are the colors? What is the composition?

metals. In this way it occurs at Cheshire, Conn.; Hatfield, Mass.; Rossie and Hammond, New York; Perkiomen, Pennsylvania, and the lead mines of the west. At Scoharie and Pillar Point, near Sackett's harbor, are other localities. Also near Fredericksburg and elsewhere, Virginia. The variety from Pillar Point receives a fine polish and looks like marble, the colors being in bands or clouds.

Uses. Heavy spar is ground up and used as white paint, and in adulterating white lead. When white lead is mixed in equal parts with sulphate of barytes it is sometimes called *Venice white*, and another quality with twice its weight of barytes is called *Hamburgh white*, and another, one-third white lead, is called *Dutch white*. When the barytes is very white, a proportion of it gives greater opacity to the color, and protects the lead from being speedily blackened by sulphureous vapors; and these mixtures are therefore preferred for certain kinds of painting. There are establishments for grinding barytes near New Haven, Ct., where the spar from Cheshire, Ct., Hatfield, Mass., and Virginia, is used. The iron ore or ferruginous clay usually mixed with it, is separated by digestion in large vats of dilute sulphuric acid.

WITHERITE.—*Carbonate of Baryta.*

Trimetric. In modified rhombic prisms, (fig. 8, p. 26.) M : M=118° 30′; M : ĕ=149° 15′. Also in six-sided prisms terminated with pyramids. Cleavage imperfect. Also in globular or botryoidal forms: often massive, and either fibrous or granular. The massive varieties have usually a yellowish or grayish white color, with a luster a little resinous, and are translucent. The crystals are often white and nearly transparent. H=3—3·75. Gr=4·29—4·35. Brittle.

Composition: baryta 77·6, carbonic acid 22·4. Decrepitates before the blowpipe and fuses easily to a translucent globule, opaque on cooling. Effervesces in nitric acid.

Dif. Distinguished by its specific gravity and fusibility from calcareous spar and arragonite; by its action with acids from allied minerals that are not carbonates; by yielding no metal from white lead ore, and by not tinging the flame red, from strontianite.

What are the uses of heavy spar? How is witherite distinguished from other minerals?

Obs. The most important foreign localities of witherite are at Alstonmoor in Cumberland, and Anglezark in Lancashire.

Uses. This mineral is poisonous, and is used in the north of England for killing rats. The salts of baryta are made from this species: these salts are much used in chemical analysis; the nitrate affords a yellow light in pyrotechny; the prepared carbonate is a common water color.

Barytocalcite occurs at Alstonmoor in Cumberland, England, in whitish oblique rhombic crystals, M : M=106° 54′. H=4. G=3·6—3·7. Consists of the carbonates of lime and baryta.

Bromlite is a mineral of the same composition from Bromley Hill near Alston, and from Northumberland, England. Its crystals are *right* rhombic prisms.

Dreelite is a compound of the sulphates of baryta and lime, occurring in small white crystals in France.

Sulphato-carbonate of Baryta occurs in six-sided prisms.

5. STRONTIA.

The salts of strontia have a high *specific gravity*, it ranging from 3·6 to 4·0. In this respect they most resemble the salts of baryta, and they are distinguished by the same characters as the baryta salts from the salts of the metals. *Hardness* below 4.

CELESTINE.—*Sulphate of Strontia.*

Trimetric. In modified rhombic prisms. M : M=104° to 104° 30′. Crystals sometimes flattened; often long and slender. *a* : *a*=103° 58′. Cleavage distinct parallel with M. Massive varieties: columnar or fibrous, forming layers half an inch or more thick with a pearly luster; rarely granular. Color generally a tinge of blue, but sometimes clear white. Luster vitreous or a little pearly; transparent to translucent. H=3—3·5. Gr=3·9—4. Very brittle.

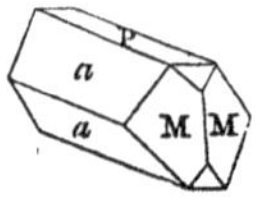

Composition: sulphuric acid 43·6, strontia 56·4. Decrepitates before the blowpipe, and on charcoal fuses rather easily to a milk white alkaline globule, tinging the flame red. Phosphoresces when heated.

How is witherite distinguished from strontianite? What are its uses? What is said of the salts of strontia? What is the usual color and appearance of celestine? What is the composition?

Dif. The long slender crystals are distinguished at once from heavy spar, as the latter does not occur in such elongated forms. From all the varieties of heavy spar, it differs in a lower specific gravity and blowpipe characters; from the carbonates it is distinguished by not effervescing with the acids.

Obs. A bluish celestine, in long slender crystals, occurs at Strontian island, Lake Erie; Scoharie, Lockport and Rossie, N. Y., are other localities. A handsome fibrous variety occurs at Franktown, Huntington county, Pennsylvania. Sicily affords very splendid crystallizations associated with sulphur: the preceding figure represents one of the crystals. The prisms are attached by one end, and being crowded over the surface, they are in beautiful contrast with the yellow sulphur beneath.

The pale *sky-blue* tint so common with the mineral, gave origin to the name celestine.

Uses. Celestine is used in the arts for making the nitrate of strontia, which is employed for producing a red color in fire-works. Celestine is changed to sulphuret of strontium by heating with charcoal, and then by means of nitric acid the nitrate is obtained.

STRONTIANITE.—*Carbonate of Strontia.*

Trimetric. In modified rhombic prisms. M : M=117° 19′. Cleavage parallel to M, nearly perfect. Occurs also fibrous and granular, and sometimes in globular shapes with a radiated structure within.

Color usually a light tinge of green; also white, gray and yellowish-brown. Luster vitreous, or somewhat resinous. Transparent to translucent. H=3·5—4. Gr=3·6—3·72. Brittle.

Composition: strontia 70·2 carbonic acid 29·8. Fuses before the blowpipe on thin edges, tinging the flame red; becomes alkaline in a strong heat; effervesces with the acids.

Dif. Its effervescence with acids distinguishes it from minerals that are not carbonates; the color of the flame before the blowpipe, from witherite; and this character and the

---

For what is celestine used? How do strontianite and celestine differ in composition? What are distinguishing characters of strontianite?

fusibility, although difficult, from calc spar. Calc spar some times reddens the flame, but not so deeply.

Obs. Strontianite occurs in limestone at Scoharie, New York, in crystals, and also fibrous and massive. Strontian in Argyleshire, England, was the first locality known, and gave the name to the mineral and the earth strontia. It occurs there with galena in stellated and fibrous groups and in crystals.

Uses. This mineral is used for preparing the nitrate of strontia, which is extensively employed for giving a red color to fire-works.

6. LIME.

With the exception of the nitrate of lime, none of the native salts of lime are soluble, unless in minute proportions. They give no odor, and no metallic reaction before the blowpipe, except such as may arise from mixture with iron or manganese. The *specific gravity* is below 3·2, and *hardness* not above 5. The few metallic salts of lime (arsenate of lime, tungstate of lime, &c.) are arranged with the metallic ores.

GYPSUM.—*Sulphate of Lime.*

2

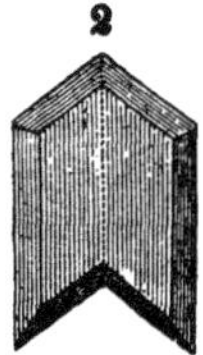

Monoclinic. Usually in right rhomboidal prisms, with beveled sides. $M : T = 111°14'$ $a : a = 143° 42'$; $e : e = 111° 42'$. Figure 2 represents a common twin (or *arrow head*) crystal. Eminently foliated in one direction and cleaving easily, affording laminæ that are flexible but not elastic. Occurs also in laminated masses, often of large size; in fibrous masses, with a satin luster; in stellated or radiating forms consisting of narrow laminæ; also granular and compact.

When pure and crystallized it is as clear and pellucid as glass, and has a pearly luster. Other varieties are gray, yellow, reddish, brownish, and even black, and opaque.

Whence the name of the mineral and earth strontia? For what is it used? What is said of the salts of lime? What are the prominent characters of gypsum?

H=1·5—2, or so soft as to be easily cut with a knife. Gr=2·31—2·33. The plates bend in one direction and are brittle in another.

Composition: lime 32·6, sulphuric acid 46·5, water 20·9. Before the blowpipe it becomes instantly white and opaque and exfoliates, and then falls to powder or crumbles easily in the fingers. At a high heat it fuses with difficulty. No action with acids.

The principal varieties are as follows:

Selenite, including the transparent foliated gypsum, so called in allusion to its color and luster from *selene*, the Greek word for *moon.*

Radiated gypsum, having a radiated structure.

Fibrous gypsum or *satin spar*, white and delicately fibrous.

Snowy gypsum and *alabaster*, including the white or light-colored compact gypsum having a very fine grain.

Dif. The foliated gypsum resembles some varieties of Heulandite, stilbite, talc and mica; and the fibrous, looks like fibrous carbonate of lime, asbestus and some of the fibrous zeolites; but gypsum in all its varieties is readily distinguished by its softness; its becoming an opaque white powder immediately and without fusion before the blowpipe, and by not effervescing nor gelatinizing with acids.

Obs. New York, near Lockport, affords beautiful selenite and snowy gypsum in limestone. At Camillus and Manlius, N. Y., and in Davidson county, Tenn., are other localities. Fine crystals of the form represented in figure 1, come from Poland and Camfield, Ohio, and large groups of crystals from the St. Marys in Maryland. Troy, N. Y., also affords crystals in clay. In the mammoth cave, Kentucky, alabaster occurs in singularly beautiful imitation of flowers, leaves, shrubbery and vines. Alabaster comes mostly from Castelino in Italy, 35 miles from Leghorn. Massive gypsum occurs abundantly in New York, from Syracuse westward to the western extremity of Genesee county, accompanying the rocks which afford the brine springs; also in Ohio, Illinois, Virginia, Tennessee, Arkansas and Nova Scotia. It is abundant also in Europe.

Uses. Gypsum when burnt and ground up forms a white

What is the composition of gypsum? What is alabaster? What effect is produced by heat? How is gypsum distinguished from talc, mica and other minerals?

powder, which, after being mixed with a little water, becomes on drying, hard and compact. This ground gypsum is *plaster of Paris*, and is used for taking casts, making models, and for giving a *hard finish* to walls. *Alabaster* is cut into vases and various ornaments, statues, &c. It owes its beauty for this purpose to its snowy whiteness, translucency and fine texture. It is moreover so soft as to be cut or carved with common cutting instruments. Gypsum is ground up and used for improving soils.

ANHYDRITE.—*Anhydrous Sulphate of Lime.*

Trimetric. In rectangular prisms, cleaving easily in three directions, and readily breaking into square blocks. The figure is a side view of a crystal; M : à=124° 10′; M : a=153° 50 ; M : *e*=135° 35′. Occurs also fibrous and lamellar, often contorted; also coarse and fine granular and compact.

Color white or tinged with gray, red, or blue. Luster more or less pearly. Transparent to subtranslucent. H=2·5—3·5. Gr=2·9—3.

The crystallized varieties have been called *muriacite.* *Vulpinite* is a siliceous variety containing 8 per cent. of silex, and a little above the usual hardness, (3·5.)

Composition: lime 41 2, sulphuric acid 58·8. It is a *sulphate of lime* like gypsum, but differs in containing no water. Whitens before the blowpipe, but does not exfoliate like gypsum, and finally with some difficulty becomes covered with a friable enamel. No action with acids.

Dif. Differs from gypsum in being harder and not exfoliating when heated; from carbonate of lime and the zeolites which it sometimes resembles, in the non-action of acids, and its action before the blowpipe. Its square forms of crystallization and cleavage are also good distinguishing characters.

Obs. A fine blue crystallized anhydrite occurs with gypsum and calcareous spar in a black limestone at Lockport. Foreign localities are at the salt mines of Bex in Swit-

---

What is plaster of Paris, and how is it used? For what is alabaster used? How is gypsum employed in agriculture? How does anhydrite differ in composition from gypsum? Mention other distinguishing characters.

zerland, at Hall in the Tyrol, at Ischil in Upper Austria, Wieliczka in Poland and elsewhere.

Uses. The vulpinite variety is sometimes cut and polished for ornamental purposes.

CALCITE—*Calcareous Spar—Carbonate of Lime.*

Rhombohedral, (fig. 1.) R : R=105° 5′. Cleavage easy parallel with the faces of the fundamental rhombohedron.

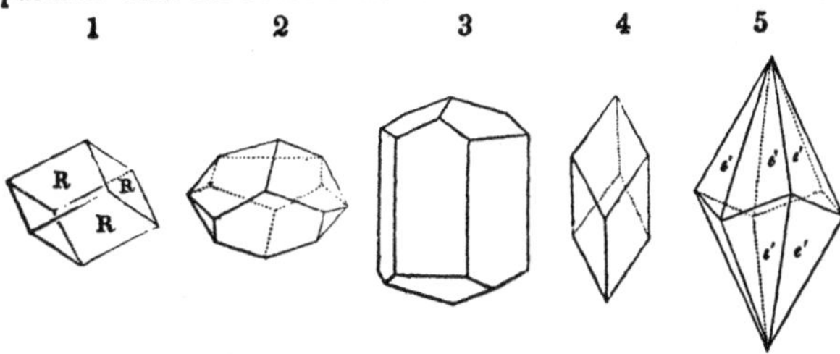

Figure 1, is the fundamental rhombohedron; figure 2, is a flat rhombohedron with the lateral angles removed, sometimes called *nail-head spar;* figure 3, is a six-sided prism; figure 4, an acute rhombohedron; figure 5, a scalene dodecahedron, the form of the variety called *dog-tooth spar.* Figures 28, 28*a*, 30, 31, page 32; 62, 63, page 39; and 66, page 40, are other forms. Calcareous spar also occurs fibrous with a silky luster, sometimes lamellar, and often coarse or fine granular and compact.

The purest crystals are transparent with a vitreous luster; the impure massive varieties are often opaque, and without luster, or even earthy. The colors of the crystals are either white or some light grayish, reddish or yellowish tint, rarely deep red; occasionally topaz yellow, rose or violet. The massive varieties are of various shades from white to black, generally dull unless polished. H=3. Gr=2·5—2·8.

Composition: lime 56·0, carbonic acid 44·0: sometimes impure from mixture with iron, silica, clay, bitumen and other minerals. Infusible before the blowpipe, but gives out an intense light, and is ultimately reduced to quicklime. Effervesces with the acids. Many varieties phosphoresce when heated.

What is the fundamental form of calcite or calc spar? What are its colors and appearance? What is its composition?

This species takes on a great variety of forms and colors, and has received names for the more prominent varieties.

Iceland spar.—Transparent crystalline calc spar, first brought from Iceland. Shows well double refraction.

Satin spar.—A finely fibrous variety with a satin luster Receives a handsome polish. Occurs usually in veins traversing rocks of different kinds.

Chalk.—White and earthy, without luster, and so soft as to leave a trace on a board. Forms mountain beds.

Rock milk.—White and earthy like chalk, but still softer, and very fragile. It is deposited from waters containing lime in solution.

Calcareous tufa.—Formed by deposition from waters like rock milk, but more cellular or porous and not so soft.

Stalactite, Stalagmite.—The name stalactite is explained on page 54. The deposits of the same origin that cover the floor of a cavern, are called stalagmite. They generally consist of different colored layers, and appear banded or striped when broken. The so-called "Gibraltar rock" is stalagmite from a cavern in the rock of Gibraltar.

Limestone is a general name for all the massive varieties occurring in extensive beds.

Oolite, Pisolite.—Oolite is a compact limestone, consisting of small round grains, looking like the spawn of a fish; the name is derived from the Greek *öon, an egg*. Pisolite, a name derived from *pisum*, the Latin for *pea*, differs from oolite in consisting of larger particles.

Argentine.—A white shining limestone consisting of laminæ a little waving, and containing a small proportion of silica.

Fontainbleau limestone.—This name is applied to crystals, of the form in figure 4, containing a large proportion of sand, and occurring in groups. They were formerly obtained at Fontainbleau, France, but the locality is exhausted.

Granular limestone.—A limestone consisting of crystalline grains. It is called also primary limestone. The coarser varieties when polished constitute the common *white and clouded marbles*, and the material of which marble buildings are made. The finer are used for statuary, and

---

What is Iceland spar? What is chalk? How does satin spar under this species differ from that which is a variety of gypsum? What is calcareous tufa? How are stalactites and stalagmite formed? What is limestone? What is oolite? What is said of granular limestone?

called *statuary marble.* The best is as clear and fine grained as loaf sugar, which it much resembles.

Compact limestone.—The common secondary limestones, breaking with a smooth surface, without any appearance of grains. The rock is very variously colored, sometimes of a uniform tint, and frequently in bands, blotches or veinings, and always nearly dull until polished. The varieties form marbles of as many kinds.

Stinkstone, Anthraconite.—A limestone, either columnar or compact, which gives out a fetid odor when struck.

Plumbocalcite, from Cornwall, contains 2·34 per cent. of carbonate of lead.

Dif. The varieties of this species are easily distinguished by their being scratched easily with a knife, in connection with their strongly effervescing with acids, and their complete infusibility. Calc spar is not so hard as aragonite, and differs entirely in its cleavage.

Obs. Crystallized calcareous spar occurs in magnificent forms in the vicinity of Rossie, New York. One crystal from there now at New Haven weighs 165 pounds. Some rose and purple varieties from this region are very beautiful. Splendid geodes of the dog-tooth spar variety occur in limestone at Lockport, along with gypsum and pearl spar. Leyden and Lowville, N. Y., are other localities. Bergen Hill, N. J., affords beautiful wine-yellow crystals in amygdaloid. *Argentine* occurs near Williamsburg and Southampton, Mass. *Rock milk* covers the sides of a cave at Watertown, N. Y., and is now forming. Stalactites of great beauty occur in Weir's and other caves in Virginia and the Western States; also in Ball's cave at Scoharie, N. Y. Chalk occurs in England and Europe, but has not been met with in the United States. Granular limestones are common in the Eastern and Atlantic States, and compact limestones in the middle and Western, and some beds of the former afford excellent marble for building and some of good quality for statuary.

Uses. Any of the varieties of this mineral when burnt, form *quicklime.* Heat drives off the carbonic acid and leaves the lime in a pure or caustic state. Some limestones contain a portion of clay disseminated throughout it, and these burn often to *hydraulic lime*, a kind of lime, of which a

---

What is said of compact limestone? How is this species distinguished from other species? What are the uses of limestone?

cement or plaster is made that "sets" under water. See further, the chapter on Rocks, for the uses of limestone.

ARAGONITE.

Trimetric. In rhombic prisms, (see fig. 8, page 26); M : M=116° 10′. Cleavage parallel with M. Usually in compound crystals having the form of a hexagonal prism, with uneven or striated sides, or in stellated forms consisting of two or three flat crystals crossing one another. Also in globular and coralloidal shapes; also in fibrous seams in different rocks.

Color white or with light tinges of gray, yellow, green and violet. Luster vitreous. Transparent to translucent. H=3·5—4. Gr=2·931.

In *composition*, it is identical with calcareous spar, and in its action before the blowpipe it differs only in falling to powder readily when heated. Effervesces also with the acids. Phosphoresces when heated. Some varieties contain a few per cent. of carbonate of strontia, but this is not an essential ingredient.

Dif. The same distinctive characters as calcareous spar, except its crystalline form and superior hardness, and its falling to powder before the blowpipe.

Obs. Aragonite occurs mostly in gypsum beds and deposits of iron ore; also in basalt and other rocks. The coralloidal forms are found in iron ore beds, and are called *flos-ferri, flowers of iron.* They look like a loosely intertwined or tangled white cord.

The *flos-ferri* variety occurs at Lockport with gypsum; also at Edenville, at the Parish iron ore bed in Rossie, and in Chester county, Pennsylvania. Aragon in Spain affords six-sided prisms of aragonite, associated with gypsum. This locality gave the name to the species.

6. DOLOMITE—*Magnesian Carbonate of Lime.*

Rhombohedral. R : R=106° 15′. Cleavage perfect parallel to the primary faces. Faces of rhombodedrons sometimes curved, as in the annexed figure. Often granular and massive, constituting extensive beds.

Color white or tinged with yellow, red, green,

What are the usual forms of arragonite? Does it differ in composition from calcite? What are its colors and luster? What effect is produced by the blowpipe?

brown, and sometimes black. Luster vitreous, o. a little pearly. Nearly transparent to translucent. Brittle. H= 3·5—4. Gr=2·8—2·9.

Composition. Dolomite is a compound of carbonate of magnesia and carbonate of lime. The common variety consists of 54·4 of the latter to 45·6 of the former. Infusible before the blowpipe. Effervesces with acids, but more slowly than calc spar.

The principal varieties of this species are as follows:

Dolomite.—White crystalline granular, often not distinguishable in external characters from granular limestone, except that it crumbles more readily.

Pearl spar.—This variety occurs in pearly rhombohedrons with curved faces.

Rhomb spar, Brown spar.—In rhombohedrons, which become brown on exposure, owing to their containing 5 to 10 per cent. of oxyd of iron or manganese.

Miemite.—A yellowish brown fibrous variety from Miemo in Tuscany.

Gurhofite.—A compact white rock, looking like porcelain and containing a few per cent. of silica.

Dif. Distinctive characters, nearly the same as for calcareous spar. It is harder than that species, and differs in the angles of its crystals, and effervesces less freely; but chemical analysis is often required to distinguish them.

Obs. Massive dolomite is common in the Eastern States, and constitutes much of the coarse white marble used for building. Crystallized specimens are obtained at the Quarantine, Richmond county, N. Y. Rhomb spar occurs in talc at Smithfield, R. I., Marlboro, Vt., Middlefield, Mass.; pearl spar in crystals of the above form at Lockport, Rochester, Glen's Falls; gurhofite on Hustis's farm, Phillipstown, N. Y.

Dolomite was named in honor of the geologist and traveler, Dolomieu.

Uses. Dolomite burns to quicklime like calc spaı, and affords a stronger cement. The white massive variety is used extensively as marble. The magnesian lime has been supposed to injure soils; but this is believed not to be the case if it is air-slaked before being used. It is also employed in the manufacture of Epsom salts or sulphate of magnesia.

What is the composition of dolomite? How does it differ from calcite? What are its uses?

The mineral is subjected to the action of sulphuric acid ; the sulphate of lime being insoluble is deposited, leaving the sulphate of magnesia in solution. A more economical method is to boil the calcined stone in proper proportions in bittern ; the muriatic acid of the bittern takes up the lime.

Ankerite. This species resembles brown spar, and like that becomes brown on exposure. The primary is a rhombohedron of 106° 12′. It consists of the carbonates of lime, magnesia, iron, and manganese. The Styrian iron ore beds and Saltzburg are some of its foreign localities. It is said to occur in veins at Quebec and at West Springfield, Mass.

7, APATITE.—*Phosphate of Lime.*

In hexagonal prisms. The annexed figure represents a crystal from St. Lawrence county, New York. Cleavage imperfect.

P
M M M

Usually occurs in crystals; but occasionally massive ; sometimes mammillary with a compact fibrous structure. Small crystals are occasionally transparent and colorless, but the usual color is green, often yellowish-green, bluish-green, and grayish-green ; sometimes yellow, blue, reddish or brownish. Coarse crystals nearly opaque. Luster resinous, or a little oily. H=5. Gr=3—3·25. Brittle. Some varieties phosphoresce when heated, and some become electric by friction.

Composition: phosphate of lime 92·1, fluorid of calcium 7·0, chlorid of calcium 0·9. Infusible before the blowpipe except on the edges. Dissolves slowly in nitric acid without effervescence. Its constituents are contained in the bones and ligaments of animals, and the mineral has probably been derived in many cases from animal fossils.*

Asparagus stone is a translucent wine-yellow variety occurring in talc at Zillerthal in the Tyrol. *Phosphorite* is a massive variety from Estremadura in Spain, and Schlackenwald in Bohemia. *Moroxite* is a greenish-blue variety from Arendal. *Eupyrchroite* (Emmons) is a fibrous mammillary variety from Crown Point, Essex county, N. Y.

What is the common form of apatite ? is colors and appearance ? Is it harder than calc spar ? What is the principal constituent in its composition ? What is a probable origin of this mineral in many cases ?

---

* Bones contain 55 per cent. of phosphate of lime, with some fluorid of calcium, 3 to 12 per cent. of carbonate of lime, some phosphate of magnesia and chlorid of sodium, besides 33 per cent. of animal matter.

Dif. Distinguished by its inferior hardness from beryl, it being easily scratched with a knife ; by dissolving in acids without effervescence from carbonate of lime and other carbonates ; by its difficult fusibility, and giving no metallic reaction before the blowpipe from phosphate of lead and other metallic species. Its phosphorescence is also an important characteristic.

Obs. Apatite occurs in gneiss and mica slate, granular limestone, and occasionally in ancient volcanic rocks. The finest localities in the United States occur in granular limestone. The crystals from the limestone of St. Lawrence county, N. Y., are among the largest yet discovered in any part of the world. One from Robinson's farm measured a foot in length and weighed 18 pounds. But they are nearly opaque and the edges are usually rounded. They occur with scapolite, sphene, &c. Edenville and Amity, Orange county, N. Y., afford fine crystals from half an inch to twelve inches long. At Westmoreland, N. H., fine crystals are obtained in a vein of feldspar and quartz ; also at Blue Hill bay in Maine. Bolton, Chesterfield, Chester, Mass., are other localities. A beautiful blue variety is obtained at Dixon's quarry, Wilmington, Delaware.

The name apatite, from the Greek *apatao, to deceive,* was given in allusion to the mistake of early mineralogists respecting the nature of some of its varieties.

8. FLUOR SPAR—*Fluorid of Calcium, Fluate of Lime.*

Monometric. Cleavage octahedral, perfect. Secondary forms, the following :

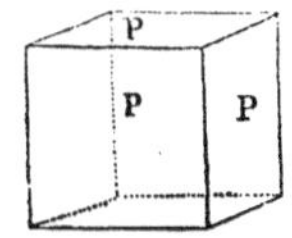

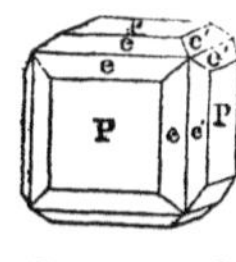

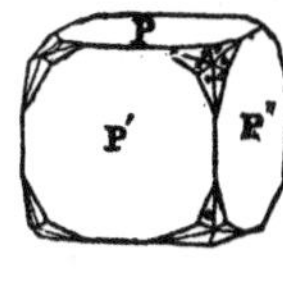

Rarely occurs fibrous ; often compact, coarse or fine granular. Colors usually bright ; white, or some shade of light green, purple, or clear yellow are most common ; rarely rose-red and sky-blue ; colors of massive varieties often

How is apatite distinguished from beryl? how from carbonates? how from phosphate of lead? What is said of the crystalline form and cleavage of fluor spar? What is said of its colors and appearance?

banded. The crystals are transparent or translucent. H=4. Gr=3·14—3·18. Brittle.

Composition: fluorine 48·7, calcium 51·3. Phosphoresces on a hot iron, giving out a bright light of different colors; in some varieties the light is emerald green; in others, purple, blue, rose-red, pink, or an orange shade. Before the blowpipe it decrepitates, and ultimately fuses to an enamel. Pulverised and moistened with sulphuric acid, a gas is given off which corrodes glass.

The name *chlorophane* has been given to the variety that affords a green phosphorescence.

Dif. In its bright colors, fluor resembles some of the gems, but its softness at once distinguishes it. Its strong phosphorescence is a striking characteristic; and also its affording easily, with sulphuric acid and heat, a gas that corrodes glass.

Obs. Fluor spar occurs in veins in gneiss, mica slate, clay slate, limestone, and sparingly in beds of coal. It is the gangue in some lead mines.

Cubic crystals of a greenish color, over a foot each way, have been obtained at Muscolonge Lake, St. Lawrence county, N. Y. Near Shawneetown on the Ohio, a beautiful purple fluor in grouped cubes of large size is obtained from limestone and the soil of the region. At Westmoreland, N. H., at the Notch in the White Mountains, Blue Hill Bay, Maine, Putney, Vt., and Lockport, N. Y., are other localities. The *chlorophane* variety is found with topaz at Huntington, Conn.

In Derbyshire, England, fluor spar is abundant, and hence it has received the name of *Derbyshire spar.* It is a common mineral in the mining districts of Saxony.

Fluorid of calcium is also found in the enamel of teeth, in bones and some other parts of animals; also in certain parts of many plants; and by vegetable or animal decomposition it is afforded to the soil, to rocks, and also to coal beds in which it has been detected.

Uses. Massive fluor receives a high polish and is worked into vases, candlesticks and various ornaments, in Derbyshire, England. Some of the varieties from this locality, consisting of rich purple shades banded with yellowish white, are very

---

What is said of the phosphorescence of calc spar? Of what does it consist? What is chlorophane? How is fluor spar distinguished from the gems? What are its uses?

beautiful. The mineral is difficult to work on account of being brittle. It is usually turned in a lathe, and worked down first with a fine steel tool; then with a coarse stone, and afterwards with pumice and emery. The crevices which occur in the masses are sometimes concealed by filling them with galena, a mineral often found with the fluor. Fluor spar is also used for obtaining fluoric acid, which is employed in etching. To etch glass, a picture, or whatever design it is desired to etch, is traced in the thin coating of wax* with which the glass is first covered; a very small quantity of the liquid fluoric acid is then washed over it; on removing the wax, in a few minutes, the picture is found to be engraved on the glass. The same process is used for etching seals, and any siliceous stone will be attacked with equal facility. Fluor spar is also used as a flux to aid in reducing copper and other ores, and hence the name *fluor*.

Hayesine or *Hydrous Borate of Lime*. Occurs in snowy white interwoven fibers, with gypsum and alum on the plains of Iquique, S. A.

Hydroboracite. A hydrous borate of lime and magnesia resembling somewhat a white fibrous gypsum. It is of Caucasian origin.

Oxalate of Lime. Observed on calc spar in small oblique crystals. Locality unknown.

Nitrate of Lime. In white delicate efflorescences; deliquescent. Also in solution in some waters. The salt is formed in calcareous caverns and covered spots of earth where the soil is calcareous. It is extensively used in the manufacture of saltpeter, (nitrate of potash.) Occurs in the caverns of Kentucky and other Western States.

7. MAGNESIA.

The sulphates and nitrate of magnesia are soluble, and are distinguished by their bitter taste. The other native magnesian salts are insoluble. The presence of magnesia when no metallic oxyds are present is indicated by a blowpipe experiment: after heating a fragment, moisten it with a solution of *nitrate of cobalt*, and then subject it again to the heat

---

How is glass etched by means of fluor spar? What is the origin of the name fluor? What is said of the occurrence and uses of nitrate of lime? What is the taste of soluble salts of magnesia? What blowpipe test distinguishes them?

---

* The best material is a mixture of bees wax and turpentine resin melted together.

of the blowpipe, and it will become pale-red, and deepen in color by fusion.

Specific gravity of the species in this family, below 3. Hardness of some species as high as 7.

EPSOM SALT.—*Sulphate of Magnesia.*

Trimetric. In modified rhombic prisms, (fig. 8, page 26.) M : M = 90° 34′. Cleavage perfect parallel with the shorter diagonal. Usually in fibrous crusts, or botryoidal masses, of a white color. Luster vitreous—earthy. Very soluble, and taste bitter and saline.

Composition : magnesia 16·3, sulphuric acid 32·5, water 50·2. Deliquesces before the blowpipe. Does not effervesce with acids.

Dif. The fine spicula-like crystalline grains of Epsom salt, as it appears in the shops, distinguish it from Glauber salt, which occurs usually in thick crystals.

Obs. The floors of the limestone caves of the West often contain Epsom salt in minute crystals mingled with the earth. In the Mammoth Cave, Ky., it adheres to the roof in loose masses like snow-balls. It occurs as an efflorescence on the east face of the Helderberg, 10 miles from Coeymans. The fine efflorescences suggested the old name *hair salt.*

At Epsom in Surrey, England, it occurs dissolved in mineral springs, and from this place the salt derived the name it bears. It occurs at Sedlitz, Aragon, and other places in Europe ; also in the Cordilleras of Chili ; and in a grotto in Southern Africa, where it forms a layer an inch and a half thick.

Uses. Its medical uses are well known. It is obtained for the arts from the bittern of sea-salt works, and quite largely from magnesian carbonate of lime, by decomposing it with sulphuric acid. The sulphuric acid takes the lime and magnesia, expelling the carbonic acid ; and the sulphate of magnesia remaining in solution is poured off from the sulphate of lime, which is insoluble. It is then crystallized by evaporation.

MAGNESITE.—*Carbonate of Magnesia.*

Rhombohedral ; R : R = 107° 29. Cleavage rhombohedral, perfect. Often in fibrous plates the surface of which

---

Of what does Epsom salt consist? Where does it occur? Whence the name Epsom?

frequently consists of minute acicular crystals; also granular and compact and in tuberous forms. Color white, yellowish or grayish-white or brown. Luster vitreous; fibrous varieties often silky. Transparent to opaque. H=3—4½. Gr=2.8—3.

Composition: carbonic acid 52·4, and magnesia 47·6. Infusible before the blowpipe. Dissolves slowly with little effervescence in nitric or sulphuric acid.

Dif. Resembles some varieties of carbonate of lime and dolomite; but effervesces more feebly in acids, does not burn to quicklime, and the light before the blowpipe is less intense. The fibrous variety is distinguished from amianthus and other fibrous minerals associated with it, by its greater hardness and more vitreous luster, and from siliceous minerals generally by its complete solubility in acids.

Obs. Magnesite is usually associated with magnesian rocks, especially serpentine. At Hoboken, N. J., it occurs in this rock in fibrous seams; similarly at Lynnfield, Mass.; and at Bolton, imperfectly fibrous, traversing white limestone.

Uses. When abundant it is a convenient material for the manufacture of sulphate of magnesia or Epsom salt, to make which, requires simply treatment with sulphuric acid.

BRUCITE.—*Hydrate of Magnesia.*

In foliated hexagonal prisms and plates. Structure thin foliated, and thin laminæ easily separated and translucent; flexible but not elastic. Color white and pearly, often grayish or greenish. H=1·5. Gr=2·35.

Composition: magnesia 69·9, water 31·0. Infusible before the blowpipe, but becomes opaque and friable. Entirely soluble in the acids without effervescence.

Dif. It resembles talc and gypsum, but is soluble in acids; it differs from heulandite and stilbite, also by its infusibility.

Obs. Occurs in serpentine at Hoboken, N. J., and Richmond Co., N. Y., also at Swinaness in Unst, one of the Shetland Isles.

Nemalite is a fibrous hydrate of magnesia or brucite. The following are its characters;

Of what does magnesite consist? How is it distinguished from most earthy minerals? How from calc spar? For what use is it fitted? What is the appearance of nemalite? its composition? its locality?

Neatly fibrous and silky; fibres brittle and easily seperable. Color whitish, grayish or bluish white; transparent, but becomes opaque and crumbling on exposure. H=2. Gr=2·35—2·4.

Composition: magnesia 62·0; protoxyd of iron 4·6; water 28·4; carbonic acid 4·1;—(Whitney.) In the flame of a candle the fibres become opaque, brownish and rigid, and in this state easily crumble in the fingers. Phosphoresces with a yellow light when rubbed with a piece of iron.

Dif. Resembles abestus or amianthus, but differs in becoming brittle before the blowpipe.

Obs. Occurs in serpentine at Hoboken, N. J., in greenstone at Piermont, Rockland Co., N. Y., and Bergen Hill, N. J.

Hydromagnesite. A pearly crystalline, or earthy white pulverulent hydrous carbonate of magnesia, from Hoboken, N. J., and Texas, Pa.

BORACITE.—*Borate of Magnesia.*

Monometric. Cleavage octahedral; but only in traces.

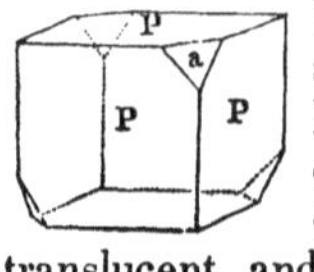

Usual in cubes with only the alternate angles replaced; or having all replaced, but four of them different from the other four. The crystals are translucent and seldom more than a quarter of an inch through. Color white or grayish; sometimes yellowish or greenish. Luster vitreous. H=7. Gr=2·97. Becomes electric when heated, the opposite angles of the cube becoming of opposite poles, one *north* and the other *south.*

Composition: boracic acid 70·0, magnesia 30·0. Intumesces before the blowpipe and forms a glassy globule, which becomes crystalline and opaque on cooling.

Dif. Distinguished readily by its form, high hardness, and pyro-electric properties.

Obs. Boracite is found only with gypsum and common salt. It occurs near Luneberg in Lower Saxony, and near Kiel in the adjoining dutchy of Holstein.

Nitrate of Magnesia. Occurs in white deliquescent efflorescences, having a bitter taste, associated with nitrate of lime, in limestone cav-

What is Brucite? What is its appearance? How is it distinguished from talc, gypsum, and other minerals? What is said of the crystals of boracite? What is stated of its electric properties? What is its composition? What is its mode of occurrence?

erns. It is used, like its associate, in the manufacture of saltpeter (see page 102.)

Polyhalite. A brick-red saline mineral, with a weak bitter taste, occurring in masses which have a somewhat fibrous appearance. Consists of the sulphates of lime, potash and magnesia, with six per cent. of water.

Wagnerite. A fluo-phosphate of magnesia, occurring in yellowish or grayish oblique rhombic prisms. Insoluble. H=5—5·5. Gr=3·1. From Salzberg, Germany.

Rhodizite. Resembles boracite in its crystals, but tinges the blowpipe flame deep red. Occurs with the red tourmaline of Siberia.

8. ALUMINA.

The compounds of alumina may often be distinguished by a blowpipe experiment. If a fragment of alumina after having been heated to redness be moistened with a solution of nitrate of cobalt and again heated, it assumes before fusion a blue color. This is a good test, and distinguishes aluminous from magnesian minerals, except when the oxyds of the metals are present.

The sulphates, fluorids and some of the phosphates, (the salts included in this family,) are soluble with more or less difficulty, in the acids; and some of the sulphates (the various alums) dissolve readily in water.

The solution in acids takes place without effervescence, and without forming a jelly like many *silicates* of alumina (the zeolites, &c.)

Specific gravities of the species below 3·1. Hardness of some species as high as 6.

NATIVE ALUM.

Monometric. Cleavage octahedral. Occurs in octahedrons; but usually in silky fibrous masses, or in efflorescent crusts. Taste sweetish astringent.

There are several kinds of native alum, differing in one of the ingredients in their constitution, but resembling one another in crystallizing in octahedrons, and in containing the ingredients in exactly the same proportions. They all contain

What blowpipe experiment distinguishes alumina ? What is said of the sulphates of alumina ? What is the composition of the alums ?

24 parts of water to 1 part of sulphate of alumina, and 1 part of some other sulphate. In *potash-alum*, this sulphate is a sulphate of potash. This is the common alum of the shops.

The corresponding sulphate in the other alums is as follows :—

Soda-alum, sulphate of soda ;
Magnesia-alum, sulphate of magnesia ;
Ammonia-alum, sulphate of ammonia ;
Iron-alum, sulphate of iron ;
Manganese-alum, sulphate of manganese.

Besides these there is also a *hydrous sulphate of alumina* without any other sulphate ; it is called *feather-alum*, and is even of more common occurrence than any of the true alums.

These alums are formed from the decomposition of pyrites, in contact with clay. Iron pyrites is a compound of sulphur and iron; in decomposition, its sulphur and iron unite with oxygen derived from the moisture present, and it then becomes sulphate of iron, or a compound of sulphuric acid and oxyd of iron. This sulphuric acid, or part of it, by uniting with the alumina of the clay rock, produces a sulphate of alumina. To form a true alum, a little potash, or soda, &c. must be present in the clay. The iron of the iron alum proceeds from the pyrites which undergoes the decomposition.

These compounds differ but little in taste and appearance.

Obs. Potash alum and more abundantly the sulphate of alumina (or feather alum), and sulphate of alumina and iron, impregnate frequently clay-slates, which are then called *aluminous slates* or *shales.* These alum rocks are often quarried and lixiviated for the alum they contain. The rock is first slowly heated after piling it in heaps, in order to decompose the remaining pyrites and transfer the sulphuric acid of any sulphate of iron to the alumina and thus produce the largest amount possible of sulphate of alumina. It is next lixiviated in stone cisterns. The lye containing this sulphate is afterwards concentrated by evaporation, and then the requisite proportion of potash (sulphate or muriate, alum containing potash as well as alumina) is added to the lix-

---

What is the composition of common potash alum? What of a soda alum? What are alum shales? Whence the alum or sulphate of alumina they contain? How is alum obtain from alum shale?

ivium. A precipitate of alum falls which is afterwards washed and re-crystallized. The mother liquor left after the precipitation is also treated for more alum. This process is carried on extensively in Germany, France, at Whitby in Yorkshire, Hurlett and Campsie, near Glasgow, in Scotland. Cape Sable in Maryland, affords large quantities of alum annually. The slates of coal beds are often used to advantage in this manufacture, owing to the decomposing pyrites present. At Whitby, 130 tons of calcined schist give one ton of alum. In France, ammoniacal salts are used instead of potash, and an ammoniacal alum is formed.

Soda alum has been observed at the Solfataras in Italy, near Mendoza in South America, on the island of Milo in the Grecian Archipelago. *Magnesia alum* forms large fibrous masses, delicately silky, near Iquique, S. A. This is the *Pickeringite* of Mr. A. A. Hayes. Ammonia alum occurs at Tschermig in Bohemia.

Alunite.—*Alum Stone.*

Rhombohedral, with a perfect cleavage parallel with a, (fig. 62, p. 39.) R : R=89° 10′. Also massive. Color white, grayish or reddish. Luster of crystals vitreous, or a little pearly on a. Transparent to translucent. H=4. Gr=2·58—2·75.

Composition: sulphuric acid 38·5, alumina 37·1, potash 11·4, water 13·0=100. Decrepitates in the blowpipe flame and is infusible both alone and with soda. In powder, soluble in sulphuric acid.

Dif. Distinguished by its infusibility, in connection with its complete solubility in sulphuric acid without forming a jelly.

Obs. Found in rocks of volcanic origin at Tolfa, near Rome, and also at Beregh and elsewhere in Hungary.

Uses. At Tolfa, alum is obtained from it by repeatedly roasting and lixiviating it and finally crystallizing by evaporation. The variety found in Hungary is so hard as to admit of being used for millstones.

Websterite. Another sulphate of alumina, in compact reniform masses and tasteless. From Newhaven in Sussex, Epernay in France, and Halle in Prussia. It is called also *aluminite.*

What is the color and appearance of alum stone? What its composition? What its use, and where is it extensively employed?

WAVELLITE.

Trimetric. Usually in small hemispheres a third or half an inch across, attached to the surface of rocks, and having a finely radiated structure within; when broken off they leave a stellate circle on the rock. Sometimes in rhombic crystals.

Color white or yellowish and brownish, with a somewhat pearly or resinous luster. Sometimes green, gray or black. Translucent. H=3·5—4. Gr=2·23—2·37.

Composition: alumina 33·8, phosphoric acid 34·9, water 26·6, fluorid of aluminium, 4·6. Whitens before the blowpipe but does not fuse.

Dif. Distinguished from the zeolites, some of which it resembles, by giving the reaction of phosphorus and also by dissolving in acids without gelatinizing. Cacoxene, to which it is allied, becomes dark reddish-brown before the blowpipe, and gives the reaction of iron.

Obs. Near Saxton's River, Bellows Falls, Vt., and also at Washington mine, Davidson Co., N. C. It was first discovered by Dr. Wavel, in clay slate in Devonshire. Occurs also in Bohemia and Bavaria.

Fischerite is another hydrous phosphate of alumina containing less phosphoric acid. Gr=2·46. Color dull green. Translucent. Sometimes in six-sided prisms. From the Ural.

TURQUOIS.

In opaque reniform masses without cleavage, of a bluish green color and somewhat waxy luster. H=6. Gr=2·6—3.

Composition: phosphoric acid 30·9, alumina 44·5, oxyd of copper 3·7, protoxyd of iron 1·8, water 19·0=99·9. Before the blowpipe it is infusible, but colors the flame green and in the inner cone becomes brown. Loses its blue color in muriatic acid.

Dif. Distinguished from bluish green feldspar, which it resembles, by its infusibility and the reaction of phosphorus.

Obs. Turquois is brought from a mountainous district in

What is the usual appearance of Wavellite? What is its composition? What distinguishes it from the zeolites? What is the color and appearance of turquois? Its constituents? How is it distinguished from a variety of feldspar? Where is it found?

Persia, not far from Nichabour, and according to Agaphi occurs in veins, that traverse the mountain in every direction.

The *callais* of Pliny was probably turquois. Pliny, in his description of it, mentions the fable that it was found in Asia, projecting from the surface of inaccessible rocks, whence it was obtained by means of slings.

Uses. Turquois receives a fine polish and is highly esteemed as a gem. In Persia it is much admired, and the Persian king is said to retain for himself, all the large and more finely tinted specimens. The *occidental* or *bone* Turquois, a much inferior and softer stone, is fossil teeth or bones, colored with a little phosphate of iron. Green *malachite* is sometimes substituted for turquois, but it is much softer and has a different tint of color. The stone is so well imitated by art as scarcely to be detected except by chemical tests. The imitation is much sofier than true turquois.

GIBBSITE.—*Hydrate of Alumina.*

In small stalactitic shapes or mammillary and incrusting. Color grayish or greenish white ; surface smooth but nearly dull. Structure sometimes nearly fibrous. Rarely in hexagonal crystals. H=3—3·5. Gr=2·3—2·4.

Composition : alumina 65·6, water 34·4.—(Torrey.) Recent examinations have shown that the mineral contains phosphoric acid only in traces. Prof. B. Silliman, Jr., has also found, in specimens examined by him, as impurity a proportion of silica without phosphoric acid. The mineral has resulted from the decomposition of feldspar or some aluminous mineral, and probably varies in composition. It whitens but does not fuse before the blowpipe.

Dif. Resembles chalcedony but is softer.

Obs. Occurs in a bed of brown iron ore at Richmond, Mass., and at Unionvale, Dutchess county, N. Y. This species was named in honor of Col. George Gibbs.

Lazulite. In compact masses; rarely in oblique crystals. Color fine azure blue, and nearly opaque, with a vitreous luster. H=5—6. Gr=3·057. Brittle. Contains phosphoric acid 41·8, alumina 35·7 magnesia 9·3, silica 2·1, protoxyd of iron 2·6, water 6·1=97·7. It in-

---

What is said of its use? How is it distinguished from false or artificial turquois? What is the appearance of Gibbsite? What is said of its composition? How is it distinguished from chalcedony? What is the constitution of lazulite? its color?

tumesces before the blowpipe without fusing. Occurs in veins in clay slate at Salzberg and in Styria; in the United States, near Crowder Mountain, Lincoln county, N. C.

Mellite or *Honey stone.* In square octahedrons, looking like a honey-yellow resin; may be cut with a knife. It is mellate of alumine. Found in Prussia and Austria.

Cryolite. In snow white masses, having rectangular cleavages, and remarkable for melting easily in the flame of a candle, to which its name (from the Greek *kruos, ice,*) alludes. H=2·25—2·5. Gr=2·95. It is a fluorid of aluminium and sodium. From Greenland.

Chiolite is near cryolite in composition and characters. H=3·5. Gr=2·6—2·90. From Siberia.

Fluellite. From Cornwall, in minute white rhombic octahedrons. Contains fluorine and aluminium.

Childrenite. Found in Derbyshire, Eng., in minute yellowish brown crystals coating spathic iron. Consists of phosphoric acid, alumina and iron, with water.

Amblygonite. A compound of phosphoric acid, alumina and lithia. Found in Saxony, in pale green crystals.

Diaspore, or *Dihydrate of Alumina.* Occurs in irregular lamellar prisms, having a brilliant cleavage; color greenish gray or hair brown. H=6—7·0. Gr=3·43. It decrepitates with violence before the blowpipe. From the Urals, in granular limestone. At Trumbull, Ct.

CLASS VI.—EARTHY MINERALS.

1. SILICA.

QUARTZ.

Rhombohedral. Occurs usually in six-sided prisms, more or less modified, terminated with six-sided pyramids: R; R=94° 15′. No cleavage apparent, seldom even in traces; but sometimes obtained by heating the crystal and plunging it into cold water. The following are some of its forms:

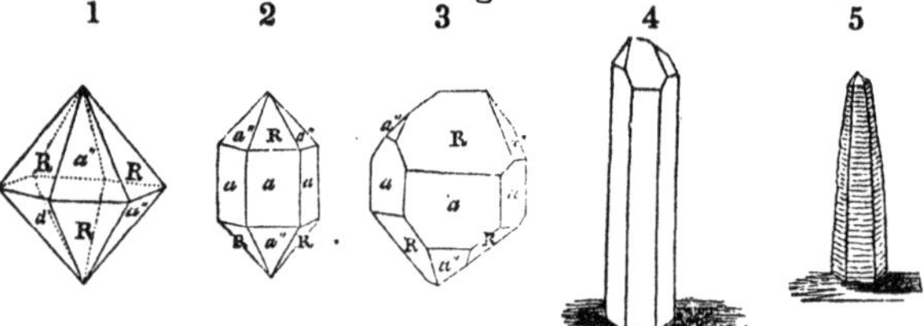

Occurs sometimes in coarse radiated forms; also coarse and fine granular; also compact, either amorphous or presenting stalactitic and mamillary shapes.

Crystals are often as pellucid as glass, and usually color-

What is the usual form of quartz crystals?

less; but sometimes present topaz-yellow, amethystine, rose or smoky tints. Also of all degrees of transparency to opacity, and of various shades of yellow, red, green, blue and brown colors, to black. In some varieties the colors are in bands, stripes, or clouds. H=7. Gr=2·6—2·7.

Composition: quartz is pure silica. Opaque varieties often contain oxyd of iron, clay, chlorite or some other mineral disseminated through them. Alone before the blowpipe infusible, but with soda melts readily with a brisk effervescence.

Dif. Quartz is a constituent of many rocks, and composes most of the pebbles of the soil or gravel beds. There is no mineral which takes on so many forms and colors, yet none is more easily distinguished. A few simple trials are all that is required.

1. *Hardness*—scratches glass with facility.
2. *Infusibility*—not melting in any heat obtained with the blowpipe.
3. *Insolubility*—not being attacked, like limestone, in any way, by the three acids.
4. *Absence of any thing like cleavage.* One variety appears to be laminated, but it consists merely of apposed plates, which are the result of having been formed or deposited in successive layers, and cannot be mistaken for cleavage plates.

To these characteristics, its action with soda might be added. In the crystallized varieties, the form alone is sufficient to distinguish it.

VARIETIES.—The varieties of quartz owe their peculiarities either to crystallization, mode of formation, or impurities, and they fall naturally into three series.

I. The *vitreous varieties*, distinguished by their glassy fracture.

II. The *chalcedonic varieties*, having a subvitreous or a waxy luster, and generally translucent.

III. The *jaspery varieties*, having barely a glimmering luster and opaque.

I. VITREOUS VARIETIES.

Rock Crystal. Pure pellucid quartz.

This is the mineral to which the word *crystal* was first applied by the ancients; it is derived from the Greek *krus-*

What is said of the color and appearance of quartz? How is it distinguished? What are the three classes of varieties? What is the origin of the word crystal?

tallos, meaning *ice*. The pure specimens are often cut and used in jewelry, under the name of "white stone."

It is often used for optical instruments and spectacle glass, and even in ancient times was made into cups and vases. Nero is said to have dashed to pieces two cups of this kind on hearing of the revolt that caused his ruin, one of which cost him a sum equal to $3000.

Amethyst. A purple or bluish-violet variety of quartz-crystal, often of great beauty. The color is owing to a trace of oxyd of manganese. It was so called on account of its supposed preservative powers against intoxication. The amethyst, especially when large and finely colored, is highly esteemed as a gem. It is always set in gold.

Rose Quartz. A pink or rose-colored quartz. It seldom occurs in crystals, but generally in masses much fractured, and imperfectly transparent. The color fades on exposure to the light, and on this account it is little used as an ornamental stone, yet is sometimes cut into cups and vases. The color may be restored by leaving it in a moist place.

False Topaz. This name is applied to the light yellow pellucid crystals. They are often cut and set for topazes. The absence of cleavage distinguishes it from true topaz. The name *citrine*, often applied to this variety, alludes to its yellow color.

Smoky Quartz. A smoky-tinted quartz crystal. The color is sometimes so dark as to be nearly black and opaque except in splinters. Crystals of the lighter shades are often extremely beautiful and are used for seals and the less delicate kinds of jewelry. It is the *cairngorum* stone.

Milky quartz. A milk-white, nearly opaque, massive quartz, of very common occurrence. It has often a greasy luster, and is then called *greasy quartz.*

Prase. A leek-green massive quartz, resembling some shades of beryl in tint, but easily distinguished by the absence of cleavage and its infusibility. It is supposed to be colored by a trace of iron.

Aventurine Quartz. Common quartz spangled throughout with scales of golden-yellow mica. It is usually translucent, and gray, brown, or reddish brown, in color. The artificial

---

What use is made of rock crystal? What is the color of amethyst? Why was it so called? What is rose quartz? What is said of its color? What is false topaz? How is it used? What is smoky quartz? What is milky quartz? What is prase? What is aventurine quartz?

imitations of this stone are more beautiful than the natural aventurine.

Ferruginous Quartz. Includes opaque, yellow, brownish-yellow, and red crystals. The color is due to oxyd of iron. These crystals are usually very regular in their forms, (figure 2,) and not distorted like the limpid crystals. They are sometimes minute and aggregated like the grains of sand in a sandstone.

II. Chalcedonic Varieties.

Chalcedony. A translucent massive variety, with a glistening and somewhat waxy luster; usually of a pale grayish, bluish, or light brownish shade. It often occurs lining or filling cavities in amygdaloid and other rocks.

These cavities are nothing but little caverns, into which siliciceous waters have filtrated at some period. The stalactites are "icicles" of chalcedony, hung from the roof of the cavity. Some of these chalcedony grottos are several feet in diameter.

Chrysoprase. An apple-green chalcedony. It is colored by nickel.

Carnelian. A bright red chalcedony, generally of a clear rich tint. It is cut and polished and much used in the more common jewelry. The colors are deepened by exposure of several weeks to the sun's rays. It is often cut for seals and beads. The Japanese cut great numbers into beads of the form of the fruit of the olive.

Sard. A deep-brownish red chalcedony, of a blood-red color by transmitted light.

Agate. A variegated chalcedony. The colors are distributed in clouds, spots, or concentric lines. These lines take straight, circular, or zigzag forms; and when the latter, it is called *fortification agate*, so named from the resemblance to the angular outlines of a fortification. These lines are the edges of layers of chalcedony, and these layers are the successive deposits during the process of its formation. *Mocha stone* or *Moss agate* is a brownish agate, consisting of chalcedony with dendritic or moss-like delineations, of an opaque yellowish brown color. They arise from disseminated oxyd of iron; all the varieties of agate are beau-

---

What is ferruginous quartz? Describe chalcedony. What is said of its formation? What is chrysoprase? What is carnelian? How is its color deepened? For what is it used? What is sard? Describe agate.

tiful stones when polished, but are not much used in fine jewelry. The colors may be darkened by boiling the stone in oil, and then dropping it into sulphuric acid. A little oil is absorbed by some of the layers, which becomes blackened or charred by the acid.

Onyx. This is a kind of agate with the colors arranged in flat horizontal layers. They are usually light clear brown and an opaque white. When the stone consists of sard and white chalcedony in alternate layers, it is called *sardonyx.*

Onyx is the material used for cameos, and is well fitted for this kind of miniature sculpture. The figure is carved out of one layer and stands in relief on another. The most noted of the ancient cameos is the Mantuan vase at Brunswick. It was cut from a single stone, and has the form of a creampot, about 7 inches high and 2½ broad. On its outside, which is of a brown color, there are white and yellow groups of raised figures, representing Ceres and Triptolemus in search of Proserpine. The Museo Borbonico contains an onyx measuring eleven inches by nine, representing the apotheosis of Augustus ; and another exhibiting the apotheosis of Ptolemy on one side and the head of Medusa on the other. Both are splendid specimens of the art, and the former is supposed to be the largest in existence.

Cat's eye. This is a greenish-gray translucent chalcedony, having a peculiar opalescence, or glaring internal reflections, like the eye of a cat, when cut with a spheroidal surface. The effect is owing to filaments of asbestus. It comes from Ceylon and Malabar, ready cut and polished, and is a gem of considerable value.

Flint, Hornstone. Flint is massive compact silica, of dark shades of smoky gray, brown, or even black, and feebly translucent. It breaks with sharp cutting edges and a conchoidal surface. It is well known as the material of gun-flints. It occurs in nodules in chalk : not unfrequently the nodules are in part chalcedonic. *Hornstone* resembles flint, but is more brittle, and therefore unfit for making into flints. It is found in limestone, and one of these rocks is called *cherty* limestone, from the abundance of it.

Plasma. This is a faintly translucent variety of chalce-

How may the colors of agate be deepened ? What is onyx ? For what is it used ? What are some of the remarkable cameos ? What is cat's eye ? What is flint ? How does it differ from hornstone.

dony approaching jasper, of a greenish color, sprinkled with yellow and whitish dots.

III. Jaspery Varieties.

Jasper. A dull red or yellow siliceous rock, containing some clay and yellow or red oxyd of iron. The yellow jasper becomes red by heat, owing to its rendering the iron anhydrous. It also occurs of green and other shades. *Riband jasper* is a jasper consisting of broad stripes of green, yellow, gray, red or brown. *Egyptian jasper* consists of these colors in irregular concentric zones, and occurs in nodules, which are usually sawn across and polished. *Ruin jasper* is a variety with delineations like ruins, of some brownish or yellowish shade on a darker ground. *Porcelain jasper* is nothing but a baked clay, and differs from jasper in being fusible before the blowpipe. Red porphyry resembles red jasper; but this is also fusible, and consists almost purely of feldspar.

Jasper admits of a high polish, and is a handsome stone for inlaid work, but is not used as a gem.

Bloodstone or *Heliotrope.* This is a deep green stone, slightly translucent, containing spots of red, which have some resemblance to drops of blood. It contains a few per cent. of clay and oxyd of iron mechanically combined with the silica. The red spots are colored with iron. There is a bust of Christ in the royal collection at Paris, cut in this stone, in which the red spots are so managed as to represent drops of blood.

Lydian stone, Touchstone, Basanite. A velvet-black siliceous stone or flinty jasper, used on account of its hardness and black color for trying the purity of the precious metals; this was done by comparing the color of the tracing left on it with that of an alloy of known character.

Besides the above there are also two or three other varieties, arising from structure.

Float stone. This variety consists of fibres or filaments, aggregated in a spongy form, and so light as to float in water. It comes from the chalk formations of Menil Montant, near Paris.

Tabular quartz. Consists of thin plates, either parallel or crossing one another and leaving large open cells.

Granular quartz. A rock consisting of quartz grains compactly cemented. The colors are white, gray, flesh-red,

---

What is plasma? What is jasper? What is bloodstone? Lydian stone?

yellowish or reddish brown. Sandstone often consists of nearly pure quartz.

Silicified wood. Petrified wood often consists of quartz. Some specimens, petrified with chalcedony or agate, are remarkably beautiful when sawn across and polished, retaining all the texture or grain as perfect as in the original wood.

Penetrating substances. Quartz crystals are sometimes penetrated by other minerals. Rutile, asbestus, actinolite, topaz, tourmaline, chlorite and anthracite, are some of these substances. The rutile often looks like needles or fine hairs of a brown color passing through in every direction. They are cut for jewelry, and in France pass by the name of *Flèches d'amour*, (love's arrows.) The crystals of Herkimer county, N. Y., often contain anthracite. Other crystals contain cavities filled with some fluid, as water, naphtha or some mineral solution.

Loc. Fine quartz crystals occur in Herkimer county, New York, at Middlefield, Little Falls, Salisbury and Newport, in the soil and in cavities in a sandstone. The beds of iron ore at Fowler and Hermon, St. Lawrence county, afford dodecahedral crystals. Diamond rock near Lansingburg is an old locality, but not affording at present good specimens. Diamond Island, Lake George, Pelham and Chesterfield, Mass., Paris and Perry, Me., and Meadow Mt., Md., are other localities. Small unpolished rhombohedrons, the primary form, have been found at Chesterfield, Mass. *Rose quartz* is found at Albany and Paris, Me., Acworth, N. H., and Southbury, Conn.; *smoky quartz* at Goshen, Mass., Paris, Me., and elsewhere; *amethyst* at Bristol, R. I., and Kewenaw Point, Lake Superior; *chalcedony* and *agates* of moderate beauty near Northampton, and along the trap of the Connecticut valley—but finer near Lake Superior, upon some of the Western rivers, and in Oregon; *chryroprase* occurs at Belmont's lead mine, St. Lawrence county, N. Y., and a green quartz (often called chryroprase) at New Fane, Vt., along with fine drusy quartz; *red jasper* occurs on the banks of the Hudson at Troy, and at Saugus near Boston, Mass.; *yellow jasper* is found with chalcedony at Chester, Mass.; *Heliotrope* occupies veins in slate at Blooomingrove, Orange county, N. Y.

What is granular quartz? What is said of silicified wood? What are common penetrating substances?

OPAL.

Compact and amorphous; also in reniform and stalactitic shapes. Presents internal reflections, often of several colors, and the finest opals exhibit a rich play of colors of delicate shades when turned in the hand. White, yellow, red, brown, green and gray are some of the shades that occur, and impure varieties are dark and opaque. Luster subvitreous. H=5·5—6·5. Gr.=2·21.

Composition: opal consists of soluble silica and 5 to 12 per cent of water.

VARIETIES.

Precious opal, Noble opal. External color usually milky, but within there is a rich play of delicate tints. *Composition,* silica 90, water 10, (Klaproth.) This variety forms a gem of rare beauty. It is cut with a convex surface. The largest mass of which we have any knowledge is in the imperial cabinet of Vienna; it weighs 17 ounces, and is nearly as large as a man's fist, but contains numerous fissures and is not entirely disengaged from the matrix. This stone was well known to the ancients and highly valued by them. They called it *paideros,* or *child beautiful as Love.* The noble opal is found near Cashau in Hungary, and in Honduras, South America; also on the Faroe Islands.

Fire opal, Girasol. An opal with yellow and bright hyacinth or fire-red reflections. It comes from Mexico and the Faroe Islands.

Common opal, Semiopal. Common opal has the hardness of opal and is easily scratched by quartz, a character which distinguishes it from some silicious stones often called semiopal. It has sometimes a milky opalescence, but does not reflect a play of colors. The luster is slightly resinous, and the colors are white, gray, yellow, bluish, greenish to dark grayish green. Translucent to nearly opaque. Phillips found nearly 8 per cent. of water in one specimen.

Hydrophane. This variety is opaque white or yellowish when dry, but becomes translucent and opalescent when immersed in water.

Cacholong. Opaque white, or bluish white, and usually

Describe opal. How does it differ from quartz in composition? What is said of the appearance and value of noble opal? What is fire opal? common opal?

associated with chalcedony. Much of what is so called is nothing but chalcedony; but other specimens contain water, and are allied to hydrophane. It contains also a little alumina and adheres to the tongue. It was first brought from the river Cach in Bucharia.

Hyalite, Muller's glass. A glassy transparent variety, occurring in small concretions and occasionally stalactitic. It resembles somewhat a transparent gum arabic. Composition, silica 92·00, water 6·33, (Bucholz.)

Menilite. A brown opaque variety, in compact reniform masses, occasionally slaty. Composition, silica 85·5, water 11·0, (Klaproth.) It is found in slate at Menil Montant, near Paris.

Wood opal. This is an impure opal, of a gray, brown or black color, having the structure of wood, and looking much like common silicified wood. It is wood petrified with a hydrated silica, (or opal,) instead of pure silica, and is distinguished by its lightness and inferior hardness. Specific gravity, 2.

Opal jasper. Resembles jasper in appearance, and contains a few per cent. of iron; but it is not so hard owing to the water it contains.

Siliceous sinter has often the composition of opal, though sometimes simply silica. The name is given to a loose porous siliceous rock usually of a grayish color. It is deposited around the Geysers of Iceland in cellular or compact masses, sometimes in fibrous, stalactitic or cauliflower-like shapes. *Pearl sinter*, or *fiorite* occurs in volcanic tufa in smooth and shining globular or botryoidal masses, having a pearly luster.

Tabasheer is a siliceous aggregation found in the joints of the bamboo in India. It contains several per cent. of water, and has nearly the appearance of hyalite.

Dif. Infusibility before the blowpipe is the best character for distinguishing opal from pitchstone, pearlstone, and other species it resembles. The absence of anything like cleavage or crystalline structure is another characteristic. Its inferior hardness separates it from quartz.

Obs. Hyalite is the only variety of opal that has yet been found in the United States. It occurs sparingly at the

---

What is hyalite? wood opal? siliceous sinter? tabasheer? How is opal distinguished from pitchstone and quartz?

Phillips ore bed, Putnam county, N. Y., and in Burke and Scriven counties, Georgia. The Suanna spring in Georgia affords small quantities of siliceous sinter.

2. LIME.

The silicates and borosilicate of lime gelatinize readily and perfectly with muriatic acid. In hardness they are not above feldspar, (6,) and their specific gravities do not exceed 3. They fuse before the blowpipe with different degrees of facility, affording no metallic reaction.

WOLLASTONITE.—*Tabular Spar.*

Monoclinic. Rarely in oblique flattened prisms. Usually massive, cleaving easily in one direction, and showing a lined or indistinctly columnar surface, with a vitreous luster inclining to pearly.

Usually white, but sometimes tinged with yellow, red, or brown. Translucent, or rarely subtransparent. Brittle. H=4—5. Gr=2·75—2·9.

Composition: silica 52, lime 48. Fuses with difficulty to a subtransparent, colorless glass; forms with borax a clear glass.

Dif. Differs from any carbonates in not effervescing with acids; from asbestus and nemalite in its more vitreous appearance and fracture; and from these and tremolite in its forming a jelly with acids; from natrolite, scolecite and dysclasite in its very broad *sub*-fibrous cleavage surface and more difficult fusibility; from feldspar in the lined appearance of a cleavage surface and the action of acids.

Obs. Usually found in granite or granular limestone; occasially in basalt or lava.

At Willsboro', Lewis, Diana, and Roger's Rock, N. Y., it is abundant, of a white color, along with garnet. At Boonville, it is found in boulders with garnet and pyroxene. Grenville, Lower Canada, and Bucks county, Pennsylvania, are other localities. Occurs also at Kewenaw Point, Lake Superior.

What are the prominent characters of the silicates and borosilicate of lime? What is the color and appearance of tabular spar? Of what does it consist? How does it differ from the carbonates? how from asbestus, tremolite, and feldspar?

DATHOLITE—*Borosilicate of Lime.*

Trimetric. In hemihedral rhombic prisms. M : M=115° 26′. Crystals without distinct cleavage ; small and glassy. Also botryoidal, with a columnar structure, and then called *botryolite.* Color white, occasionally grayish, greenish, yellowish or reddish. Translucent. H=5—5·5. Gr=2·9—3.

Composition: silica 37·4, lime 35·7, boracic acid 21·3, water 5·7. Botryolite contains twice the proportion of water. Rendered friable in the flame of a candle. Before the blowpipe becomes opaque, intumesces and melts to a glassy globule coloring the flame green. Forms a jelly easily with nitric acid.

Dif. Its small glassy complex crystallizations without cleavage are unlike any other mineral that gelatinizes with acid, except some chabazites, from which it is distinguished by tinging the blowpipe flame green, and having greater hardness.

Obs. Occurs in amygdaloid and gneiss. In Connecticut, the finest come from Roaring brook, 14 miles from New Haven. The Rocky Hill quarry near Hartford, Berlin, Middlefield Falls, Conn., and Bergen Hill and Patterson in New Jersey, are other localities ; also in great abundance at Eagle Harbor in the copper region, Lake Superior.

Uses. Where abundant, as near Lake Superior, it may be profitably employed in the manufacture of boracic acid. It is suggested by Dr. C. T. Jackson as a good flux for the copper ores.

Okenite. In white fibrous seams or masses, consisting of delicate fibers, and singularly tough under the hammer ; color whitish, yellowish or bluish. H=4·5. Gr=2·28—2·36. *Composition,* silica 57·0, lime 26·4, water 16·6. Fuses on the edges. Gelatinizes easily in muriatic acid. From the Faroe Islands in trap ; also from Greenland. *Dysclasite* is this species.

Pectolite. Divergent, fibrous and resembling dysclasite. Luster weak pearly. H=4—5. Gr=2·69. *Composition,* silica 52·5, alumina, 36·1, soda 8.0, water 3·4. Fuses to a white transparent glass. From the Tyrol and Fassa-thal. Also from Bergen Hill and Isle Royale, Lake Superior. The Bergen Hill mineral has been called stellite.

What is said of the crystals of datholite ? How much boracic acid does datholite contain ? How is it distinguished ?

Danburite. A silicate of lime and boracic acid, of a yellowish white color. H=7. Gr=2·96. Occurs at Danbury, Ct., with oligoclase.

3. MAGNESIA.

The blowpipe test for distinguishing magnesia when not disguised by the presence of a metallic oxyd, is given on page 123. None of the silicates of magnesia gelatinize with acids. The species vary in hardness from 1 to 8.*

1. *Hydrous Silicates of Magnesia.*

TALC.

Trimetric. In right rhombic or hexagonal prisms. M : M=120° Usually in pearly foliated masses, separating easily into thin translucent folia. Sometimes stellate, or divergent, consisting of radiating laminæ; often massive, consisting of minute pearly scales; also crystalline granular, or of a fine impalpable texture.

Luster eminently pearly, and feel unctuous. Color some shade of light green or greenish white; occasionally silvery white; also grayish green and dark olive green. H=1—1·5; easily impressed with the nail. Gr=2·5—2·9. Laminæ flexible, but not elastic.

VARIETIES.

Foliated talc. The purest talc, occurring in foliated masses, of a white or greenish white color, and having an unctuous feel.

Soapstone, or *Steatite.* A gray or grayish green massive talc, showing often when broken a fine crystalline texture, occasionally yellowish or reddish. The Briançon variety is milk-white, with a pearly lustre, very greasy to the feel, or like soap.

Potstone, or *Lapis ollaris.* An impure talc, of grayish green and dark green colors and slaty structure. Feel unctuous.

Do any silicates of magnesia gelatinize with acids? Describe talc. What is steatite? What is potstone?

---

* The base magnesia is replaceable by protoxyd of iron, protoxyd of manganese, or lime, as illustrated in the species pyroxene, and consequently this group embraces compounds which are not purely silicates of magnesia.

Indurated talc. A slaty talc, of compact texture, and above the usual hardness, owing to impurities. Feel somewhat unctuous. This passes into *talcose slate*, still less pure and less unctuous in its feel, and coarser in its slaty structure.

Rensselaerite. This name has been given by Professor Emmons to a kind of soapstone from St. Lawrence, Jefferson county, N. Y., which has a very compact structure, a soapy feel, slight translucency, and hardness 3 to 4. It occurs of white, yellow, or grayish white colors, and even black. It works up with a very smooth and handsome surface, and is made into inkstands.

Composition of foliated talc, silica 62·8, magnesia 32·4, with protoxyd of iron 1·6, alumina 1·0, water 2·3. Water is considered by some chemists an essential ingredient, and 4 per cent. have been detected in some talcs.

Composition of steatite, silica 62·2, magnesia, 30·5, protoxyd of iron 2·5, water 5·0. Before the blow-pipe talc loses its color and fuses with great difficulty.

Dif. The unctuous feel, foliated structure, and pearly uster of talc are good characteristics. It differs from mica also in being inelastic, although flexible; from chlorite, saponite and serpentine in yielding no water when heated in a glass tube. Only the massive varieties resemble the last mentioned species, and chlorite has a dark olive-green color.

Obs. Handsome foliated talc occurs at Bridgewater, Vt.; Smithfield, R. I.; Dexter, Me.; Lockwood, Newton and Sparta, N. J., and Amity, N. Y. On Staten Island, near the quarantine, both the common and indurated are obtained; at Cooptown, Md., green, blue and rose colored talc occur. *Steatite* or soapstone is abundant, and is quarried at Grafton, Vt., and an adjacent town; at Francestown and Orford, N. H. It also occurs at Keene and Richmond, N. H.; at Marlboro and New Fane, Vt.; at Middlefield, Mass.; in Loudon county, Va., and at many other places.

Uses. Steatite may be sawn into slabs and turned in a lathe. It is used for fire stones in furnaces and stoves, and for jambs for fire-places. It receives a polish after being heated, and has then a deep olive-green color. It is bored out for conveying water, in place of lead tubes. Steatite is

---

How does talc differ from mica? Of what does talc consist? Why is it useful for fire stones? What other uses has it?

also used in the manufacture of porcelain; it makes the biscuit semi-transparent, but brittle and apt to break with slight changes of heat. It forms a polishing material for serpentine, alabaster and glass, and removes grease spots from cloth. When ground up, it is employed for diminishing the friction of machinery. Potstone is worked into vessels for culinary purposes, at Como in Lombardy.

CHLORITE.

Usually in dark olive-green masses, having a granular texture: rarely in hexagonal crystals, foliated like talc and in radiated forms. Luster a little pearly. Rarely subtransparent; subtranslucent to opaque. Laminæ inelastic. H=1·5. Gr=2·65—2·85. Feel scarcely unctuous.

Composition: silica 30·4, alumina 17, magnesia 34·0, protoxyd of iron 4·4, water 12·6. Fuses with difficulty on the thinnest edges. Yields water when heated in a glass tube.

This species has lately been subdivided on chemical grounds, and the name *Ripidolite* applied to the new species instituted.

Dif. Its olive green color and granular texture when massive are characteristic, and the latter character will distinguish it from serpentine and potstone. From talc and its varieties it is distinguished also by yielding water in a glass tube; from green iron earth in its difficult fusibility.

Obs. Chlorite and chlorite slate, the latter an impure slaty variety, form extensive deposits in many regions, and the latter often contains crystals of magnetic iron, hornblende or tourmaline.

Saponite. Soft and almost like butter, but brittle on drying; color white, or tinged with yellow, blue or red. *Composition,* silica 45·0, magnesia 24·7, alumina 9·3, peroxyd of iron 1·0, potash, 0·7, water 18·0=98·7. From Lizard's Point, Cornwall, and the north shore of Lake Superior. It may be kneaded like dough when first extracted.

SERPENTINE.

Rarely in right rectangular prisms. Cleavage indistinct. Usually massive and compact in texture, of a dark oil green, olive-green, or blackish-green color. Occurs also fibrous

---

What effect has it in porcelain? What is the color and usual appearance of chlorite? How is chlorite distinguished from green iron earth? What is the color and appearance of serpentine?

and lamellar. The lamellar varieties consist of thin folia, sometimes separable, but brittle; colors greenish-white, and light to dark-green.

Luster weak; resinous, inclining to greasy. Finer varieties translucent; also opaque. H=2½—4. May be cut with a knife. Gr=2·5—2·6. Becomes yellowish-gray on exposure. Feel sometimes a little unctuous.

VARIETIES AND COMPOSITION.

Precious serpentine. Purer specimens of a rich oil green color, and translucent, breaking with a splintery fracture. It is a beautiful stone when polished. *Composition:* silica 42·3, magnesia 44·2, protoxyd of iron 0·2, carbonic acid 0·9, water 12·4. Gives off water when heated; becomes brownish-red before the blowpipe, but fuses only on the edges.

Common serpentine. Opaque of dark green shades of color.

Picrolite, Schiller asbestus. A fibrous serpentine, of an olive-green color, constituting seams in serpentine. The fibers are coarse or fine, and brittle. Resembles some forms of asbestus, but differs in its difficult fusibility. Thomson's *Baltimorite* belongs here. *Amianthus* is a silky variety.

Marmolite. A foliated serpentine, of greenish white and light green shades of color, and pearly luster, consisting of thin folia rather easily separable. The folia are brittle, and the variety is thus distinguished from talc and brucite. *Composition:* silica 40·1, magnesia 41·4, protoxyd of iron 2·7, water 15·7, (Shepard.)

Kerolite. Near marmolite, but folia not separable.

Dif. Precious and common serpentine are easily distinguished from other green minerals by their dull resinous luster and compact structure, in connection with their softness, being easily cut with a knife, and their low specific gravity,

Obs. Serpentine occurs as a rock, and the several varieties mentioned either constitute the rock or occur in it. Occasionally it is disseminated through granular limestone, giving the latter a clouded green color: this is the *verd antique* marble.

Good Serpentine is found in the United States at Phil-

What is the hardness of serpentine? Of what does it consist? What is precious serpentine? What are the peculiarities of marmolite and kerolite? How is serpentine distinguished? How does serpentine occur?

lipstown, Port Henry, Gouverneur, Warwick, N. Y.; Newburyport, Westfield, and Blandford, Mass.; at Kellyvale and New Fane, Vt.; Deer Isle, Maine; New Haven, Conn.; Bare Hills, Md., &c. Marmolite and kerolite, at Hoboken, N. J., and Blandford, Mass. The quarries of Milford and New Haven, Ct., afford a beautiful verd-antique, and have been wrought; but the works are now suspended.

Uses. Serpentine forms a handsome marble when polished, especially when mixed with limestone, constituting *verd-antique* marble. Its colors are often beautifully clouded, and it is much sought for, as a material for tables, jambs for fire-places, and ornamental in-door work. Exposed to the weather, it wears uneven, and soon loses its polish. Chromic iron is usually disseminated through it, and increases the variety of its shades. Dr. C. T. Jackson of Boston has lately shown that Epsom salts (sulphate of magnesia) may be profitably manufactured from serpentine.

NEPHRITE.—*Jade.*

Massive, and very tough and compact; greenish or bluish to white. Translucent to subtranslucent. Luster vitreous. H=6·5—7·5. Gr=2·9—3·03.

Composition: contains silica, magnesia, and some water, with or without alumina, oxyd of iron, and lime. It varies in constitution, and has been lately considered a massive tremolite. Infusible alone before the blowpipe.

Dif. Differs from beryl in having no cleavage; and from quartz by its finely uneven surface of fracture, instead of smooth and glassy.

Obs. A greenish and reddish-gray variety is found at Easton, Pa., and Stoneham, Mass. The so-called nephrite from Smithfield, R. I., has the composition of serpentine.

Nephrite is made into images, and was formerly worn as a charm. It was supposed to be a cure for diseases of the kidney, whence the name, from the Greek *nephros*, kidney. In New Zealand, China and Western America, it is carved by the inhabitants or polished down into various fanciful shapes. Much of the mineral from China called jade is prehnite.

What is verd-antique? What are the uses of serpentine? What are the characters of nephrite? What is the origin of the name?

MEERSCHAUM.—*Sea Froth.*

Dull white, opaque and earthy, nearly like clay. H=2 Gr=2·6—3·4.

Composition of a variety from Anatolia : silica 60·9, magnesia 27·8, water 11·3, oxyd of iron and alumina 0·1. When heated it gives out water and a fetid odor, and becomes hard and perfectly white. When first dug up it is soft, has a greasy feel and lathers like soap; and on this account it is used by the Tartars in washing their linen. It is used for making the bowls of Turkish pipes, by a process like that for pottery ware. When imported into Germany, the bowls of the pipes are prepared for sale by softening them first in tallow, then in wax, and finally polishing them.

Aphrodite is another meerschaum from Longbanshyttan.

Quincite is a variety or related species of a reddish color.

SCHILLER SPAR.

Triclinic. Occurs massive, with cleavage in two directions, producing a thin foliated structure. Folia brittle and separable. Color olive and blackish-green, inclining on the cleavage face to pinchbeck-brown. Luster metallic-pearly on a cleavage face; vitreous in other directions. H=3·5—4. Sectile. Gr=2·5—2·7.

Composition: silica 43·9, magnesia 25·9, oxyd of iron and chromium 13·0, water 12·4, alumina 1·3, lime 2·6, protoxyd of manganese 0·5. Gives off water, and becomes pinchbeck-brown and magnetic before the blowpipe, but fuses only on the thinnest edges.

Dif. Distinguished from diallage, which also occurs in serpentine, and is the only species with which it can be confounded, by its yielding water before the blowpipe. Marmolite is much softer. Talc and mica are flexible.

Obs. Occurs imbedded in serpentine. Baste in the Hartz is a foreign locality. Blandford and Westfield, Mass., and Amity, N. Y., are given as American localities.

Clintonite. In oblique crystals: but usually massive, thin foliated, and brittle, with a submetallic luster, and reddish or yellowish-brown, or copper-red color. Streak yellowish-gray. *Composition,* silica 17·0, alumina 37·6, magnesia 24·3, lime 10·7, protoxyd of iron 5·0, water

---

What is meerschaum? its appearance? What is the structure of Schiller spar? its luster? What does it occur with? How does it differ from diallage?

3·6, (Clemson.) Infusible. Affords a transparent bead with borax. Acted on by the acids when pulverised. Occurs in limestone with serpentine at Amity, N. Y. It was named in honor of De Witt Clinton. It has also been called *Seybertite*.

Xanthophyllite is considered by Rose, its describer, as identical with Clintonite.

Pennine. Near chlorite; occurs in hexagonal tables, secondary to a rhombohedron of 118°. From the Pennine Alps.

Picrosmine. A green or greenish-white mineral, either fibrous like asbestus, or in rectangular prisms. H=2·5—3. Gr=2·59—2·7. Gives out water when heated, and has an argillaceous odor when moistened with the breath. Near serpentine in composition. From an iron mine in Bohemia.

Monradite is a cleavable yellowish mineral near picrosmine in composition.

Retinalite. A massive mineral, having a resinous appearance, found with and allied to serpentine. From Granville, Upper Canada.

Dermatine. Occurs massive, reniform or in crusts on serpentine, of a resinous luster and green color. Feel greasy. Odor when moistened argillaceous.

Villarsite. Occurs in yellowish rhombic octahedrons in dolomite at Traversella, in Piedmont. Allied in composition to serpentine.

Antigorite. A brownish or leek green mineral, in foliated masses and resembling Schiller spar.

Spadaïte. A flesh-red mineral, near Schiller spar.

Pyrallolite. A white or greenish cleavable mineral, dull and a little resinous in luster. Becomes black and then white again before the blowpipe, whence the name, from the Greek *pyr*, fire, *allos*, other, and *lithos*, stone. From Pargas, Finland. It is altered augite.

Pyrosclerite. A hydrous silicate of magnesia and alumina, of a light green, violet or grayish color, soft, and often foliated or micaceous. *Kæmmererite* is a violet variety of this mineral. Occurs in the Urals, at Unst in the Shetlands, at Texas in Pennsylvania.

Pyrophyllite. Foliated and pearly like talc; plates more or less radiating; very soft. Color white or greenish. It swells up and spreads out in fan-like shapes before the blowpipe. Occurs in the Urals.

Vermiculite is probably identical with pyrosclerite. It looks and feels like steatite; but when heated before the blowpipe, worm-like projections shoot out, owing to a separation of the thin leaves composing the grains, arising from the vaporization of the water present. Occurs at Milbury, Massachusets.

Periclase. Occurs at Vesuvius in small transparent octahedrons, and is pure magnesia. Luster vitreous; nearly as hard as feldspar. Gr=3·75.

Steatitic pseudomorphs. Pseudomorphous crystals often consist of a kind of steatite. A pseudomorph of this kind from Warwick, N. Y., having the form of hornblende, but so soft as to be easily cut with a knife, afforded Beck, silica 34·7, alumina 25·3, lime 5·1, magnesia 25·2, water 9·1. These crystals have been produced by a change of the original hornblende. Others have the form of spinel, &c.

The *Rensselaerite* of Emmons is believed to be a steatitic pseudomorph, or altered pyroxene.

2. *Anhydrous Silicates of Magnesia, and Compounds Isormorphous with them.*

PYROXENE.

Monoclinic. In modified oblique rhombic prisms; M : M=87° 5′. Cleavage perfect parallel with the sides of the prisms, and also distinct parallel with the diagonals. Usually in thick and stout prisms, of 6 or 8 sides, terminating in two faces meeting at an edge; *a* ; *a*= 120° 32′, M : *ĕ*=133° 33′, M : *ĕ*=136° 27′. Occurs also in oblique octahedrons, much modified. Massive varieties of a coarse lamellar structure; also fibrous, usually very fine and often long capillary; also granular, usually in coarse angular grains and friable, sometimes round; sometimes fine and compact.

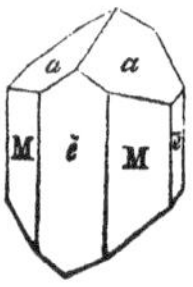

Colors green of various shades, verging to white on one side and brown and black on the other, passing through *blue* shades, but *not yellow*. Luster vitreous, inclining to resinous or pearly; the latter especially in fibrous varieties. Transparent to opaque. H=5—6. Brittle. Gr=3·2—3·5.

Pyroxene consists of silica and magnesia, combined with one or more of the bases, lime, protoxyd of iron, or protoxyd of manganese. These bases replace one another in a compound without changing the crystalline form, and have the same form nearly in their own crystallizations, as explained on page 74. The varieties of pyroxene arise from the variations in composition dependent on this isomorphism, and they differ much in appearance.

Varieties and Composition. The varieties may be divided into three sections—the light colored, the dark colored, and the thin foliated.

I. *White malacolite* or *white augite*—includes white or grayish-white crystals or crystalline masses. *Diopside*, in greenish-white or grayish-green crystals, and cleavable masses cleaving with a bright smooth surface. *Sahlite*; of a more dingy green color, less luster and coarser structure than diopside, but otherwise similar; named from the place

---

What is the character of the crystals of pyroxene? What is a common form? What is said of its massive varieties? its colors and luster? What are the constituents of pyroxene?

Sahla, where it occurs. *Fassaite;* in crystals of rich green shades and smooth and lustrous exterior. The name is derived from the foreign locality Fassa. *Alalite;* a diopside from Piedmont. *Coccolite* is a general name for granular varieties, derived from the Greek *coccos, grain.* The green is called *green coccolite,* the white, *white coccolite.* The specific gravity of these varieties varies from 3·25 to 3·3.

Composition: silica 55·3, lime 27·0, magnesia 17·0, protoxyd of manganese 1·6, protoxyd of iron 2·2. Fuse before the blowpipe to a colorless glass; with borax or soda form a transparent glass.

Asbestus. This name includes *fibrous* varieties of both pyroxene and hornblende; it is more particularly noticed under the latter species.

II. *Augite* includes black and greenish-black crystals, mostly presenting the form figured above. Specific gravity 3·3—3·4. *Hedenbergite* is a greenish-black opaque variety, in cleavable masses affording a greenish-brown streak. Specific gravity 3·5. *Polylite, Hudsonite,* and *Jeffersonite* fall here.

The varieties in this section contain a large proportion of iron, or iron and manganese. *Composition* of one variety, silica 54·1, lime 23·5, magnesia 11·5, protoxyd of iron 10·0, protoxyd of manganese 0·6=99·7. Fuse like the preceding, but the globule obtained is colored with iron.

III. *Diallage* is a thin-foliated, clear green variety, occurring imbedded in serpentine; folia thin, brittle, translucent. *Bronzite* occurs in serpentine and greenstone, and is similarly foliated; its colors are dark green, or greenish brown, with a metallic-pearly luster, or like bronze. Specific gravity 3·25. *Hypersthene* is less thinly foliated than bronzite, but cleaves readily; color grayish or greenish black, and luster metallic-pearly, Gr=3·39. The *Labrador hornblende,* and *Metalloidal diallage* are here included.

Composition of hypersthene, silica 54·25, lime 1·5, magnesia 14·0, protoxyd of iron 24·5, protoxyd of manganese *a trace,* alumina 2·25, water 1·0. The edges fuse with difficulty to a grayish green semi-opaque glass; some varieties wholly fuse. Other hypersthenes contain much less *iron* and a large proportion of lime.

Dif. Resembles hornblende, but is distinct in cleavage

What is coccolite? What is the appearance of asbestus? What is diallage? What is hypersthene?

and in the angles of its crystals. Moreover, the crystals are usually stout and thick, and never have the slender bladed form common with hornblende. Some fibrous varieties, however, can scarcely be distinguished except by analysis; yet it is a general fact, that asbestus occurring where pyroxene abounds, belongs to this species, and that with hornblende pertains to hornblende. White crystals of scapolite may be mistaken for this species, especially where two of the pyramidal faces in a crystal of scapolite are enlarged so as to resemble the oblique roof-like termination of crystals of pyroxene; but the angle between these faces in the former is 136° 7′, while it is 120° 32′ in pyroxene. Their relations to schiller spar and serpentine have already been stated. The species is never yellowish green like epidote.

Obs. Pyroxene is one of the most common minerals. It occurs in granite, granular limestone, serpentine, basalt and lavas. In basalt and lavas the crystals are generally small and black or greenish black. In the other rocks, they occur of all the shades of color given, and of all sizes to a foot or more in length. One crystal from Orange county, measured 6 inches in length, and 10 in circumference. White crystals occur at Canaan, Conn., Kingsbridge, New York county, and the Singsing quarries, Westchester county, N. Y., in Orange county at several localities; green crystals at Trumbull, Ct., at various places in Orange county, N. Y., Roger's Rock and other localities in Essex, Lewis, and St. Lawrence Co's. Dark green or black crystals are met with near Edenville, N. Y., Diana, Lewis county. Green coccolite is found at Roger's Rock, Long Pond, and Willsboro, N. Y.; black coccolite, in the forest of Dean, Orange county, N. Y. Diopside, at Raymond and Rumford, Me., Hustis's farm, Phillipstown, N. Y.

Pyroxene was thus named by *Haüy* from the Greek *pur* fire, and *xenos* stranger, in allusion to its occurring in lavas, where, according to a mistake of Haüy, it did not belong. The name *augite* is from the Greek *auge*, luster.

HORNBLENDE.

Monoclinic. In oblique rhombic prisms more or

What is said of the occurrence of pyroxene? How does it differ from hornblende? how from scapolite? What is the derivation of the names *pyroxene* and *augite*.

less modified; M : M=124° 30′. Cleavage perfect parallel with the sides of the prism. Often in long slender flat rhombic prisms, (fig. 3) breaking easily transversely; also 4, 6, and 8 sided prisms with oblique extremities. ě : ě=148° 30′. Occurs also frequently columnar, with a bladed structure; often fibrous, the fibers coarse or fine and frequently like flax, with a pearly or silky luster; also lamellar; also granular, either coarse or fine; generally firmly compact; rarely friable.

Colors from white to black passing through bluish green, grayish green, green, and brownish green shades, to black. Luster vitreous, with the cleavage face inclining to pearly. Nearly transparent to opaque. H=5—6. Gr=2·9—3.4.

Varieties and Composition. This species, like pyroxene, has numerous varieties, differing much in external appearance, and arising from the same causes—isomorphism and crystallization. Alumina enters into the constitution of some and replaces part of the other ingredients. The following are the most important:

1. Light Colored Varieties.

Tremolite, Grammatite. Tremolite comprises the white, grayish, and light greenish slender crystallizations, usually in blades or long crystals, penetrating the gangue or aggregated into coarse columnar forms. Sometimes nearly transparent. Gr=2·93. The name is from the foreign locality, *Tremola* in Switzerland.

Actinolite. The light green varieties. *Glassy actinolite* includes the bright glassy crystals, of a rich green color, usually long and slender (fig. 3) and penetrating the gangue like tremolite. *Radiated actinolite* includes olive green masses, consisting of aggregations of coarse acicular fibers, radiating or divergent. *Asbestiform actinolite* resembles the radiated, but the fibers are more delicate. *Massive actinolite* consists of angular grains instead of fibers. Gr=3·02—3·03. The name actinolite alludes to the radiated struc-

What is the crystallization of hornblende? What are common forms? What is said of the columnar and fibrous varieties? What are its colors? On what do the characters of its varieties depend? What is tremolite? what actinolite? Mention the characters of the varieties of actinolite?

ture of some varieties, and is derived from the Greek *aktin*, a ray of the sun. It is often mispelt *actynolite*.

Asbestus. In slender fibers easily separable, and sometimes like flax. Either green or white. *Amianthus* occurs in narrow seams, with a rich satin luster: much so-called is serpentine. *Ligniform asbestus* is compact and hard; it occurs of brownish and yellowish colors, and looks somewhat like petrified wood. *Mountain leather* occurs in thin tough sheets, looking and feeling a little like kid leather. It consists of interlaced fibers of asbestus, and forms thin seams between layers or in fissures of rocks. *Mountain cork* is similar, but is in thicker masses; it has the elasticity of cork, and is usually white or grayish-white.

The preceding light colored varieties contain little or no alumina or iron. *Composition* of glassy actinolite, silica 59·75, magnesia 21·1, lime 14·25, protoxyd of iron 3·9, protoxyd of manganese 0·3, hydrofluoric acid 0·8, (Bonsdorf.)

2. Dark Colored Varieties.

Pargasite. This name is applied to dark green crystals, short and stout, (resembling fig. 1,) with bright luster, of which Pargas in Finland is a noted locality. Gr = 3·11.

Hornblende. The black and greenish-black crystals and massive specimens. Often in slender crystallizations like actinolite; also short and stout like figures 1 and 2, the latter more especially. It contains a large per-centage of oxyd of iron, and to this owes its dark color. It is a tough mineral, as is implied in the name it bears. This character however is best seen in the massive specimens. Pargasite and hornblende contain both alumina and iron.

Composition of hornblende, silica 48·8, magnesia 13·6, lime 10·2, alumina 7·5, protoxyd of iron 18·75, protoxyd of manganese 1·15, hydrofluoric acid and water 0.9, (Bonsdorf.)

Composition of pargasite, silica 46·3, magnesia 19·0, lime 14·0, alumina 11·5, protoxyd of iron 3·5, protoxyd of manganese 0·4, hydrofluoric acid and water 2·2.

Amphibole is a name often given to this species.

The varieties of hornblende fuse easily with some ebullition, the white varieties forming a colorless glass and the green a globule more or less colored by iron.

What is asbestus and amianthus? mountain leather and mountain cork? What is the peculiarity in composition of the light colored varieties of hornblende? what of the dark varieties?

Dif. Distinguished from pyroxene as stated under that species; the black variety from black tourmaline by its perfect cleavage, (tourmaline having none,) and also by the form of its crystals; the fibrous varieties from picrosmine, nemalite, and tabular spar, as stated under those species; from the fibrous zeolites by not gelatinizing, and, when in limestone or serpentine, by its gangue.

Obs. Hornblende is an essential constituent of certain rocks, as syenite, trap and hornblende slate. Actinolite is usually found in magnesian rocks, as talc, steatite or serpentine; tremolite in granular limestone and dolomite; asbestus in the above rocks and also in serpentine. Black crystals of hornblende occur at Franconia, N. H., Chester, Mass., Thomaston, Me., Willsboro', N. Y. in Orange county, N. Y., and elsewhere. Pargasite occurs at Phipsburg and Parsonsfield, Me.; glassy actinolite, in steatite or talc, at Windham, Readsboro', and New Fane, Vt., Middlefield and Blandford, Mass.; and radiated varieties at the same localites and many others. Tremolite and gray hornblende occur at Canaan, Ct., Lee, Newburgh, Mass., in Thomaston and Raymond, Me., Lee and Great Barrington, Mass., Dover, Kingsbridge, and in St. Lawrence county, N. Y., at Chesnut Hill, Penn., at the Bare Hills, Md. Asbestus at many of the above localities; also at Milford, Conn., Brighton and Sheffield, Mass., Cotton Rock and Hustis's farm, Phillipstown, N. Y., near the quarantine, Richmond county, N. Y. Mountain leather is met with at the Milford quarries, and also at Brunswick, N. J.

Uses. Asbestus is the only variety of this species of any use in the arts. The flax-like variety is sometimes woven into cloth; it has been proposed of late to use clothes of it for firemen, and patents have been taken out. Its incombustibility and slow conduction of heat, render it a complete protection against the flames. It is often made into gloves. A garment when dirty, need only be thrown into the fire for a few minutes to be white again. The ancients, who were acquainted with its properties, are said to have used it for napkins, on account of the ease with which it was cleaned. It was also the wicks of the lamps in the ancient temples; and because it maintained a perpetual flame

---

How does the species hornblende differ from tourmaline and other minerals mentioned? What is said of the occurrence of hornblende? What are the uses of asbestus? Why was it so called?

without being consumed, they named it *asbestos*, unconsumed. It is now used for the same purpose by the natives of Greenland. The name amianthus alludes to the ease of cleaning it, and is derived from *amiantos*, undefiled. Asbestus is now extensively used for lining iron safes. The best locality for collecting asbestus in the United States, is that near the quarantine, in Richmond County, N. Y.

Anthophyllite. In oblong grayish, greenish or brownish crystals, or in needles, imbedded in mica slate, or penetrating it. Brittle; fibers sharp. Gr=2·9—3·16. It is a variety of hornblende. Occurs at Haddam and Guilford, Conn., and Chesterfield, Chester and Blandford, Mass.

Cummingtonite. Fibrous; the fibers divergent, stellular or scopiform; ash-gray; a little silky. A variety of hornblende. From Cummington and Plainfield, Mass., in mica slate.

Acmite. In long highly polished prisms, of a dark brown or reddish-brown color, with a pointed extremity, penetrating granite, near Kongsberg in Norway. M : M=86° 56′. Resembles pyroxene. Fuses easily before the blowpipe.

Babingtonite. Resembles some varieties of pyroxene. It occurs in greenish-black splendent crystals in quartz at Arendal in Norway.

SPODUMENE.

Monoclinic. Crystals like those of pyroxene. Surface of cleavage pearly. Color grayish or greenish. Translucent to subtranslucent. H=6·5—7. Gr=3·1—3·19.

Composition: silica 64·5, alumina 2·3, lithia 6·2. Intumesces before the blowpipe, and fuses to a transparent glass. In fine powder mixed with bisulphate of potash and fluor, and fused on platinum foil, it tinges the flame red, owing to the lithia contained.

Dif. Resembles somewhat feldspar and scapolite, but has a higher specific gravity and a more pearly luster, and affords rhombic prisms by cleavage.

Obs. Occurs in granite at Goshen; also at Chesterfield, Norwich and Sterling, Mass.; at Windham, Me.; at Brookfield, Ct. It is found at Uton, in Sweden, Sterzing in the Tyrol; and at Killiney bay, near Dublin.

Triphane is another common name of this mineral.

Uses. This mineral is remarkable for the *lithia* it contains, and has been used for obtaining this rare earth.

Mention the characters of spodumene. How much lithia does it contain? How does it differ from feldspar and scapolite?

CHRYSOLITE.—*Olivine.*

Trimetric. In right rectangular prisms, having perfect cleavage parallel with the smaller lateral plane. Usually in imbedded grains of an olive green color, looking like green bottle glass. Also yellowish-green. Transparent to translucent. H=6·5—7. Gr=3·3—3·5. Looks much like glass in the fracture, except in the direction of the cleavage.

Composition : silica 38·5, magnesia 48·4, protoxyd of iron 11·2, oxyd of manganese 0·3, alumina 0·2. Darkens before the blowpipe but (except certain varieties) does not fuse. Forms a green glass with borax.

Dif. Distinguished from green quartz by its occurring disseminated in basaltic rocks, which never so occurs ; also in its cleavage. On account of its gangue it cannot be mistaken for beryl. From obsidian or volcanic glass it differs in its infusibility.

Obs. Occurs disseminated through basalt and lavas, and is a characteristic mineral of some varieties of these rocks. Has been found in New Hampshire. *Boltonite,* from limestone at Bolton, Mass., is a variety of chrysolite.

Uses. Sometimes used as a gem, but it is too soft to be valued, and is not delicate in its shade of color.

CHONDRODITE.

Usually in imbedded grains or small rounded or flattened kernels or nodules in limestone, and apearing brittle. Structure finely granular without cleavage. Color brownish yellow or brown, sometimes reddish or white, and occasionally black. Luster vitreous, inclining a little to resinous. Streak rarely colored. Translucent or subtranslucent. Fracture uneven. H=6—6 5. Gr=3·1—3·2.

Composition : silica 33·1, magnesia 55·5, protoxyd of iron 3·6, fluorine 7·6. From New Jersey. Fuses with difficulty on the edges. With borax fuses easily to a yellowish-green glass.

Dif. As it occurs only in limestone it will hardly be confounded with any species resembling it in color when the gangue is present. The specific gravity is less than that of tourmaline or garnet, some brownish yellow varieties of which it approaches in appearance ; moreover, it is seldom in crystals, and when so, the faces are not polished. This

---

What is the crystallization of chrysolite ? describe its character ; its blowpipe action ; composition ; occurrence ; differences. Describe chondrodite.

mineral has been called *Brucite ;* but chondrodite is of prior authority ; it is from the Greek *chondros*, a grain.

Obs. It is abundant in the adjoining counties, Sussex, N. J., and Orange, N. Y., occurring at Sparta, and Bryam, N. J., and in Warwick and other places in New York. At Vesuvius it occurs in small crystals, called Humite.

4. ALUMINA.

1. *Uncombined.*

CORUNDUM.

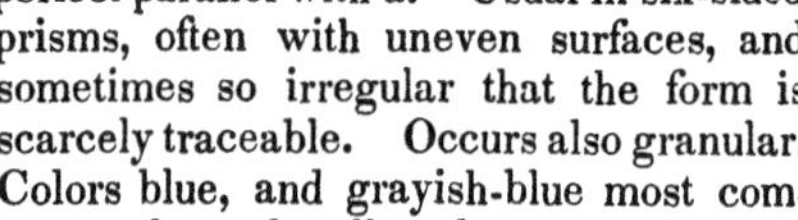

Rhombohedral. $R : R = 86° 4'$. Cleavage sometimes perfect parallel with a. Usual in six-sided prisms, often with uneven surfaces, and sometimes so irregular that the form is scarcely traceable. Occurs also granular. Colors blue, and grayish-blue most common ; also red, yellow, brown, and nearly black ; often bright. When polished on the surface a, a star of six rays, corresponding with the six-sided form of the prism, is sometimes seen within the crystal. Transparent to translucent. $H=9$, or next to the diamond. Exceedingly tough, when compact. $Gr=3{\cdot}9$—$4{\cdot}16$.

Composition : pure alumina. It remains unaltered before the blowpipe both alone and with soda. Fuses with difficulty with borax.

Varieties. The name *sapphire* is usually restricted in common language to clear crystals of bright colors, used as gems ; while dull, dingy-colored crystals and masses are called *corundum*, and the granular variety of bluish-gray and blackish colors is called *emery.*

Blue is the true sapphire color. When of other bright tints, it receives other names ; as *oriental ruby*, when red ; *oriental topaz*, when yellow ; *oriental emerald*, when green ; *oriental amethyst*, when violet ; and *adamantine spar*, when hair-brown. Crystals with a radiate chatoyant interior are often very beautiful, and are called *asteria*, or *asteriated sapphire.*

What is the usual form of crystals of corundum ? What are their colors ? hardness ? Of what does sapphire consist ? What are the red, yellow and green varieties called ? What the hair-brown variety ? What are corundum and emery ? What is asteriated sapphire ?

Dif. Distinguished readily by its hardness, exceeding all species except the diamond, and scratching quartz crystals with great facility.

Obs. The sapphire is usually found loose in the soil: primitive rocks, and especially gneissoid mica slate, talcose rock and granular limestone, appear to be its usual matrix. It is met with in several localities in the United States, but seldom sufficiently fine for a gem. A blue variety occurs at Newton, N. J., in crystals sometimes several inches long; bluish and pink, at Warwick, N. Y.; white, blue and reddish crystals, at Amity, N. Y.; grayish, in large crystals in Delaware and Chester counties, Pennsylvania; pale blue crystals have been found in boulders at West Farms and Litchfield, Ct. It occurs also in considerable quantities in North Carolina; also in Chester county, Georgia, where a fine red sapphire has been obtained.

The principal foreign localities are as follows: blue, from Ceylon; the finest red from the Capelan Mountains in the kingdom of Ava, and smaller crystals from Saxony, Bohemia and Auvergne; corundum, from the Carnatic, on the Malabar coast, and elsewhere in the East Indies; adamantine spar, from the Malabar coast; emery, in large boulders from near Smyrna, and also at Naxos and several of the Grecian islands.

The name sapphire is from the Greek word *sappheiros*, the name of a blue gem. It is doubted whether it included the sapphire of the present day.

Uses. Next to the diamond, the sapphire in some of its varieties is the most costly of gems. The red sapphire is much more highly esteemed than those of other colors. A crystal weighing 3½ carats, perfect in transparency and color, has been valued at the price of a diamond of the same size. They seldom exceed half an inch in their dimensions. Two splendid red crystals, as long as the little finger and about an inch in diameter, are said to be in the possession of the king of Arracan.

Blue sapphires occur of much larger size. According to Allan, Sir Abram Hume possesses a crystal which is three inches long; and in Mr. Hope's collection of precious stones

---

How is the species sapphire distinguished? In what rocks does the sapphire occur? What are some of the American localities? what are the principal foreign? What is said of the value of sapphires?

there is one crystal formerly belonging to the Jardin des Plantes of Paris, for which he gave £3000 sterling.

The largest oriental ruby known was brought from China to Prince Gargarin, governor of Siberia; it afterwards came into the possession of Prince Menzikoff, and constitutes now a jewel in the imperial crown of Russia.

2. *Combined with other oxyds.*

SPINEL.

Monometric. In octahedrons, more or less modified, and dodecahedrons. Figure 1, is the octahedron with truncated

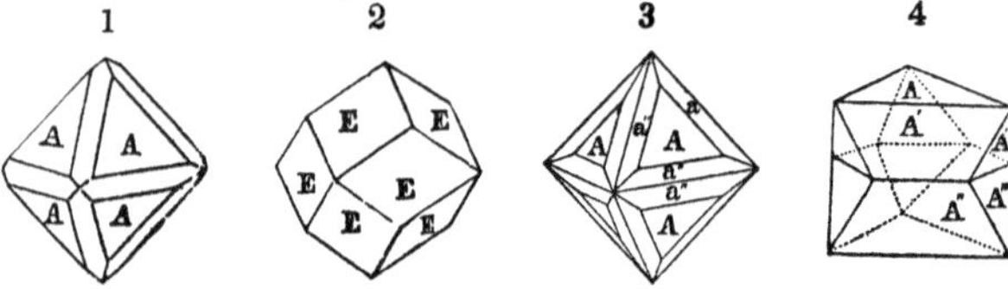

edges; figure 3, the same with beveled edges; figure 2, the dodecahedron. Occurs only in crystals; cleavage octahedral, but difficult. Figure 4 represents a twin crystal.

Color red, passing into blue, green, yellow, brown and black. The red shades often transparent and bright; the dark shades usually opaque. Luster vitreous. H=8. Gr=3·5—3·6.

Composition: of a red spinel, from Ceylon, alumina 69·0, magnesia 26·2, protoxyd of iron 0·7, silica 2·0, chromic acid 1·1. Essentially alumina and magnesia. Infusible alone, and with difficulty with borax.

Varieties. The following are the varieties of this species that have received distinct names: The scarlet or bright red crystals, *spinel ruby;* the rose-red, *balas-ruby;* the orange-red, *rubicelle;* the violet, *almandine-ruby;* the green, *chlorospinel;* while the black varieties are called *pleonaste.* Pleonaste crystals contain sometimes 8 to 20 per cent. of oxyd of iron.

Dif. The form of the crystals and their hardness distinguish the species. Garnet is fusible. Magnetic iron ore

---

What is the usual crystalline form of spinel? What is its hardness? What are its colors? Of what does it essentially consist? Mention the colors and names of some of the varieties?

is attracted by the magnet. Zircon has a high specific gravity and is not so hard.

Obs. Occurs in granular limestone; also in gneiss and volcanic rocks. At numerous places in the adjoining counties of Sussex in New Jersey, and Orange county, of various colors from red to brown and black; especially at Franklin, Newton and Sparta, in the former, and in Warwick, Amity and Edenville, in the latter. The crystals are octahedrons, and often grouped or disseminated singly in granular limestone. One crystal found at Amity by Dr. Heron, weighs 49 pounds. The limestone quarries of Bolton, Boxborough, Chelmsford and Littleton, Mass., afford a few crystals.

Crystals of spinel are occasionally soft, having undergone a change of composition, and approaching steatite in all characters except form. They are true *pseudomorphs*. They are met with in Sussex and Orange counties.

Uses. The fine colored spinels are much used as gems. The red is the common ruby of jewelry, the *oriental* rubies being sapphire. Crystals weighing 4 carats have been valued at half the price of a diamond of the same size.

Automolite. A variety of spinel, containing 20—35 per ct. of oxyd of zinc. Color dark green or black. H=7·5—8. Gr=4—4·6. With soda it forms at first a dark scoria, and when fused again with more soda, a ring of oxyd of zinc is deposited on the charcoal. Infusible alone, and nearly so with borax.

Occurs in granite at Haddam with beryl, chrysoberyl, garnet, &c. In Sweden, near Fahlun, in talcose slate.

Dysluite. A variety of the species spinel, containing oxyd of iron and zinc. Color yellowish or grayish-brown. H=7·5—8. G=4·55. *Composition,* alumina 30·5, oxyd of zinc 16·8, peroxyd of iron 41·9, protoxyd of manganese 7·6, silica 3, moisture 0·4. Becomes red before the blowpipe, but loses the color on cooling. Infusible alone with borax affords a translucent bead of a deep garnet-red color. The name *dysluite* is from the Greek *dus*, with difficulty, and *luo*, to dissolve. From Sterling, N. J., with Franklinite and Troostite.

Hercinite. A spinel consisting of alumina and protoxyd of iron, with only 2·9 per cent. of magnesia.

3. *Hydrous combinations with Silica.*

HALLOYSITE.—*Hydrous Silicate of Alumina.*

Massive and earthy, resembling a compact steatite. Yields to the nail, and may be polished by it.

How is spinel distinguished from magnetic iron? from garnet? from zircon? For what are spinels used? What is automolite? What is the appearance of halloylite?

Color white or bluish. Adheres to the tongue, and small pieces become transparent in water. Gr=1·8—2·1.

Composition: silica 39·5, alumina 34·0, water 26·5. Dissolves in sulphuric acid, yielding a jelly. Becomes milk-white before the blowpipe.

Obs. From Liege and Bayonne, France. Named in honor of the geologist, Omalius d' Holly.

NOTE.—There are several other hydrous silicates of alumina allied to halloylite, having the following names: *Pholerite, kollyrite, cimolite, bole, fettbol, rock soap, rosite, groppite, malthacite,* and *smelite.* They are in general soft and earthy, often clay-like, and are distinguished from similar magnesian species by the blowpipe test for alumina.

There are also stalactitic hydrous silicates, found in volcanic and other igneous rocks, and formed by the decomposition of feldspar or other ingredients. Such silico-aluminous stalactites are not uncommon in the Pacific Islands. They are of mixed composition, as necessarily results from their mode of origin. *Gibbsite* is in some cases of this character. When containing an alkali they become zeolites.

Allophane. Reniform and massive, occasionally with traces of crystallization; sometimes almost pulverulent. Color pale blue; sometimes green, brown or yellow. Luster vitreous or resinous. Splendent and waxy internally. Streak white. H=3. Gr=1·85—1·90. *Composition,* alumina 29·2, silica 21·9, water 44·2, mixed clay 4·7. Becomes opaque, colorless and pulverulent before the blowpipe, intumesces a little and tinges the flame green. Forms a jelly with acids. In marl in Thuringia and Saxony, and in chalk at Beauvais in France.

The name allophane is from the Greek *allos,* other, and *phaino,* to appear, alluding to its changes of appearance before the blowpipe.

Schrœtterite, or *opal allophane,* resembles allophane; it consists of silica 12·0, alumina 46·3, water 36·2, with some iron, copper and lime.

FAHLUNITE.—*Chlorophyllite—Pinite.*

In six and twelve-sided prisms, usually foliated, parallel to the base. Folia soft and brittle, of a grayish-green to dark olive-green color, and pearly luster. Gr=2·7.

Composition: of *Fahlunite,* silica 44·9, alumina 30·7, peroxyd of iron 7·22, potash 1·38, magnesia 6·04, water 8·65, protoxyd of manganese 1·90, lime 0·95. (Wachtmeister.) Of *Chlorophyllite,* silica 45·2, alumina 27·6, magnesia 9·6, protoxyd of iron 8·2, protoxyd of manganese 4·1, water 3·6, (Jackson.) Yields water before the blowpipe and becomes bluish-gray, but fuses only on the edges.

Dif. It is distinguished from talc by affording water be-

Of what does halloylite consist?

fore the blowpipe, and readily by its association with iolite, and its large hexagonal forms, with brittle folia.

Obs. Occurs with iolite in granite at Haddam, Ct., and at Unity, N. H. The iolite and chlorophyllite are often interlaminated, and the latter appears to result from the alteration of the former, in which the principal change is the addition of water. A variety from Brevig, in Norway, has been called *esmarkite*. The *fahlunite* is from Fahlun, Sweden.

The name chlorophyllite, given to this species by Dr. Jackson, is derived from the Greek *chlôros*, green, and *phullon*, leaf.

Pinite includes the alkaline varieties of altered iolite. The cleavage is often indistinct. Color gray to grayish green. Occurs in Auvergne, in decomposed feldspar-porphyry, and at Schneeberg, in Saxony.

The following species, like chlorophyllite in crystallization, appear also to have proceeded from the alteration of iolite.

Gigantolite. Color greenish to dull steel gray. Gr=2.85—2·88. From Tamela, Finland. *Iberite* is near gigantolite. Color pale grayish green. Gr=2·89. *Hydrous iolite* of Bonsdorf, differs from chlorophyllite in containing one per cent. more of water.

Aspasiolite is another hydrous mineral allied to the above, and found associated with iolite. It usually resembles a light green serpentine, and occurs in six-sided prisms.

ZEOLITE FAMILY.

NOTE.—The following species from heulandite to chabazite, inclusive, constitute what has been called the *zeolite* family, so named because the species generally melt and intumesce before the blowpipe, the term being derived from the Greek *zeo*, to boil. They consist essentially of silica, alumina and some alkali, with more or less water. The most of them gelatinize in acids, owing to the separation of the silica in a gelatinous state.

They occur filling cavities in rocks, constituting narrow seams, or implanted on the surface, and rarely in imbedded crystals; and never disseminated through the body of a rock like crystals of garnet or tourmaline. All occur

---

What is the meaning of the word zeolite? What is the constitution of the zeolites? their mode of occurrence?

in amygdaloid, and some of them occasionally in granite or gneiss. The first four, *heulandite, laumonite, apophyllite, stilbite*, have a *strong pearly cleavage*, and do not occur in *fine fibrous* crystallizations; when columnar, the structure is thin lamellar. Excepting laumonite, these species dissolve in the strong acids, but do not gelatinize. The species *natrolite, scolecite, stellite*, and *thomsonite*, are *often fibrous*, and the crystallizations generally *slender*. The remaining species, *harmotome, analcime, sodalite, hauyne, lapis lazuli*, and *chabazite*, occur in *short or stout glassy crystals*, and are seldom *fibrous*. To the *second* division above given might be added the species *dysclasite* and *pectolite*, described under *Lime*. They have a more pearly or silky luster than natrolite.

HEULANDITE.

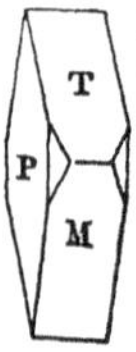

T P M

Monoclinic. In right rhomboidal prisms and their modifications. P on M or T=90°. M : T=129° 40′. Cleavage highly perfect, parallel to P. Luster of cleavage face pearly, of other faces vitreous. Color white; sometimes reddish, gray, brown. Transparent to subtranslucent. Folia brittle. H=3·5—4. Gr=2·2.

Composition: silica 59·3, alumina 16·8. lime 9·2, water 14·7. Intumesces and fuses, and becomes phosphorescent. Dissolves in acid without gelatizing.

Dif. Distinguished from gypsum by its hardness and the action of acids and the blowpipe; from apophyllite and stilbite by its crystals.

Obs. Found in amygdaloid; occasionally in gneiss, and in some metalliferous veins.

Occurs at Bergen Hill, N. J., in trap; at Hadlyme, Ct., and Chester, Massachusetts, on gneiss; near Baltimore, on a syenitic schist; at Peter's Point and Cape Blomidon, Nova Scotia, in trap.

The species was named by Brooke in honor of Mr. Heuland, of London. *Lincolnite* is here included.

Brewsterite. Crystals right rhomboidal prisms, with a perfect pearly cleavage like heulandite; but M : T=93° 40′. H=4½—5. Gr=2·1—2·5. From Argyleshire and the Giant's Causeway.

What is the appearance and structure of heulandite? How is it distinguished from gypsum? how from apophyllite and stilbite?

STILBITE.

In right rectangular prisms, more or less modified; cleavage perfect parallel with M̆. The prism is usually flattened parallel with the cleavage face, (annexed figure,) and terminates in a pyramid; a : a = 119°. Also in sheath-like aggregations and thin columnar.

Color white; sometimes yellow, brown or red. Luster of cleavage face pearly, of other faces vitreous. Subtransparent to translucent. H = 3·5—4. Gr = 2·13—2·15.

Composition: silica 57·6, alumina 16·3, lime 8·9, water 16·3. Before the blowpipe fuses with intumescence to a colorless glass. Does not gelatinize except after long boiling in nitric acid.

Dif. Distinguished from gypsum like heulandite; and from heulandite by its crystals, which are usually thin, elongated rectangular prisms, with pyramidal terminations, often uneven in surface.

Obs. Occurs mostly in amygdaloid; also on gneiss and granite.

It is found sparingly at the Chester and Charlestown syenite quarries, Mass., at Thatchersville and Hadlyme, Ct., at Phillipstown, N. Y., at Bergen Hill, N. J., in trap, in the copper region of Lake Superior, in amygdaloid. In beautiful crystallizations at Partridge Island, Nova Scotia.

The name stilbite is derived from the Greek *stilbè*, luster.

APOPHYLLITE.

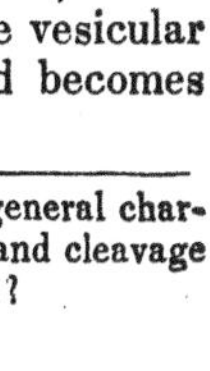

Dimetric. In right square prisms or octahedrons. Cleavage parallel with the base highly perfect. Prisms often terminate in a sharp pyramid, (annexed figure,) a : a = 104° 2′ and 121°. Massive and foliated. Color white or grayish; sometimes with a shade of green, yellow, or red. Luster of P pearly: of the other faces vitreous. Transparent to opaque. H = 4·5—5. Gr = 2·3—2·4.

Composition: silica 51·9, lime 25·2, potash 5·1, water 16·0. Exfoliates and ultimately fuses to a white vesicular glass. In nitric acid separates into flakes and becomes somewhat gelatinous and subtransparent.

What is the crystallization of stilbite? What are its general characteristics? How is it distinguished? What is the form and cleavage of crystals of apophyllite? What are its other characters?

Dif. The acute pyramidal terminations of its glassy crystals at once distinguish it from the preceding, as also its cleavage *across* the prism.

The name alludes to its exfoliation before the blowpipe.

Obs. Found in amygdaloidal trap and basalt.

Occurs in fine crystallizations at Peter's Point and Partridge Island, Nova Scotia, and at Bergen Hill, N. J.

LAUMONITE.

Monoclinic. In oblique rhombic prisms; M : M=86° 15′, P : M=68° 40′. Cleavage parallel to the acute lateral edge; also massive, with a radiating or divergent structure.

Color white, passing into yellow or gray. Luster vitreous, inclining to pearly on the cleavage face. Transparent to translucent. H=3·5—4. Gr=2·3. Becomes opaque on exposure, and readily crumbles.

Composition: silica 51·1, alumina 21·8, lime 11·9, water 15·2. Intumesces and fuses to a white frothy mass. Gelatinizes with nitric or muriatic acid, but is not affected by sulphuric unless heated.

Dif. The alteration this species undergoes on exposure to the air, at once distinguishes it. This result may be prevented with cabinet specimens, by dipping them into a solution of gum arabic.

Obs. Found in amygdaloid and also in gneiss, porphyry, and clay slate. Peter's Point, Nova Scotia, is a fine locality of this species. Occurs also at Phipsburg, Me.; Charlestown syenite quarries, Mass.; Bergen Hill, N. J.; in the amygdaloid of the copper region, Lake Superior.

Leonhardite resembles laumonite; it contains silica 55, alumina 24·1, lime 10·5, water and loss 12·30.

NATROLITE.

Trimetric. In right rhombic prisms, usually slender and terminated by a short pyramid; M : M=91° 10′; e : e=143° 14′, M : e=116° 37′. Cleavage perfect parallel with M. Also in globular, stellated, and divergent groups, consisting of delicate acicular fibers, the fibers often terminating in acicular prismatic crystals.

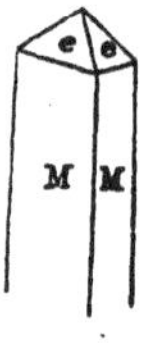

Color white, or inclining to yellow, gray, or red.

How is apophyllite distinguished? What are the characters of laumonite? What takes place when it is exposed to the air? What is the crystallization of natrolite? mention other characters.

Luster vitreous. Transparent to translucent. H=4·5—5·5. Brittle. Gr=2·14—2·23.

Composition: silica 47·4, alumina 26·9, soda 16·2, water 9·5 Becomes opaque before the blowpipe and fuses to a glassy globule. Forms a thick jelly in the acids, *after* heating as well as *before*.

Dif. Distinguished from scolecite by its action before the blowpipe.

Obs. Found in amygdaloidal trap, basalt and volcanic rocks. The name natrolite is from *natron*, soda.

Occurs in the trap of Nova Scotia and Bergen Hill, N. J.

Scolecite resembles natrolite, and differs in containing *lime* in place of *soda*. The luster is vitreous or a little pearly. Before the blowpipe it curls up like a worm (whence the name from the Greek *skolex* a *worm*) and then melts. From Staffa, Iceland, Finland, Hindostan.

Poohnahlite is a related species, from Poohnah, Hindostan. M : M= 91° 49′.

Mesole is another related species, occurring usually in implanted globules, having a flat columnar or lamellar radiated structure, with a pearly or silky luster. Gr=2·35—2·4. Fuses easily before the blowpipe and gelatinizes readily with acids. From the Faroe islands and Greenland. *Harringtonite* from the north of Ireland, and *Brevicite* from Brevig, Norway, appear to be identical with mesole.

Natrolite, scolecite, mesole, and some other zeolites, together correspond to the old species *mesotype*.

THOMSONITE.

Trimetric. In right rectangular prisms. Usually in masses, having a radiated structure within, and consisting of long fibers or acicular crystals; also amorphous.

Color snow-white. Luster vitreous, inclining to pearly. Transparent to translucent. H=5—5½. Brittle. Gr= 2·3—2·4.

Composition: silica 37·4, alumina 31·8, lime 13·0, soda 4·8, water 13·0. Intumesces and becomes opaque; but the edges merely are rounded at a high heat. When pulverized, it gelatinizes with nitric or muriatic acids.

Dif. Distinguished from natrolite and other zeolites by its difficult fusibility.

Obs. Occurs in amygdaloid, near Kilpatrick, Scotland; in lavas at Vesuvius; in clinkstone in Bohemia. Also at Peter's Point, Nova Scotia, in trap.

The species was named in honor of Dr. Thomas Thomson, of Glasgow.

The species *comptonite* and *ozarkite* are identical with thomsonite.

HARMOTOME.

Trimetric. In modified rectangular prisms; and very commonly twin crystals similar to the annexed figure.

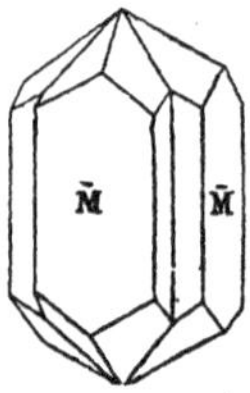

Color white; sometimes grayish, yellowish, or brownish. Subtransparent to translucent. Luster vitreous. H=4·—4·5. Brittle. Gr=2·39—2·5.

Composition: silica 44·0, alumina 16·6, baryta 24·8, water 14·6. Fuses without intumescence to a clear globule. Phosphoresces with a yellow light when heated. Scarcely attacked by the acids unless they are heated.

Dif. Its twin crystals, when distinct, cannot be mistaken for any other species except phillipsite. It is much more fusible than glassy feldspar or scapolite; it does not gelatinize in cold acids like thomsonite.

Obs. Occurs in amygdaloid, gneiss, and metalliferous veins. Fine crystallizations are found at Strontian in Argyleshire, Andreasberg in the Hartz, and Kongsberg in Norway.

The name harmotome is from the Greek *harmos* a joint, and *temno* to cleave.

Phillipsite. Near harmotome in its cruciform crystals and other characters; but differing in containing lime in place of baryta. It differs also in gelatinizing with acids and in fusing with some intumescence. It also occurs in sheaf-like aggregations and in radiated crystallizations. From the Giant's Causeway, Capo di Bove, and Vesuvius.

Zeagonite, from the last two localities mentioned, is identical with Phillipsite.

ANALCIME.

Monometric. Occurs usually in trapezohedrons, (fig. 1,) also fig. 2; cleavage cubic and only in traces.

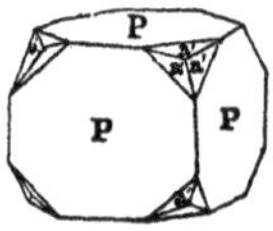

Often colorless and transparent, also milk-white, grayish and reddish-white, and sometimes opaque. The appearance sometimes seen in polarized light is shown in figure 96, page 61. Luster vitreous. H=5—5·5. Gr=2·07—2·28.

What is the common form of harmotome? what its color and appearance? What are its distinguishing characters? What is the form of crystals of analcime? the color and other characters?

Composition: silica 54·6, alumina 23·2, soda 14·0, water 8·1. Fuses before the blowpipe on charcoal without intumescence to a clear glassy globule. Gelatinizes in muriatic acid, with difficulty.

Dif. Characterized by its crystallization, without cleavage. Distinguished from quartz and leucite by its inferior hardness; from calc spar by its fusibility, and by not effervescing with acids; from chabazite and its varieties by fusing *without* intumescence to a *glassy* globule, and by the crystalline form.

Obs. Found in amygdaloid and lavas; also in gneiss.

Occurs in fine crystallizations in Nova Scotia; also at Bergen Hill, N. J.; Perry, Me.; and in the amygdaloid of the copper region, Lake Superior. The Faroe Ids., Iceland, Vicentine, the Hartz, Sicily, and Vesuvius are some of the foreign localities.

The name *analcime* is from the Greek *analkis, weak,* alluding to its weak electric power when heated or rubbed.

CHABAZITE.

Rhombohedral. Often in rhombohedrons, much resembling cubes. (Fig. 1.) R : R=94° 46′. Cleavage paral-

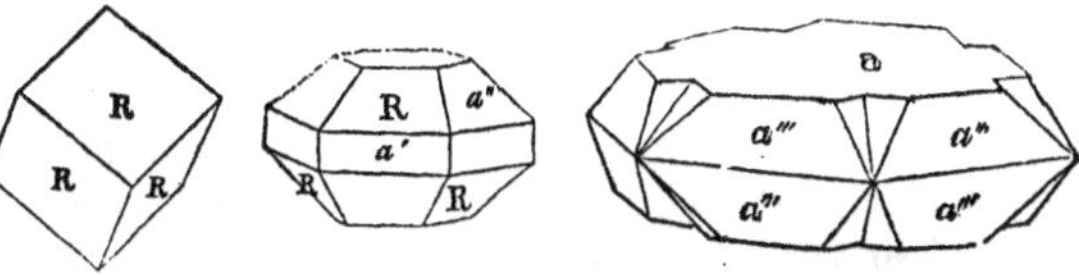

lel to the primary faces. Also in complex modifications of this form, and double six-sided pyramids or short six-sided prisms terminating in truncated pyramids. (Fig. 2.) Also in compound crystals, (fig. 3.) Never massive or fibrous.

Color white, also yellowish and red. Luster vitreous. Transparent to translucent. H=4—4·5. Gr=2·06—2·17.

Composition: silica 48·4, alumina 19·3, lime 8·7, potash 2·5, water 21·1.

This species includes *gmelinite,* occurring in small glassy crystals of the form in figure 2; also *levyne,* occurring in compound crystals (fig. 3;) also *ledererite,* which has the form

---

Mention some of the distinctive characters of analcime. What is said of the crystallization of chabazite? mention other characters.

of gmelinite, but appears to differ in containing just one third the proportion of water; also *phacolite*, occurring in small glassy crystals having the form of double six-sided pyramids. The *acadiolite* is a red variety from Nova Scotia. *Herschelite* is another variety in small hexagonal tables.

The varieties intumesce and whiten before the blowpipe. Gmelinite forms a jelly with acids.

Dif. The nearly cubical form often presented by the crystals of chabazite is a striking character. It is distinguished from analcime as stated under that species; from calc spar by its hardness and action with acids; from fluor spar by its form and cleavage, and its showing no phosphorescence.

Obs. Found in trap, gneiss, and syenite. Chabazite is met with in the trap of the Connecticut valley, but in poor specimens; also at Hadlyme, and Stonington, Ct., at Charlestown, Mass., Bergen Hill, N. J., Piermont, N. Y. Nova Scotia affords common chabazite and also the ledererite. The Faroe Islands, Iceland, and Giant's Causeway are some of the foreign localities. *Gmelinite* comes from the Vicentine; also the county of Antrim, Ireland; *levyne* from Glenarm, Scotland; also Iceland, Faroe, &c.

Haydenite. Resembles chabazite in the appearance of its crystals and is probably the same species. Occurs with heulandite at Jones's Falls, near Baltimore.

PREHNITE.

Primary form a right rhombic prism; M : M=99° 56. Cleavage, basal. Usually in six-sided prisms, rounded so as to be barrel-shaped, and composed of a series of united plates; also in thin rhombic or hexagonal plates. Often reniform and botryoidal; texture compact.

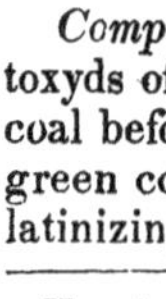

Color light green to colorless. Luster vitreous, except the face P, which is somewhat pearly. Subtranspa-rent to translucent. H=6—6·5. Gr=2·8—2·96.

Composition: silica 43·0, alumina 23·25, lime 26·0, protoxyds of iron and manganese 2·25, water 4·0. On charcoal before the blowpipe froths and melts to a slag of a light green color. Dissolves slowly in muriatic acid without gelatinizing, leaving a flaky residue.

How is chabazite distinguished from calc spar? how from fluor spar? What is the usual form and structure of prehnite? What is its color? luster? hardness?

Dif. Distinguished from beryl, green quartz, and chalcedony by fusing before the blowpipe, and from the zeolites by its superior hardness. The ordinary broken appearance of its crystals is quite characteristic.

Obs. Found in trap, gneiss, and granite.

Occurs in the trap of Farmington, and Woodbury, Ct., West Springfield, Mass., and Patterson and Bergen Hill, N. J.; in gneiss at Bellows Falls, Vt.; in syenite at Charlestown, Mass.; and very abundant, forming a large vein, in the copper region of Lake Superior, three miles south of Cat harbor, and elsewhere.

The Fassa valley in the Tyrol, St. Crystophe in Dauphiny, and the Salisbury Crag, near Edinburgh, are some of the foreign localities.

Uses. Prehnite receives a handsome polish and is sometimes used for inlaid work. In China it is polished for ornaments, and large slabs have been cut from masses brought from there.

Epistilbite. A hydrous silicate of alumina and lime. Occurs in thin rhombic prisms, of a white color, with a perfect pearly cleavage like stilbite. H=3½—4. Gr=2·25. Before the blowpipe froths and forms a vesicular enamel. Does not gelatinize. From Iceland and Hindostan, and sparingly at Bergen Hill, N. J.

Antrimolite. A stalactitic zeolite, from Antrim, Ireland.

Edingtonite. In small right square prisms, with lateral cleavage. Nearly colorless; luster vitreous. H=4—4·5. Gr=2·7—2·75. Occurs with thomsonite at Dumbartonshire.

Carpholite. In minute radiated and stellate tufts of a straw yellow color, and silky luster. From the tin mines of Schlackenwald, Austria, with fluor.

Chlorastrolite. Light bluish-green, with a radiated structure, and somewhat like prehnite. H=5·5—6. Gr=3·18. Occurs on the shores of Isle Royale, Lake Superior. Named from the Greek *chloros*, green, *astron*, star, *lithos*, stone.

Faujasite. A hydrous silicate of alumina, lime and soda. Crystals square octahedrons. A : A=111° 30′ and 105° 30′. Scratches glass. Occurs with augite, at Kaiserstuhl.

Glottalite. A hydrous silicate of alumina and lime, said to be monometric in crystallization. H=3·5. Gr=2·18. Color white. Luster vitreous. Translucent. From Scotland.

Margarite, a mineral resembling a pearly mica, but hardly elastic, and *Euphyllite*, are hydrous species, somewhat related both to chlorite and to mica. They are mentioned on page 193. *Emerylite* is identical with margarite; *diphanite* may also be the same species.

Where does prehnite occur? How is it distinguished from the zeolites and quartz? What are its uses?

Damourite. Occurs in lamellar pearly crystals, a little harder than talc. Gr=2·7—2·82. It is a hydrous silicate of alumina and potash. Reported from Leiperville, Penn., and Chesterfield, Mass. It may be only a hydrous mica.

Chloritoid. A coarsely foliated mineral, folia bent, brittle; color greenish-black. H=5·5—Gr=3·55. Infusible before the blowpipe, but becomes finally black and magnetic. From the Ural. *Sismondine* is a related mineral from St. Marcal.

Masonite is chloritoid. Occurs coarsely foliated or tabular; color dark gray; luster nearly pearly; folia brittle and often curved. H=6. Gr=3·45. Fuses with difficulty on the edges. From the vicinity of Natic village, Rhode Island.

4. *Anhydrous combinations with Silica.*

SILLIMANITE.

In long, slender rhombic prisms, often much flattened, penetrating the gangue. M : M=110°—98°. A brilliant and easy cleavage, parallel to the longer diagonal. Also in masses, consisting of aggregated crystals or fibers.

Color hair-brown or grayish-brown. Luster vitreous, inclining to pearly. Translucent crystals break easily. H=6—7·5. Gr=3·2—3·3.

Composition: silica 37·0, alumina 63·0. Identical therefore with kyanite. Infusible alone and with borax.

Dif. Distinguished from tremolite and the varieties generally of hornblende by its brilliant diagonal cleavage, and its infusibility; from kyanite by its brilliant cleavage, and a rhombic, instead of flat-bladed crystallization.

Obs. Found in gneiss at Chester, Ct., and the Falls of the Yantic, near Norwich, Ct. The long, slender prisms penetrate the gangue in every direction. Also in Yorktown, Westchester county, N. Y.

This species was named by Bowen in honor of Prof. B. Silliman, of Yale College.

Bucholzite is supposed to be a variety of Sillimanite. *Composition,* silica 46·4, alumina 52·9, (Thomson.) A specimen from Chester, Penn., gave Erdmann, silica 40·1, alumina 58·9, protoxyd of manganese. From Fassa, Tyrol; also from Chester, Penn.; Munroe, Orange county, N. Y.; Worcester, Mass.; and Humphreysville, Conn.

What is the crystallization and appearance of Sillimanite? What is its hardness? How is it distinguished from tremolite and kyanite?

The analyses of *bucholzite* and *sillimanite* give varying results, and still they make but one species. *Fibrolite* is another variety of this mineral from the Carnatic.

KYANITE.

Triclinic. Usually in long thin-bladed crystals aggregated together, or penetrating the gangue. The annexed figure is a portion of one of these crystals. Crystals sometimes short and stout. Lateral cleavage, distinct. Sometimes fine fibrous.

Color usually light blue, sometimes white, or a blue center with a white margin; sometimes gray, green, or even black. Luster of flat face a little pearly. H=5—7. Rather brittle, but less so than Sillimanite. Gr=3·6—3.7.

Composition: silica 37·0, alumina 63·0. Unaltered alone before the blowpipe. With borax forms slowly a transparent colorless glass.

Dif. Distinguished by its infusibility from varieties of the hornblende family. The short crystals have some resemblance to staurotide, but their sides and terminations are usually irregular; they differ also in their cleavage and luster.

Obs. Found in gneiss and mica slate, and often accompanied by garnet and staurotide.

Occurs in long-bladed crystallizations at Chesterfield and Worthington, Mass.; at Litchfield and Washington, Conn.; near Philadelphia; near Wilmington, Delaware; and in Buckingham and Spotsylvania counties, Va. Short crystals (sometimes called improperly *fibrolite*) occur in gneiss at Bellows Falls, Vt., and at Westfield and Lancaster, Mass.

In Europe, transparent crystals are met with at St. Gothard in Switzerland, and in Styria, Carinthia, and Bohemia. Villa Rica in South America, affords fine specimens.

The name kyanite is from the Greek *kuanos*, sky-blue. It is also called *sappar*, a corruption of sapphire; also *disthene*, and when white, *rhætizite*.

Uses. Kyanite is sometimes used as a gem, and has some resemblance to sapphire.

Wœrthite. Resembles kyanite, but gives off water before the blowpipe. It may be an altered kyanite. From St. Petersburg.

Describe kyanite? What is the origin of the name? For what is it used?

ANDALUSITE.

Trimetric. In right rhombic prisms. M : M=90° 44′. Cleavage lateral, distinct; also massive and indistinctly coarse columnar, but never fine fibrous.

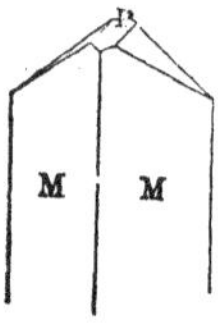

Colors gray and flesh-red. Luster vitreous, or inclining to pearly. Translucent to opaque. Tough. H=7·5. Gr=3·1—3·32.

Composition: silica 37·0, alumina 63·0. Infusible. With borax fuses with extreme difficulty.

Varieties. *Chiastolite* and *macle* are names given to crystals of andalusite which show a tesselated or cruciform structure when broken across and polished. The annexed figure represents one from Lancaster, Mass. The structure is owing to impurities, (usually the material of the gangue,) disseminated by the powers of crystallization in a regular manner along the sides, edges and diagonals of the crystal. Their hardness is sometimes as low as 3. The same structure has been observed by Dr. Jackson in staurotide crystals.

Dif. Distinguished from pyroxene, scapolite, spodumene and feldspar, by its infusibility, hardness and form.

Obs. Found in granite and gneiss.

Westford, Mass.; Litchfield and Washington, Ct.; Bangor, Me.; Chester, Penn., are some of its American localities. Chiastolite occurs at Sterling and Lancaster, Mass., and near Bellows Falls, Vermont. This species was first found at Andalusia in Spain.

STAUROTIDE.

Trimetric. In right rhombic and six-sided prisms. M : M=129° 20′. Cleavage imperfect. P : a=124° 38′, M : ě=115° 20′. Figure 2 is a common cruciform crystal, (consisting of two prisms crossing one another.) Never in massive forms or slender crystallizations.

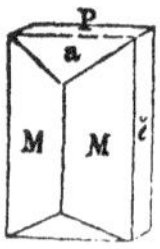

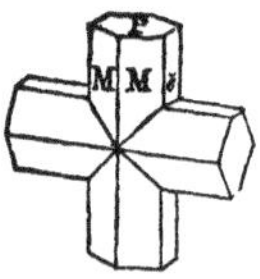

What is the appearance of andalusite? What is chiastolite or macle? How is andalusite distinguished from pyroxene and spodumene? What crystalline forms are presented by staurotide? Is it ever found massive?

Color dark brown or black. Luster vitreous, inclining to resinous; sometimes bright, but often dull. Translucent to opaque. H=7—7·5. Gr=3·65—3·73.

Composition: silica 29·3, alumina 53·5, peroxyd of iron 17·2. Before the blowpipe it darkens, but does not fuse.

Dif. Distinguished from tourmaline and garnet by its infusibility and form.

Obs. Found in mica slate and gneiss, in imbedded crystals.

Very abundant through the mica slate of New England. Franconia, Vt.; Windham, Me.; Lisbon, N. H.; Chesterfield, Mass.; Bolton and Tolland, Ct.; on the Wichichon, eight miles from Philadelphia, and near New York city, are some of the localities. St. Gothard in Switzerland, and the Greiner mountain, Tyrol, are noted foreign localities.

The name staurotide is from the Greek *stauros*, a cross.

LEUCITE.

Occurs only under the form of the trapezohedron, as in the annexed figure. Cleavage imperfect. Usually in dull glassy crystals, of a grayish color; sometimes opaque-white, disseminated through lava. Translucent to opaque. H=5·5—6. Brittle. Gr=2·48—2·49.

Composition: silica 55·1, alumina 23·4, potash 21·5=100. Infusible except with borax or carbonate of lime, and then with difficulty to a clear globule. A fine blue color, with cobalt solution.

Dif. Distinguished from analcime by its hardness and infusibility.

Obs. In lavas, especially those of Italy. Abundant at Vesuvius. Crystals from a pin's head to an inch in diameter.

The name leucite is from the Greek *leukos*, white.

Saccharite resembles a granular feldspar, of a white or greenish-white color, but has the constitution of leucite. Infusible alone, and with great difficulty with soda. From Silesia. Perhaps Andesine.

What are the colors and hardness of staurotide? What is its constitution? What is its mode of occurrence? How is it distinguished from tourmaline? Describe the forms and appearance of leucite. How does it differ from analcime?

ORTHOCLASE.—*Common Feldspar.**

Monoclinic. In modified oblique rhombic prisms. T : T=118° 49′, P : T=67° 15′; T : ĕ=120° 40′. Usually in thick prisms, often rectangular, (fig. 2,) and also in modified tables, (fig. 1.) Cleavage perfect parallel with ĕ, the shorter diagonal; also distinct parallel to P. Also massive, with a granular structure, or coarse lamellar. Colors light; white, gray, and flesh-red common; also greenish and bluish white and green. Luster vitreous; sometimes a little pearly on the face of perfect cleavage. Transparent to subtranslucent. H=6. Gr=2·39—2·62.

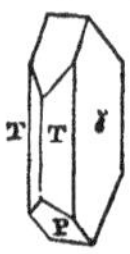

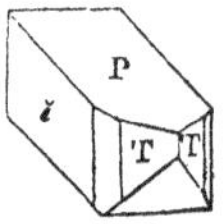

Composition: silica 64·20, alumina 18·40, potash 16·95, Fuses only on the edges. With borax forms slowly a transparent glass. Not acted upon by the acids.

Varieties. *Common feldspar* includes the common subtranslucent varieties; *adularia*, the white or colorless subtransparent specimens. The name is derived from Adula, one of the highest peaks of St. Gothard. *Glassy feldspar* and *ice-spar* include transparent vitreous crystals, found in lavas. Some crystals called by these names belong to the species anorthite. *Ryacolite* and *Loxoclase* belong here.

Moonstone is an opalescent variety of adularia, having when polished peculiar pearly reflections. *Sunstone* is similar; but contains minute scales of mica. *Aventurine feldspar* often owes its iridescence to minute crystals of specular or titanic iron, or limonite.

Dif. Distinguished from scapolite by its more difficult fusibility, and by a slight tendency to a fibrous appearance in the cleavage surface of the latter, especially in massive varieties; from spodumene by its blowpipe characters.

Obs. Feldspar is one of the constituents of granite, gneiss, mica slate, porphyry and basalt, and often occurs in these rocks in crystals. St. Lawrence county, N. Y., affords fine crystals; also Orange county, N. Y.; Haddam and

---

What is the crystallization and appearance of feldspar? What is its hardness? what its composition? Mention the principal varieties, with their peculiarities? In what rocks is feldspar an ingredient?

---

* The following species, from feldspar to nepheline inclusive, form a natural group called the *feldspar family*.

Middletown, Conn. ; South Royalston and Barre, Mass., besides numerous other localities. Green feldspar occurs at Mount Desert, Me. ; an aventurine feldspar at Leyperville, Penn. ; Adularia at Haddam and Norwich, Conn., and Parsonsfield, Me. A fetid feldspar (sometimes called *necronite*) is found at Rogers' Rock, Essex county ; at Thomson's quarry, near 196th street New York city, and 21 miles from Baltimore. Carlsbad and Elbogen in Bohemia, Baveno in Piedmont, St. Gothard, Arendal in Norway, Land's End, and the Mourne mountains, Ireland, are some of the more interesting foreign localities.

The name feldspar is from the German word *feld*, meaning *field*.

Uses. Feldspar is used extensively in the manufacture of Porcelain. *Moonstone* and *Sunstone* are often set in jewelry. They are polished with a rounded surface, and look somewhat like cat's-eye, but are much softer.

Kaolin. This name is applied to the clay that results from the decomposition of feldspar. It is the material used for making porcelain or china ware. The change the feldspar undergoes in producing kaolin consists principally in a removal of the alkali, potash, with part of the silica and the addition of water. *Composition* of a specimen from Schneeberg, silica 43·6, alumina 37·7, peroxyd of iron 1·5, water 12·6, (Berthier.) It occurs in extensive beds in granite regions, where it has been derived from the decomposition of this rock. A granite containing talc seems to be the most common source of it. See farther, the chapter on Rocks.

ALBITE.

Triclinic. In modified oblique rhomboidal prisms. M : T=122° 15′, P : T=115° 5′ ; P : M=110° 51′. The crystals are usually more or less thick and tabular. Also massive, with a granular or lamellar structure. Laminæ brittle.

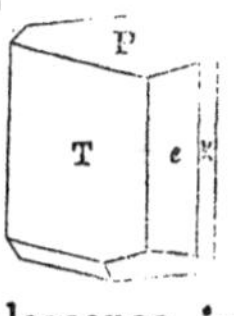

Color white ; occasionally light tints of bluish white, grayish, reddish and greenish. Luster vitreous to pearly, and sometimes a bluish opalescence is exhibited. Transparent to subtranslucent. H=6. Gr=2·6—2·7.

What are the uses of feldspar ? What is kaolin, and for what is it used ? What is the crystallization and appearance of albite ?

Composition: silica 68·5, alumina 19·3, peroxyd of iron and manganese 0·3, lime 0·7, soda 9·1. Acts like feldspar before the blowpipe, but tinges the flame yellow.

Cleavelandite is a lamellar variety occurring in wedge-shaped masses at the Chesterfield albite vein, Mass.

Dif. Albite differs from feldspar in containing a large proportion of soda. It may generally be distinguished when associated with that species by its uniform white color ; also by the form of the crystals, which are more oblique and irregular, often tabular, with two of the edges very acute ; also by the yellow tinge given the blowpipe flame.

Obs. Albite like feldspar is a constituent of many rocks, replacing feldspar. Albite granite is commonly lighter colored than feldspar granite, arising from the usual whiteness of the albite. Fine crystals occur at Middletown and Haddam, Conn., at Goshen, Mass., and Granville, N. Y.

The name albite is from the Latin *albus*, white.

Andesine. Triclinic, like albite. H=6. Gr=2·65—2·74. Color white, gray, greenish, yellowish, flesh-red. *Composition* of a specimen from the Andes, silica 59·6, alumina 24·2, peroxyd of iron 1·6, lime 5·8, magnesia 1·1, potash 1·1, soda 6·5=99·92, (Abich). Found in the Andes at Marmato ; in the Vosges, France ; in Canada.

Anorthite. Near albite. The form is an oblique rhomboidal prism, P : T=110° 57′ T : T=120° 30′. Its crystals are glassy and tabular in form. H=6. Gr=2·6—2.8. Differs from albite in not tinging the blowpipe flame deep yellow, nor affording a clear glass with soda. From Mount Somma, near Naples.

Bytownite of Thomson, is a greenish-white massive mineral from Bytown, Canada. H=6—6·5. Gr=2·7—2·8.

LABRADORITE.

Triclinic. P : M=93° 28′, P : T=114° 48′, M : T=119° 16′. Cleavage parallel with P, nearly perfect ; M distinct. Usually in cleavable massive forms.

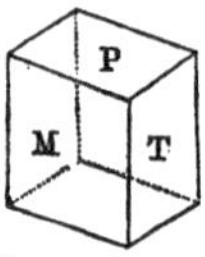

Color dark gray, brown, or greenish brown ; and usually a series of bright chatoyant colors from internal reflections, especially blue and green, with more or less of yellow, red and pearl-gray. Translucent,

How does albite differ from feldspar ? What is cleavelandite ? What is peculiar in the colors of labradorite ? Mention other characters.

subtranslucent. Luster of principal cleavage face pearly other faces vitreous. H=6. Gr=2·69—2·76.

Composition: silica 53·1, alumina 30·1, lime 12·3, soda 4·5, water 0·5. Like feldspar before the blowpipe, but fuses with a little less difficulty to a colorless glass. Entirely dissolved by muriatic acid.

Dif. Differs from feldspar and albite in containing a large percentage of lime, and it is farther distinguished by dissolving in muriatic acid, and generally by its chatoyant reflections.

Obs. A constituent of some granites, and was originally from Labrador. It is abundant in Essex county, N. Y., at Moriah, Westport and Lewis.

Uses. Labradorite receives a fine polish, and owing to the chatoyant reflections of rich and delicate colors, the specimens are often highly beautiful. It is sometimes used in jewelry.

Glaucolite. Considered by Frankenheim identical with Labradorite. Color lavender-blue, passing into green. From near Lake Baikal in Siberia.

Oligoclase. A feldspar-like mineral, with a distinct cleavage, nearly white color, of imperfectly vitreous to somewhat greasy luster. H=6. Gr=2·58—2·67. *Composition,* silica 63·5, alumina 23·1, lime 2·4, potash 2·2, soda 9·4, magnesia 0·8. Fuses with difficulty, and not attacked by acids. Occurs at Stockholm in granite, and at Arendal, Norway, and elsewhere, in granular limestone. Also found at Haddam, Ct., with iolite ; at Danbury, Ct. ; at Unionville, Penn.

Couzeranite, another allied species from the Pyrenees, of a gray or greenish gray color. Composition near that of Labradorite.

Latrobite. Resembles some reddish scapolites, but occurs in oblique rhomboidal prisms, like the feldspars, and has been referred to the species anorthite. Occurs in crystals and also in cleavable masses. H =6. Gr=2·7—2 8. *Composition,* silica 41·8, alumina 32·8, lime 9·8, oxyd of manganese with magnesia 5·8, potash 6·6. water 2·0. Fuses with some intumescence. From Labrador in granite.

Amphodelite is united with the species anorthite.

NEPHELINE.

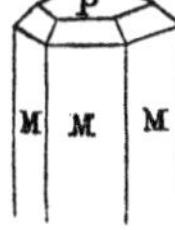

In hexagonal prisms. Also massive; sometimes thin columnar.

Color white, or gray, yellowish, greenish, bluish-red. Luster vitreous or greasy. Transparent to opaque. H=5·5—6. Gr=2·4—2·65.

Varieties and *Composition. Nepheline* includes

How does it differ from feldspar and albite? For what is it used? What is the form of crystals of nepheline? Mention its colors and luster.

glassy crystals from Vesuvius, which become clouded in nitric acid. The name is from the Greek *nephelè*, a cloud.

Elæolite (from *elaion*, oil) includes the dingy translucent or subtranslucent cleavable masses having a strong greasy luster. Altered crystals from Greenland have been called *gieseckite.*

Nepheline contains silica 43·4, alumina 33·5, peroxyd of iron 1·5, lime 0·9, soda 13·4, potash 7·1, water 1·4 Rounded on the edges before the blowpipe : some varieties fuse readily. In nitric acid, fragments become clouded and gelatinize.

Dif. Distinguished from scapolite and feldspar by the greasy luster when massive, and forming a jelly with acids ; from apatite by the same characters, and also its hardness.

Obs. *Nepheline* occurs at Vesuvius and near Rome, in lava. *Elæolite* is obtained at Brevig and other places in Norway ; also in Siberia. It is also found in the Ozark mountains in Arkansas, and at Litchfield in Maine.

SCAPOLITE.

Dimetric. In modified square prisms, often terminating in pyramids ; e : e = 136° 7′. Cleavage rather indistinct parallel with M and *e*. Also massive, sublamellar or subfibrous.

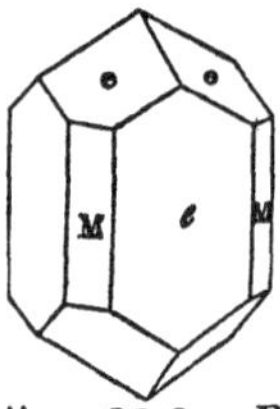

Colors light ; white, pale blue, green or red. Streak uncolored. Transparent to nearly opaque. Luster usually a little pearly. H = 5—6. Gr = 2·6—2·75.

Composition : silica 49·3 alumina 27·9 lime 22·8. Before the blowpipe it fuses slowly with intumescence. With borax dissolves with effervescence to a transparent glass.

Dif. Its square prisms and the angle of the pyramid at summit are characteristic. In cleavable masses it resembles *feldspar*, but there is a slight fibrous appearance often distinguished on the cleavage surface of scapolite, which is peculiar. It is more fusible than feldspar, and has higher specific gravity. *Spodumene* has a much higher specific gravity, and differs in its action before the blowpipe. *Tabu-*

What have specimens with a greasy luster been called? What is the effect of nitric acid? What is the usual form of scapolite crystals? What are its colors and hardness? What is its composition? How does it differ from feldspar and tabular spar?

lar spar is more fibrous in the appearance of the surface, and is less hard; it is also phosphorescent, and gelatinizes with acids.

Obs. Found mostly in the older crystalline rocks, and also in some volcanic rocks. It is especially common in granular limestone. Fine crystals occur at Gouverneur, N. Y., and at Two ponds and Amity, N. Y.; at Bolton, Boxborough and Littleton, Mass.; at Franklin and Newton, N. J. It occurs massive at Marlboro', Vt.; Westfield, Mass.; Monroe, Ct. Foreign localities are at Arendal, Norway; Wärmland, Sweden; Pargas in Finland, and also at Vesuvius, whence comes the small crystals called *meionite.*

Nuttallite, Wernerite, and *Glaucolite* are varieties of this species.
Dipyre from the Pyrenees, occurring in four or eight-sided prisms, has also been considered one of its varieties. It however contains silica 55·5, alumina 24·8, lime 9·6, with 9·4 per cent. of soda, and is more allied in composition to the feldspars. Sp. gr.=2·65. Occurs with talc and chlorite.

MEIONITE.

Dimetric. In small glassy square prisms, terminating in pyramids, and resembling scapolite; e : e=136° 11′. Cleavage rather perfect, parallel with M and *e.*

Colorless or white, and transparent to translucent. H= 5·5—6. Gr=2.5—2·75.

Composition: silica 42·1, alumina 31·9, lime 26·0. Before the blowpipe yields a colorless glass.

Dif. Differs from scapolite in the angle of the summit and in composition , from the zeolites in being anhydrous.

Obs. Found at Mt. Somma, near Naples, in small crystals in geodes in lava.

Mizzonite is closely similar. It has for the angle e : e= 135° 56′, (Scacchi.)

Sarcolite. Dimetric. Resembling somewhat analcime in appearance, being flesh-red to reddish white. Extremely brittle. Gelatinizes with acids. Of rare occurrence at Mt. Somma.

Gehlenite. Crystals square prisms like meionite: color gray; nearly opaque. H=5·5—6. Gr=2·9—3·1. *Composition,* silica 29·6, alumina 24·8, lime 35·3, protoxyd of iron 6·6, water 3·3. Infusible. With

---

In what rocks does it occur? Mention the characters of spodumene. How much lithia does it contain? How does it differ from feldspar and scapolite?

borax fuses with difficulty. Gelatinizes in muriatic acid. From the Fassa valley, Tyrol.

Humboldtilite. Crystals as above. Cleavage basal, distinct. Color brown or yellow; luster vitreous. H=5. Gr=2·9—3·2. *Composition*, silica 44·0, alumina 11·2, lime 32·0, magnesia 6·1, protoxyd of iron 2·3, soda 4·3, potash 0·4. Gelatinizes with nitric acid. From Vesuvius in lava. *Somervillite* and *mellilite* are here included.

PETALITE.

In imperfectly cleavable masses, affording a prism of 142°. Color white or gray, or with pale reddish or greenish shades. Luster vitreous to subpearly. Translucent. H=6—6·5. Gr=2·4—2·45.

Composition: silica 77·9, alumina 17·7, lithia 3·1, soda 1·3. Phosphoresces when gently heated. Fuses with difficulty on the edges. Gives the reaction of lithia like spodumene.

Dif. Its lithia reaction allies it to spodumene, but it differs from that mineral in luster, specific gravity, and greater fusibility.

Castor. Supposed to be petalite. *Zygadite* is another lithia mineral: it occurs in twins like albite. From the Hartz.

Violan is a dark violet-blue mineral, resembling glaucophane.

Glaucophane occurs in cleavable masses of a dull bluish color, and in thin prisms. Translucent. Gr=1·08. H=5·5. Fuses easily. Contains silica 56·5, alumina 12·2, protoxyd of iron 10 9, protoxyd of manganese 0·5, magnesia 8·0, lime 2 2, soda 9·3. From the Island of Syra.

Wichtine is a black mineral rectangularly cleavable in two directions. Contains silica 56·3, alumina 13·3, protoxyd of iron 13·0, peroxyd of iron 4·0, soda 3 5, lime 6·0, magnesia 3·0. From Wichty in Finland.

EPIDOTE.

Monoclinic. In right rhomboidal prisms more or less modified, often with six or more sides. M : T=115° 24′. T : ĕ=128° 19′; ă : ă=109° 27′; ĕ : ă=125° 16′

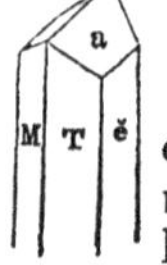

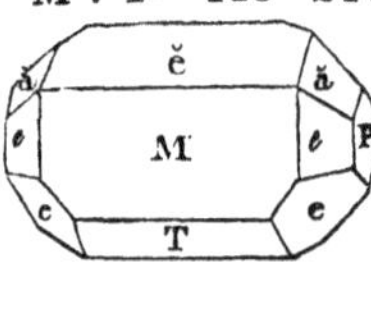

Cleavage parallel to M; less distinct parallel to T.—Also massive granular and of a columnar structure.

Describe petalite. What is the proportion of lithia in its constitution? How does it differ from spodumene? Where does it occur? What is the form of epidote?

Color yellowish-green (pistachio-green) and ash or hair brown. Streak uncolored. Translucent to opaque. Luster vitreous, a little pearly on M ; often brilliant on the faces of crystals. Brittle. H=6—7. Gr=3·25—3·46.

Varieties and *Composition*. There are three prominent varieties of this species; one of a yellowish-green color, another called *zoisite*, of a grayish-brown or hair-brown; a third of dark reddish shades, which contains 14 per cent. of oxyd of manganese, and is called *Manganesian epidote*. *Thulite* is another red variety, of paler color.

The yellowish-green epidote is sometimes called *Pistacite*. The mineral *Bucklandite* is an iron-epidote.

The green epidote consists of silica 37·0, alumina 26·6, lime 20·0, protoxyd of iron 13·0, protoxyd of manganese 0.6, water 1·8.

Zoisite consists of silica 40·2, alumina 30·3, lime 22·5, peroxyd of iron 4·5, water 2.0. Before the blowpipe, epidote and zoisite fuse on the edges and swell up, but do not liquefy. The manganesian epidote and thulite fuse readily to a black glass.

Dif. The peculiar yellowish-green color of ordinary epidote distinguishes it at once. The prisms of zoisite are often longitudinally striated or fluted, and they have not the form or brittleness of tremolite.

Obs. Occurs in crystalline rocks, and also in some sedimentary rocks that have been heated by the passage of dykes of trap or basalt. Splendid crystals, six inches long, and with brilliant faces and rich color, have been obtained at Haddam, Ct. Crystallized specimens are also found at Franconia, N. H., Hadlyme, Chester, Newbury and Athol, Mass., near Unity and Monroe, N. Y., Franklin and Warwick, N. J. Zoisite in columnar masses is found at Willsboro and Montpelier, Vt., at Chester, Goshen, Chesterfield, and elsewhere in Massachusetts; at Milford, Ct.

The name *epidote* was derived by Haüy from the Greek *epididomi*, to increase, in allusion to the fact that the base of the primary is frequently much enlarged in the crystals.

The mineral *Allanite*, p. 207, is near epidote in form and composition, although containing cerium.

What are the colors and other characters of epidote? What is the color of the variety zoisite? What is the composition of epidote? what are its distinguishing characters?

IDOCRASE.

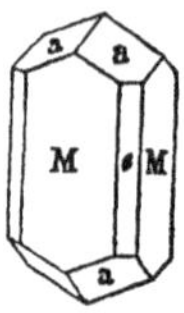

Dimetric. In square prisms usually modified. P : a=142° 53′; a : a=129° 29′, a : *e*=127° 07′. Cleavage not very distinct parallel with M. Also found massive granular and subcolumnar.

Color brown; sometimes passing into green. In some varieties the color is oil-green in the direction of the axis and yellowish-green at right angles with it. Streak uncolored. Subtransparent to nearly opaque. H=6·5. Gr=3·33—3.4.

Composition: silica 37·4, alumina 23·5, protoxyd of iron 4·0, lime 29·7, magnesia and protoxyd of manganese 5·2. Before the blowpipe fuses with effervescence to a yellow translucent globule.

Dif. Resembles some brown varieties of garnet, tourmaline and epidote, but besides its difference of crystallization, it is much more fusible.

Obs. Idocrase was first found in the lavas of Vesuvius, and hence called *Vesuvian.* It has since been obtained in Piedmont, near Christiania, Norway, in Siberia, also in the Fassa valley. Specimens of a brown color from Eger, Bohemia, have been called *egeran. Cyprine* includes blue crystals from Tellemarken, Norway; supposed to be colored by copper.

In the United States, idocrase occurs in fine crystals at Phipsburg and Rumford, Parsonsfield and Poland, Me.; Newton, N. J.; Amity, N. Y., and sparingly at Worcester, Mass. The *xanthite* of Amity is nothing but idocrase.

The name *idocrase* is from the Greek *eido*, to see, and *krasis*, mixture; because its crystalline forms have much resemblânce to those of other species.

Uses. This mineral is of little value except as a mineralogical curiosity. It is sometimes cut as a gem for rings.

GARNET.

Monometric. Common in dodecahedrons, (fig. 1,) also in trapezohedrons, (fig. 2,) and both forms are sometimes variously modified. Cleavage parallel to the faces of the dode-

---

What is the crystallization of idocrase? its color, hardness, and luster? its composition? How does it differ from garnet and tourmaline? What is the usual form of garnet?

cahedron rather distinct. Also found massive granular, and coarse lamellar.

Color deep red, prevalent; also brown, black, green,

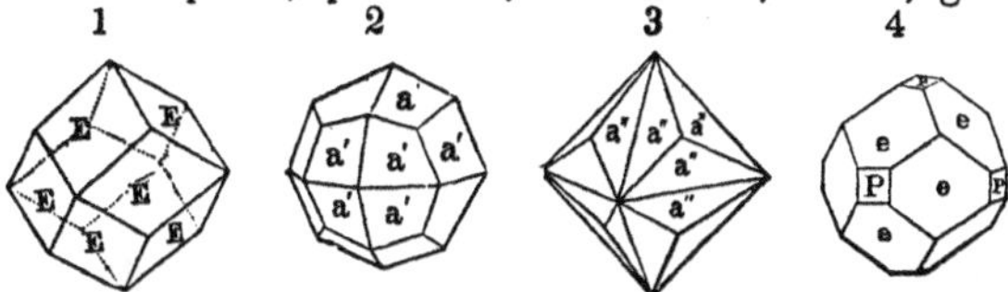

white. Transparent to opaque. Luster vitreous. Brittle. H=6·5—7·5. Gr=3·5—4·3.

Varieties and *Composition.* Garnet is a compound of three or four silicates, the silicates of alumina, lime, iron, and manganese, and the varieties of color arise from their various combinations. Oxyd of chrome is sometimes present, producing an emerald-green variety.

Precious garnet or *almandine* is a clear deep red variety, and is used much in jewelry. A specimen from New York afforded Wachtmeister, silica 42·5, alumina 19·15, protoxyd of iron 33·6, protoxyd of manganese 5·5.

Common garnet has a brownish red color. and is imperfectly translucent or opaque.

Cinnamon stone, called also *essonite,* is of a light cinnamon-yellow color and high luster. It differs from the preceding principally in containing but 5 or 6 per cent. of iron and 30 to 33 per cent. of lime. *Topazolite* is another yellow variety, approaching topaz in color, and presenting the form in figure 3.

Melanite (from the Greek *melas,* black) is a black garnet, containing 15 to 25 per cent. of the oxyds of iron and manganese. *Pyrenäite* is another name for a black variety from France.

Manganesian garnet has a deep red color, and is usually quite brittle. A Haddam specimen afforded Seybert, silica 35·8, alumina 18·1, protoxyd of iron 14·9, protoxyd of manganese 31·0.

Grossularite occurs in greenish trapezohedrons; and contains 30 to 34 per cent. of lime with but little iron.

Ouvarovite is a chrome garnet, containing 22·5 per cent. of oxyd of chromium, and having the rich color of the emerald.

What is the color and hardness of garnet? of what does it consist? what is precious garnet? What is cinnamon stone? What is ouvarovite?

Colophonite (from the Greek *kolophonia*, a resin) is a coarse granular variety, usually presenting iridescent hues and a resinous luster.

Aplome is a deep brown garnet, sometimes inclining to orange. It presents the form in figure 4, and has a cleavage parallel to the shorter diagonal of the faces. For this reason it has been separated from the species garnet, and a cube is considered its primary form.

The different varieties fuse with more or less difficulty to a dark vitreous globule.

Dif. The vitreous luster of fractured garnet, without a prismatic structure even in traces, and its usual dodecahedral forms, are easy characters for distinguishing it. Staurotide differs in being infusible ; tourmaline has less specific gravity ; idocrase fuses much more readily.

Obs. Garnet occurs abundantly in mica slate, hornblende slate, and gneiss, and somewhat less frequently in granite and granular limestone ; sometimes in serpentine and lava.

The best *precious garnets* are from Ceylon and Greenland ; *cinnamon stone* comes from Ceylon and Sweden ; *grossularite* occurs in the Wilui river, Siberia, and at Tellemarken in Norway ; *green garnets* are found at Swartzenberg, Saxony ; *melanite,* in the Vesuvian lavas ; *ouvarovite,* at Bissersk in Russia ; *topazolite,* at Mussa, Piedmont ; *aplome,* in Siberia, on the Lena, and at Swartzenberg.

In the United States, *precious garnets,* of small size, occur at Hanover, N. H. ; and a clear and deep red variety, sometimes called *pyrope,* comes from Green's creek, Delaware county, Penn. Dodecahedrons, of a *dark red color,* occur at Haverhill, N. H., some 1½ inches through ; also at New Fane, Vt., still larger ; also Lyme, Conn. ; at Unity, Brunswick, Streaked Mountain, and elsewhere, Maine ; at Monroe, Conn. ; Bedford, Chesterfield, Barre, Brookfield, and Brimfield, Mass. ; Dover, Dutchess county, Roger's rock, Crown Point, Essex county, Franklin, N. J. *Cinnamon colored crystals* occur at Carlisle, Mass., transparent, and also at Boxborough ; with idocrase at Parsonsfield, Phippsburg and Rumford, Me. ; at Amherst, N. H. ; at Amity, N. Y., and Franklin, N. J., ; at Dixon's quarry, seven miles from Wilmington, Del., in fine trapezohedral crystals. Melanite is found at Franklin, N. J., and Germantown, Penn. *Coloph-*

What is colophonite ? What is aplome ? How is garnet distinguished ?

onite is abundant at Willsborough and Lewis, Essex county, N. Y.; it occurs also at North Madison, Conn.

The garnet is the carbuncle of the ancients. The alabandic carbuncles of Pliny were so called because cut and polished at Alabanda, and hence the name *Almandine* now in use. The garnet is also supposed to have been the *hyacinth* of the ancients.

Uses. The clear deep red garnets make a rich gem, and are much used. Those of Pegu are most highly valued. They are cut quite thin, on account of their depth of color. An octagonal garnet, measuring 8½ lines by 6½ has sold for near $700. The cinnamon stone is also employed for the same purpose. Pulverized garnet is sometimes employed as a substitute for emery. When abundant, as in some parts of Germany, garnet is used as a flux to some iron ores.

Pliny describes vessels, of the capacity of a pint, formed from large carbuncles, "devoid of luster and transparency, and of a dingy color," which probably were large garnets.

Pyrope or *Bohemian garnet.* Occurs usually in rounded grains, resembling a rich garnet, but the primary form is supposed to be the cube. Cleavage none. H=7·5. Gr=3·69—3·8. *Composition:* silica 43·0, alumina 22·3, oxyd of chromium 1·8, magnesia 18·5, protoxyd of iron 8·7, lime 5·7; and, according to Apjohn, there are also 3 per cent. of yttria. From Bohemia, in trap tufa.

Helvin, a wax yellow garnet-like mineral, occurring in *tetrahedral* crystals. From Saxony and Norway.

TOURMALINE.

Rhombohedral. Usual in prisms terminating in a low pyramid. R : R=134° 03′. R : *e*=112° 59′; R : *á*=141° 30; e : e=154° 59′. The crystals are hemihedrally modified, or have unlike secondary planes at the two extremities, as shown in figure 3. They are commonly long, and often there are but three prismatic sides, which are convex and strongly furrowed.

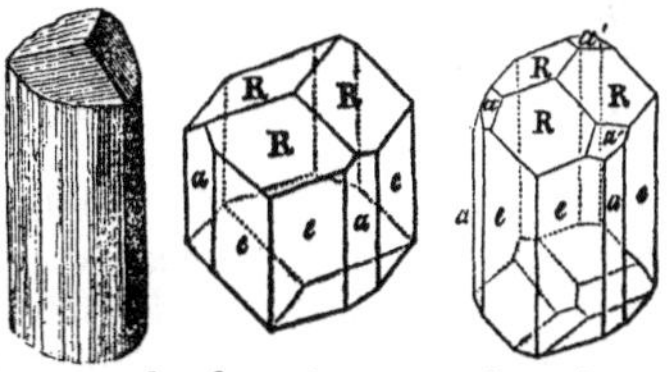

How is garnet distinguished? What are its uses? What is said of the ancient carbuncle? What is pyrope? What are the usual forms and appearance of tourmaline?

Occurs also compact massive, and coarse columnar, the columns sometimes radiating or divergent from a center.

Color black, blue-black, and dark brown, common; also bright and pale red, grass-green, cinnamon-brown, yellow, gray, and white. Sometimes red within and green externally, or one color at one extremity and another at the other. Transparent; usually translucent to nearly opaque. Luster vitreous, inclining to resinous on a surface of fracture. Streak uncolored. Brittle; the crystals often fractured across and breaking very easily. H=7·8. Gr=3—3·1. Electrically polar when heated, (page 62.)

Varieties and *Composition.* Tourmalines of different colors have been designated by different names, as follows:—

Rubellite is *red* tourmaline.

Indicolite is *blue* and *bluish-black* tourmaline.

Schorl, formerly included the common black tourmaline, but the name is not now used.

A *black* variety afforded, on analysis, silica 33·0, alumina 38·2, lime 0·8, protoxyd of iron, 23·8, soda 3·2, boracic acid 1·9.

A red variety from Siberia, silica 39·4, alumina 44·0, potash 1·3, boracic acid 4·2, lithia 2·5, peroxyd of manganese 5·0. The presence of *boracic acid* is the most remarkable point in the constitution of this mineral. It is also observed that lithia is sometimes present; over 4 per cent. have been obtained from a green tourmaline from Utön, Sweden.

Before the blowpipe the dark varieties intumesce, and fuse with difficulty; the red and light-green only become milk-white and a little slaggy on the surface.

Dif. The black and the dark varieties generally, are readily distinguished by the form and luster and absence of distinct cleavage, together with their difficult fusibility. The black when fractured often appear a little like a black resin. The brown variety resembles zoisite, though very distinct in crystallization. The light brown looks like garnet or idocrase, but is more infusible. The red, green, and yellow varieties are distinguished from any species they resemble, by the crystalline form, the prism of tourmaline always having 3, 6, 9, or 12 prismatic sides, (or some multiple of

---

What is the color and hardness of tourmaline? what has been called schorl? What is rubellite? What are the distinctive characters of tourmaline?

3.) The electric polarity of the crystals, when heated, is another remarkable character of this mineral.

Obs. Tourmalines are common in granite, gneiss, mica slate, chlorite slate, steatite, and granular limestone. They usually occur penetrating the gangue. The black crystals are often highly polished and at times a foot in length, though perhaps of no larger dimensions than a pipe-stem, or even more slender. This mineral has also been observed in sandstones near basaltic or trap dikes.

Red and green tourmalines, over an inch in diameter and transparent, have been obtained at Paris, Me., besides pink and blue crystals. These several varieties occur also, of less beauty, at Chesterfield and Goshen, Mass. Good black tourmalines are found at Norwich, New Braintree, and Carlisle, Mass.; Alsted, Acworth, and Saddleback Mountain, N. H.; Haddam, Conn.; Saratoga and Edenville, N. Y.; Franklin and Newton, N. J.

Dark brown tourmalines are obtained at Orford, N. H.; in thin black crystals in mica at Grafton, N. H.; Monroe, Ct.; Gouverneur and Amity, N. Y.; Franklin and Newton, N. J. A fine cinnamon brown variety occurs at Kingsbridge, Amity, and also south in New Jersey. A gray or bluish-gray and green variety occurs near Edenville.

The word tourmaline is a corruption of the name in Ceylon, whence it was first brought to Europe. *Lyncurium* is supposed to be the ancient name for common tourmaline; and the red variety was probably called *hyacinth*.

Uses. The red tourmalines, when transparent and free from cracks, such as have been obtained at Paris, Me., are of great value and afford gems of remarkable beauty. They have all the richness of color and luster belonging to the ruby, though measuring an inch across. A Siberian specimen of this variety, now in the British museum, is valued at £500. The yellow tourmaline, from Ceylon is but little inferior to the real topaz, and is often sold for that gem. The green specimens, when clear and fine, are also valuable for gems. A stone measuring 6 lines by 4, of a deep green color, is valued at Paris at $15 to $20. The thin crystals of Grafton, N. H. are transparent, and may be used as suggested by B. Silliman, Jr., in polarizing instruments.

Where have fine specimens of red and green tourmaline been found in the United States? What is said of yellow tourmaline? What is the value of tourmaline as a gem?

AXINITE.

Triclinic. In acute edged oblique rhomboidal prisms; P : M=134° 40′, P : T=115° 5′, M : T=135° 10′. Cleavage indistinct. Also rarely massive or lamellar.

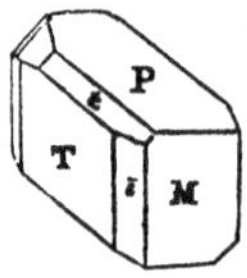

Color clove brown; differing somewhat in shade in two directions. Luster vitreous. Transparent to subtranslucent. Brittle. H=6·5—7. Gr=3·27. Pyro-electric.

Composition: silica 45, alumina 19, lime 12·5, peroxyd of iron 12·25, peroxyd of manganese 9, boracic acid 2·0, magnesia 0·2. In another specimen 5·6 per cent. of boracic acid were found. Before the blowpipe fuses readily with intumescence to a dark green glass, which becomes black in the oxydating flame.

Dif. Remarkable for the sharp thin edges of its crystals, and its glassy brilliant appearance, without cleavage. The crystals are implanted, and not disseminated like garnet. In one or all of these particulars, and also in blowpipe reaction, it differs from any of the titanium ores.

Obs. St. Cristophe in Dauphiny, is a fine locality of this mineral. It occurs also at Kongsberg in Norway, Normark in Sweden, and Cornwall, England; also Thum in Saxony, whence the name *Thummerstein* and *Thumite.*

In the United States, it has been found at Phippsburg in Maine, by Dr. C. T. Jackson.

IOLITE.—*Dichroite, Cordierite.*

Trimetric. In rhombic and hexagonal prisms. Usually occurs in six or twelve-sided prisms, or disseminated in masses without distinct form. Cleavage indistinct; but crystals often separable into layers parallel to the base.

Color various shades of blue; often deep blue in the direction of the axis, and yellowish-gray transversely. Streak uncolored. Luster and appearance much like that of glass. Transparent to translucent. Brittle. H=7—7·5. Gr=2·6—2·7.

Composition of a specimen from Haddam, Ct.: silica 48·3,

What is the form and color of axinite? What characters distinguish it? Why was it so called? What are the forms of iolite? What are its colors, appearance and hardness?

alumina 32·5, magnesia 10, protoxyd of iron 6·0, protoxyd of manganese 0·1, water (hygrometric) 3·1. Before the blowpipe fuses on the edges with difficulty to a blue glass resembling the mineral.

Dif. The glassy appearance of iolite is so peculiar that it can be confounded with nothing but blue quartz, from which it is distinguished by its fusing on the edges. It is easily scratched by sapphire. .

Obs. Found at Haddam, Conn., in granite; also in gneiss at Brimfield, Mass.; at Richmond, N. H., in talcose rock. The principal foreign localities are at Bodenmais in Bavaria; Arendal, Norway; Capo de Gata, Spain; Tunaberg, Finland; also Norway, Greenland and Ceylon.

The name iolite is from the Greek *iodes*, violet, alluding to its color; it is also called *dichroite*, from *dis*, twice, and *chroa*, color, owing to its having different colors in two directions.

Uses. Occasionally employed as an ornamental stone; when cut it presents different shades of color in different directions.

NOTE.—Iolite exposed to the air and moisture undergoes a gradual alteration, becoming a hydrate (absorbing water) and assuming a foliated micaceous structure, so as to resemble talc, though more brittle and hardly greasy in feel. *Hydrous iolite, chlorophyllite,* and *esmarkite,* are names that have been given to the altered iolite; and *fahlunite* and *gigantolite* are of the same origin. (See pages 162, 163.)

MICA.—*Muscovite.*

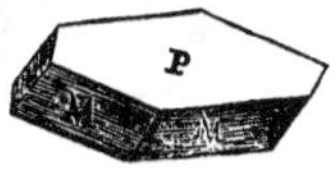

Trimetric. In oblique rhombic prisms of about 120° and 60°; but the fundamental form right rhombic. Crystals usually with the acute edge replaced. Cleavage eminent, parallel to P, yielding easily thin elastic laminæ of extreme tenuity. Usually in thinly foliated masses, plates or scales. Sometimes in radiated groups of aggregated scales or small folia.

Colors from white through green, yellowish and brownish shades to black. Luster more or less pearly. Transparent or translucent. Tough and elastic. H=2—2·5. Gr=2·8—3.

Composition: silica 46·3, alumina 36·8, potash 9·2, per-

How is iolite distinguished from quartz and sapphire? Why was it called iolite and dichroite? Describe mica. What is its composition?

oxyd of iron 4·5, fluoric acid 0·7, water 1·8. Before the blowpipe infusible, but becomes opaque white.

Varieties.—A variety in which the scales are arranged in a plumose form is called *plumose mica;* another, in which the plates have a transverse cleavage, has been termed *prismatic mica.*

Dif. Mica differs from talc in affording thinner folia and being elastic ; also in not having the greasy feel of that mineral. The same characters, excepting the last, distinguish it from gypsum ; besides, it does not crumble so readily on heating.

Obs. Mica is one of the constituents of granite, gneiss and mica slate, and gives to the latter its laminate structure. It also occurs in granular limestone. Plates two and three feet in diameter, and perfectly transparent, are obtained at Alstead, Acworth and Grafton, New Hampshire. Other good localities are Paris, Me. ; Chesterfield, Barre, Brimfield, and South Royalston, Mass. ; near Greenwood furnace, Warwick and Edenville, Orange county, and in Jefferson and St. Lawrence counties, N. Y. ; Newton and Franklin, N. J. ; near Germantown, Pa., and Jones's Falls, Maryland. Oblique prisms from near Greenwood are sometimes six or seven inches in diameter.

A green variety occurs at Unity, Maine, near Baltimore, Md., and at Chestnut Hill, Pa. Prismatic mica is found at Russel, Mass.

Uses. Mica, on account of the toughness, transparency and the thinness of its folia, has been used in Siberia for glass in windows : whence it has been called *Muscovy glass.* It was formerly employed in the Russian navy, because not liable to fracture from concussion. It is in common use for lanterns, and also for the doors of stoves. It affords a convenient material for preserving minute objects for the microscope, and is sometimes used for holding minerals before the blowpipe flame.

The best localities of the mineral in this country for the arts, are those of New Hampshire.

Lepidolite, or *Lithia mica.* Occurs in crystals or laminæ, of a purplish color, and often in masses consisting of aggregated scales. A specimen from the Ural consisted, according to Rosales, of silica 47·7,

---

How does mica differ from talc and gypsum? Of what rocks is it a constituent? What are its uses? What is the peculiarity of lepidolite?

alumina 20·3, lime 6·1, protoxyd of manganese 4·7, potash 11·0, lithia 2·8, soda 2·2, fluorine 10·2, chlorine 1·2.

Lepidolite occurs at the albite vein in Chesterfield, Mass., and at Goshen in the same state; also at Paris, Me., with red tourmalines, and near Middletown, Ct.

Fuchsite. A green mica from the Zillerthal, containing nearly 4 per cent. of oxyd of chromium.

Biotite. Resembles common mica or muscovite, but crystals usually right prisms, and angle between the optical axes only 1 or 2 degrees or less; while in muscovite the angle is 56 to 75 degrees. The form is usually regarded as hexagonal and not trimetric. Colors mostly dark green to black, sometimes white. H =2·5—3. Gr=2·7—3·1.

Composition: essentially like garnet. A variety from Monroe, N. Y., afforded Smith and Brush, silica 39·9, alumina 15·0, peroxyd of iron 7·7, magnesia 23·7, soda 1·1, potash 9·1, water 1·3, fluorine 0·9, chlorine 0·4. Biotite is a magnesia mica.

Obs. Occurs at Vesuvius, at Greenwood Furnace in Monroe, N. Y., and elsewhere. Most of the black and greenish-black micas are biotite.

Phlogopite. A mica, near biotite in the form of the crystals, but angle between the optical axes 5 to 20 degrees. Form trimetric. Color usually brown, yellowish brown, sometimes white.

Composition: a variety from Edwards, N. Y., afforded Craw, silica 40·15, alumina 17·36, magnesia 28·10, potash 10·56, soda 0·63, fluorine 4·20.

Obs. Occurs in granular limestone, being characteristic of that rock. Found at Gouverneur and other places in northern New York, Warwick, Orange Co., &c.

Margarite, or *Pearl Mica.* In hexagonal prisms, having the structure of mica; and also in intersecting laminæ. Luster pearly, approaching talc, but differing from that mineral in being a silicate of ***alumina*** instead of ***magnesia.*** Color nearly white, or gray. It intumesces and fuses before the blowpipe. From Sterzing in the Tyrol, associated with chlorite. *Emerylite* and *diphanite* belong here.

Euphyllite is a new species, related somewhat to margarite, and found associated with emerylite and corundum in Pennsylvania and elsewhere. Rather brittle.

Margarodite, or *Schistose talc of Zillerthal,* is near common mica, but contains 4 or 5 per cent. of water.

Lepidomelane. A black iron-mica, occurring in six-sided scales or tables aggregated together. It contains silica 37·4, alumina 11·6, peroxyd of iron 27·7, protoxyd of iron 12·4, magnesia and lime 0·3, potash 9·2, water 0·6. From Wärmland. *Ottrelite* (which includes the ***phyllite*** from Sterling, Mass.,) is an allied mineral occurring in black scales, disseminated through the rock.

What are other kinds of mica ?

5. *Combination of a Silicate and Fluorid.*

TOPAZ.

Trimetric. In right rhombic prisms, usually differently modified at the two extremities. Pyro-electric. M : M=124° 19'. Cleavage perfect, parallel to the base.

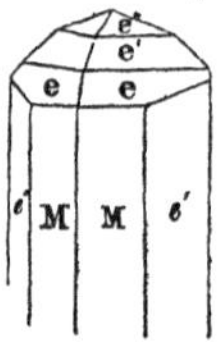

Color pale yellow; sometimes greenish, bluish, or reddish. Streak white. Luster vitreous. Transparent to subtranslucent. Fracture subconchoidal, uneven.

Composition: silica 34·2, alumina 57·5, fluorine 15·0. Infusible alone on charcoal before the blowpipe. Some varieties are changed by heat to a wine yellow or pink tinge.

Dif. Topaz is readily distinguished from tourmaline and other minerals it resembles by its brilliant transverse cleavage.

Obs. Pycnite has been separated from this species. It differs from topaz mainly in the state of aggregation of the particles, it presenting a thin columnar structure and forming masses imbedded in quartz. The *physalite* or *pyrophysalite* of Hisinger, is a coarse, nearly opaque variety, found in yellowish-white crystals of considerable dimensions This variety intumesces when heated, and hence its name from *phusao*, to blow.

Topaz is confined to granitic regions, and commonly occurs in granite, associated with tourmaline, beryl, occasionally with apatite, fluor spar and tin. With quartz, tourmaline and lithomarge, it forms the mixture called topaz rock by Werner.

Fine topazes are brought from the Uralian and Altai mountains, Siberia, and from Kamschatka, where they occur of green and blue colors. In Brazil they are found of a deep yellow color, either in veins or nests in lithomarge, or in loose crystals or pebbles. Magnificent crystals of a sky-blue color have been obtained in the district of Cairngorum, in Aberdeenshire. The tin mines of Schlackenwald, Zinnwald, and Ehrenfriedersdorf in Bohemia, St. Michael's Mount in

---

What are the forms and cleavage of topaz crystals? What are their colors? their luster and hardness? their composition? How is topaz distinguished from tourmaline and other minerals? How does topaz occur?

Cornwall, etc., afford smaller crystals. The physalite variety occurs in crystals of immense size at Finbo, Sweden, in a granite quarry, and at Broddbo, in a boulder. A well defined crystal from this locality, in the possession of the College of Mines of Stockholm, weighs eighty pounds. Altenberg in Saxony, is the principal locality of pycnite. It is there associated with quartz and mica.

Trumbull, Conn., is the principal locality of this species in the United States. It seldom affords fine transparent crystals, except of a small size: these are usually white, occasionally with a tinge of green or yellow. The large coarse crystals sometimes attain a diameter of several inches, (rarely six or seven,) but they are deficient in luster, usually of a dull yellow color, though occasionally white, and often are nearly opaque.

The ancient *topazion* was found on an island in the Red Sea, which was often surrounded with fog, and therefore difficult to find. It was hence named from *topazo*, to seek. This name, like most of the mineralogical terms of the ancients, was applied to several distinct species. Pliny describes a statue of Arsinoe, the wife of Ptolemy Philadelphus, four cubits high, which was made of topazion, or topaz, but evidently not the topaz of the present day, nor chrysolite, which has been supposed to be the ancient topaz. It has been conjectured that it was a jasper or agate; others have imagined it to be prase, or chrysoprase.

Uses. Topaz is employed in jewelry, and for this purpose its color is often altered by heat. The variety from Brazil assumes a pink or red hue, so nearly resembling the Balas ruby, that it can only be distinguished by the facility with which it becomes electric by friction. The finest crystals for the lapidary are brought from Minas Novas, in Brazil. From their peculiar limpidity, topaz pebbles are sometimes denominated *gouttes d'eau.* When cut with facets and set in rings, they are readily mistaken, if viewed by daylight, for diamonds. The coarse varieties of topaz may be employed as a substitute for emery in grinding and polishing hard substances.

Topaz is cut on a leaden wheel, and is polished on a copper wheel with rotten stone. It is usually cut in the form of the brilliant or table, and is set either with gold foil or *à jour*. The white and rose-red are most esteemed.

What are the uses of topaz? What is the effect of heat?

6. *Combination of a Silicate and Sulphate.*

LAPIS-LAZULI.—*Ultramarine.*

Monometric. In dodecahedrons. Cleavage imperfect. Also massive. Color rich Berlin or azure blue. Luster vitreous. Translucent to opaque. H=5·5. Gr=2·3—2·5.

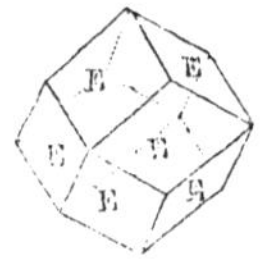

Composition: silica 45·5, alumina 31·8, soda 9·1, lime 3·5, iron 0·8, sulphuric acid 5·9, sulphur 0·9, chlorine 0·4, water 0·1. Fuses to a white translucent or opaque glass, and if calcined and reduced to powder loses its color in acids. The color of the mineral is supposed to be due to sulphuret of sodium.

Dif. Distinguished from azurite by its hardness and by giving no indications of copper before the blowpipe; and from lazulite by its fusibility, hardness, and not giving the reaction of phosphoric acid.

Obs. Found in granite and granular limestone, and is brought from Persia, China, Siberia, and Bucharia. The specimens often contain scales of mica and disseminated pyrites.

Uses. The richly-colored lapis lazuli is highly esteemed for costly vases, and for inlaid work in ornamental furniture. Magnificent slabs are contained in some of the Italian cathedrals. It is also used in the manufacture of mosaics. When powdered it constitutes the most beautiful and most durable of blue paints, called *ultramarine*, and has been one of the most costly colors. The late discovery of a mode of making an artificial ultramarine, quite equal to the native, has afforded a substitute at a comparatively cheap rate. This artificial ultramarine consists of silica 45·6, alumina 23·3, soda 21·5, potash 1·7, lime *trace*, sulphuric acid 3·8, sulphur 1·7, iron 1·1, and chlorine a small quantity undetermined. It has taken the place in the arts, entirely, of the native lapis-lazuli.

Hauyne, (including *nosean* and *spinellane*.) In dodecahedrons, and allied to the preceding. Color bright blue, occasionally greenish. Transparent to translucent. H=6. Gr=2·28—2·5. *Composition*,

---

What is the crystalline form of lapis-lazuli? What is its color? its hardness? its composition? How is it distinguished from apatite and lazulite? How does it occur? What are its uses? What is said of the artificial ultramarine?

silica 35·0, alumina 27·4, soda 9·1, lime 12·6, sulphuric acid 12·6, with traces of chlorine, sulphur and water. The *nosean* afforded silica 35·9, alumina 32·6, soda 17·8, sulphuric acid 9·2, with a small per-centage of other ingredients. A variety from Litchfield, Maine, afforded Dr. Jackson nearly the same proportions—silica 35·4, alumina 31·75, soda 17·6, sulphuric acid 6·5, with oxyd of manganese 4·4, and lime 1·8. Hauyne comes from the Vesuvian lavas and near Rome. The *nosean* is found in blocks with feldspar mica and zircon on the Rhine, near the Laacher See. Also at Litchfield, Maine.

7. *Silicate with a Chlorid.*

SODALITE.

In dodecahedrons like lapis-lazuli. Color brown, gray, or blue. H=6. Gr=2·25—2·3.

Composition: silica 37·2, alumina 31·7, soda 19·1, sodium 4·7, chlorine 7·3=100.

From Greenland, Vesuvius and Brisgau.

5. GLUCINA.

The minerals containing glucina are above quartz (7) in hardness, excepting one, (leucophane,) which contains largely of lime. The specific gravity is between 2·7 and 3·75. Excepting leucophane, they fuse before the blowpipe with extreme difficulty, or not at all.

BERYL.—*Emerald.*

Hexagonal. In hexagonal prisms. Usually in long, stout prisms, without regular terminations. Cleavage basal, not very distinct; rarely massive.

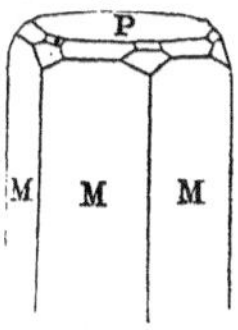

Color green, passing into blue and yellow; color rather pale, excepting the deep and rich green of the emerald. Streak uncolored. Luster vitreous; sometimes resinous. Transparent to subtranslucent. Brittle. H=7·5—8. Gr=2·65—2·75.

Varieties and *Composition.* The *emerald* includes the rich green variety; it owes its color to oxyd of chrome. *Beryl* especially includes the paler varieties, which are col-

---

What is sodalite? What is said of minerals containing glucina? What is the crystalline form of beryl? it colors and hardness?

ored by oxyd of iron. *Aquamarine* includes clear beryls of a sea-green, or pale-bluish or bluish-green tint.

The *beryl* consists of silica 66·9, alumina 19·0, glucina 14·1=100. *Emerald* contains less than one per cent. of oxyd of chromium. Before the blowpipe becomes clouded, but fuses on the edges with difficulty.

Dif. The hardness distinguishes this species from apatite; and this character, and also the form of the crystals, from green tourmaline; the imperfect cleavage, from euclase and topaz.

Obs. The finest emeralds come from Grenada, where they occur in dolomite. A crystal from this locality, 2¼ inches long and about 2 inches in diameter, is in the cabinet of the Duke of Devonshire. It weighs 8 oz. 18 dwts., and though containing numerous flaws, and therefore but partially fit for jewelry, has been valued at 150 guineas. A more splendid specimen, but weighing only 6 oz., is in the possession of Mr. Hope of London. It cost £500. Emeralds of less beauty, but of gigantic size, occur in Siberia. One specimen in the royal collection of Russia measures 4½ inches in length and 12 in breadth, and weighs 16¾ pounds troy. Another is 7 inches long and 4 broad, and weighs 6 pounds. Mount Zalora in Upper Egypt, affords a less distinct variety.

The finest beryls (*aquamarines,*) come from Siberia, Hindostan and Brazil. One specimen belonging to Don Pedro is as large as the head of a calf, and weighs 225 ounces, or more than 18½ pounds troy; it is transparent and without a flaw.

In the United States, beryls of enormous size have been obtained, but seldom transparent crystals. They occur in granite or gneiss. One hexagonal prism from Grafton, N. H., weighs 2900 pounds and measured 4 feet in length, with one diameter of 32 inches and another of 22; its color was bluish-green, excepting a part at one extremity, which was dull green and yellow. At Royalston, Mass., one crystal has been obtained a foot long, and pellucid crystals are sometimes met with. Haddam, Conn., has afforded fine crystals,

---

What is the composition of beryl? What are the different varieties and their distinctions? How is beryl distinguished from apatite and tourmaline? Where are the finest emeralds brought from? What is said of the Siberian emeralds? What of the finest beryls? What is the size of some beryls found in the United States?

(see the figure.) Other localities are Barre, Fitchburg, Goshen, Mass.; Albany, Norwich, Bowdoinham and Topham, Me.; Wilmot, N. H.; Monroe, Conn.; Leyperville, Penn.

The name beryl is from the Greek *beryllos.*

EUCLASE.

Monoclinic. In oblique rhombic prisms; M : M=115°. Cleavage in one direction highly perfect, affording smooth polished faces.

Color pale green. Luster vitreous; transparent. Very brittle. H=7·5. Gr=2·9—3·1. Pyro-electric.

Composition: silica 43·2, alumina 32·6, glucina 24·2. Before the blowpipe with a strong heat it intumesces, and finally fuses to a white enamel.

Dif. The very perfect cleavage of this glassy mineral is like that of topaz, and at once distinguishes it from tourmaline and beryl. It differs from topaz in its very oblique crystals

Obs. Occurs in Peru, and with topaz in Brazil.

Uses. The crystals of this mineral are elegant gems of themselves, but they are seldom cut for jewelry on account of their brittleness.

CHRYSOBERYL.

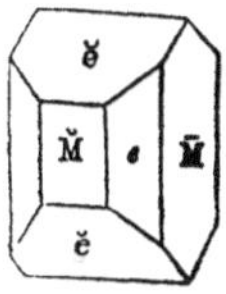

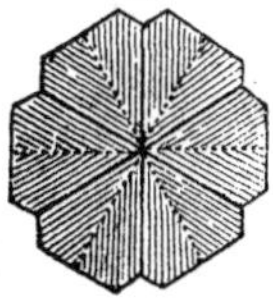

Trimetric. In modified rectangular prisms. é : ĕ=119° 46′. M̆ : *e*=125° 20′. Cleavage not very distinct, parallel to M̆. Also in compound crystals, as in fig. 2. Crystals sometimes thick; often tabular.

Color bright green, from a light shade to emerald green; rarely raspberry or columbine red by transmitted light. Streak uncolored. Luster vitreous. Transparent to translucent. H=8·5. Gr=3·5—3·8.

Composition: alumina 80·2, glucina 19·8=100. A little iron is sometimes present. Infusible and unaltered before the blow-pipe.

Alexandrite is a name given to an emerald-green variety from the Urals, which is supposed to be colored by chrome,

What is the form and cleavage of euclase? what the color and luster? How is it distinguished? What are its uses? What is the appearance of chrysoberyl? its hardness? its composition? What is alexandrite?

and to bear the same relation to ordinary chrysoberyl as emerald to beryl.

Dif. Near beryl, but distinct in its often tabular crystallizations, and its entire infusibility.

Obs. Chrysoberyl occurs in the United States in granite at Haddam, Conn., and Greenfield, near Saratoga, N. Y., associated with beryl, garnet, etc.

The name chrysoberyl is from the Greek *chrysos*, golden, and *beryllos*, beryl. *Cymophane* is another name of the species, alluding to its opalescence, and derived from the Greek *kuma*, wave, and *phaino*, to appear.

Uses. The crystals are seldom sufficiently pellucid and clear from flaws to be valued in jewelry; but when of fine quality, it forms a beautiful gem, and is often opalescent.

Phenacite. Colorless or bright wine-yellow, inclining to red, of vitreous luster and transparent to opaque. Crystals and cleavage rhombohedral. H=8. Gr=2·97. *Composition*, silica 54·3, glucina 45·7, with a trace of magnesia and alumina. Unaltered before the blowpipe. From Perm, Siberia, with emerald.

Leucophane. Resembles somewhat a light green apatite. H=3·5. Gr=2·97. Powder phosphorescent. Pyro-electric. *Composition*, silica 47·8, glucina 11·5, lime 25·0, protoxyd of manganese 1·01, potassium 0·3, sodium 7·6, fluorine 6·2. From Norway in syenite, accompanying albite and elæolite.

Helvin. Helvin occurs in Saxony and Norway in tetrahedrons of a wax yellow or brownish color. H=6—6·5. Gr=3·1—3·3. Luster vitreous. It contains silica, oxyds of iron and manganese, sulphuret of manganese, with glucina and alumina.

6. ZIRCONIA.

ZIRCON.

Dimetric. In square prisms and octahedrons. M : e = 132° 10′; e : e = 123° 19′. Cleavage parallel to M, but not strongly marked. Usually in crystals; but also granular.

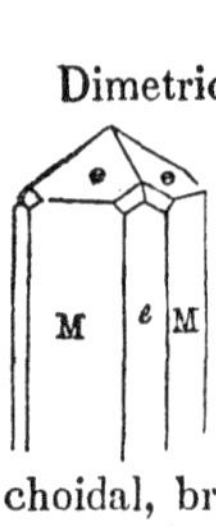

Color brownish-red, brown, and red, of clear tints; also yellow, gray and white. Streak uncolored. Luster more or less adamantine. Often transparent; also nearly opaque. Fracture conchoidal, brilliant. H = 7·5. Gr=4·0—4·8.

How does chrysoberyl differ from beryl? Where and how does it occur? What is the origin of the name chrysoberyl? What are its uses? Describe zircon?

Varieties and *Composition.* Transparent red specimens are called *hyacinth.* A colorless variety from Ceylon, having a smoky tinge, is called *jargon;* it is sold for inferior diamonds, which it resembles, though much less hard. The name *zirconite* is sometimes applied to crystals of gray or brownish tints. Consists of silica 33·2, zirconia 66·8. Infusible before the blowpipe, but loses color. Forms with borax a diaphanous glass.

Dif. The hyacinth is readily distinguished from spinel by its prismatic form and specific gravity, as well as its adamantine luster and a less clear shade of red. Its infusibility, hardness, and other characters, distinguish it from tourmaline, idocrase, staurotide, and the minerals it resembles.

Obs. The zircon is confined to the crystalline rocks, including lavas and granular limestone. Hyacinth occurs mostly in grains, and comes from Ceylon, Auvergne, Bohemia, and elsewhere in Europe. Siberia affords crystals as large as walnuts. Splendid specimens come from Greenland.

In the United States, fine crystals of zircon occur in Buncombe county, N. C.; of a cinnamon red color in Moria, Essex county, N. Y.; also at Two ponds and elsewhere, Orange county, in crystals sometimes an inch and a half long; in Hammond, St. Lawrence county, and Johnsbury, Warren county, N. Y.; at Franklin, N. J.; in Litchfield, Me.; Middlebury, Vt.; Haddam and Norwich, Conn.

The name *hyacinth* is from the Greek *huakinthos.* But it is doubtful whether it was applied by the ancients to stones of the zircon species.

Uses. The clear crystals (hyacinths) are of common use in jewelry. When heated in a crucible with lime, they lose their color, and resemble a pale straw-yellow diamond, for which they are substituted. Zircon is also used in jewelling watches. The hyacinth of commerce is to a great extent cinnamon stone, a variety of garnet.

The earth zirconia is also found in the rare minerals *eudialyte* and *wöhlerite;* also in *polymignite, æschynite, œrstedite;* also sparingly in *fergusonite.*

What is the composition of zircon? What are its varieties? How does it differ from spinel and other minerals? How does it occur? What is said of its uses? Does the earth zirconia occur in other minerals?

Eudialyte. In modified acute rhombohedrons; vitreous and of a red color. R : R=73° 30′. Transverse cleavage, perfect; opaque or nearly so. It is a silicate of zirconia, lime, soda and iron, and gelatinizes in acids. From West Greenland, in white feldspar.

Wöhlerite. In tabular crystals of light yellow and brownish shades; sometimes transparent. Consists mainly of silica, columbic acid, zirconia, (15 per cent.,) lime and soda. From Brevig, Norway.

Æschynite. A titanate of zirconia and oxyd of cerium, with some lime and oxyd of iron. Black and submetallic, or resinous in luster. H=5—6. Gr=4·9—5·2. From the Ural.

Œrstedite. A titanate and silicate of zirconia. Color brown. H=5·5. Gr=3.629. In brilliant crystals from Arendal, Norway.

Malacone. Contains silica 31·3, zirconia 63·4, with water 3. Form that of zircon. Gr=3·9. H=6. Appears to be a zircon containing water. Color bluish white, brownish, reddish. Streak colorless.

7. THORIA.

The earth Thoria has been found only in a rare mineral named from its constitution *thorite*, and in the ores *monazite*, (p. 206,) and *pyrochlore*, (p. 208.)

Thorite is a hydrous silicate of thoria. Color black to resin-yellow. Powder light orange to brown. Gr=4·6—5·3. From Norway.

CLASS VII.—METALS AND METALLIC ORES.

General condition of Metals and Metallic Ores in nature.— Metals are found either *native*, or *mineralized* by combination with other substances. The common ores are compounds of the metals with *oxygen*, *sulphur*, *arsenic*, *carbonic acid*, or *silica*. For example, the *oxyds* and *carbonate* of iron are the common workable iron ores; *sulphuret* of lead (called galena) is the lead ore of the arts; *arsenical* cobalt is the principal source of cobalt and arsenic.

Only a few of the metals occur native* in the rocks. Of these, *gold*, *platinum*, *palladium*, *iridium*, and *rhodium*, are with a rare exception, found only native. The *bismuth*

What is said of thoria? How do metals occur? What are ores? Give examples from ores of iron, lead, cobalt? What metals occur principally native?

---

* By *native* is understood either *pure*, or *alloyed* with other metals, excluding those metals, like arsenic or tellurium, which destroy the malleability of the metal and disguise its character. *Native gold* is much of it an alloy of gold and silver. But *aurotellurite*, a compound of gold and tellurium with some lead and silver, is properly mineralized gold.

of the shops is obtained from *native* bismuth. *Native silver, native mercury*, and *native copper*, are sometimes abundant, but are far from being the main sources of these metals. The other native metals are mineralogical rarities. Perhaps we should except from this remark native iron, which constitutes large meteoric masses, though very rarely if ever seen of terrestrial origin.

Their associations and impurities.—The ores of the metals are often much disguised by mixtures with one another or with earthy material. Thus a large part of the iron ore worked in England and this country is so mixed with clay or silica, that its real character might not be suspected without some experience in ores.

Occasionally ores contain phosphate of iron or some arsenical ores or certain sulphurets, scattered through them; and on account of the difficulty of separating the phosphorus, sulphur, or arsenic, the ore is rendered comparatively useless. By this intimate mixture of species, the difficulties of reducing ores are much increased.

When different ores are not intimately commingled, they are frequently closely disseminated together through the rock. We find ores of lead and zinc often thus associated; also of cobalt and nickel; of iron and manganese; the ores of silver, lead and copper, and often cobalt and antimony; platinum, iridium, palladium and rhodium.

Position in rocks.—Metals and their ores occur in the rocks in different ways:

1. In beds or layers between layers of rock, as some iron ores;
2. Disseminated through rocks in grains, nests, or crystals, or extended masses, as is the case with iron pyrites, cinnabar, or mercury ore, and much argillaceous iron;
3. In veins, intersecting different rocks, as ores of tin, lead, copper, and nearly all metallic ores;
4. Very frequently, metallic ores, instead of occurring in true veins, are found in rocks near their intersection with a mass or dike of igneous rock, as in the vicinity of a porphyry or trap dike. This is the case with much of the copper ore in Connecticut and Michigan, as well as with much

---

What is said of native iron? How are ores often disguised? Explain by example. How do they occur together? What is an effect of this mixture? What are the positions of ores in the rocks?

silver ore and mercury in South America and elsewhere, and often the igneous rock itself contains the same metals disseminated through it.

Gangue.—The rock immediately enveloping the ore is called the *gangue*. A vein often consists for the most part of the rock material called the gangue; and the ore either intersects the gangue in a continued band, or more commonly, is partly disseminated through it in some places, and is continuous for long distances in others. Often a good vein gradually loses its character, the metal disappears, and the gangue alone is left; but by following on for some distance, it will often resume its former character.

The usual *gangue* in metallic veins is either *quartz*, *calc spar*, or *heavy spar;* less frequently *fluor spar.* Calc spar is the gangue of the Rossie lead ore; heavy spar of much of the lead ore of the Mississippi valley; fluor spar in some places of the lead of Derbyshire, England.

Reduction of Ores.—In the *reduction* of an ore, the object is to obtain the metal in a pure state. It is necessary for this purpose to separate, 1, the gangue; 2, the impurities or minerals mixed with the ore; and 3, the ingredient with which the ore is mineralized—as the *sulphur*, for example, in the common ore of lead.

1. Much of the gangue will be separated in the process of mining and selecting the ore. Another portion is in many cases removed by pounding the ore coarsely, while a current of water is made to pass over it; the water carries off the lighter earthy matters and leaves the heavier ore behind. This process is called *washing*. With a fusible native metal, as bismuth, it is only necessary to heat the pounded ore in crucibles, and the metal flows out. A fusible ore, as gray antimony, is separated from the rock in the same manner. In the case of gold, which is usually in disseminated grains, mercury is mixed with the pounded rock after washing, which unites with the gold; and thus the gold is dissolved out from the gangue as water dissolves a salt; by vaporizing the solvent, mercury, the gold is afterwards obtained.

With iron ores, there is no special effort to separate the gangue beyond what is done in the process of mining.

What is the gangue? What is said of the ore in the gangue? What are the common kinds of gangue? What is meant by the *reduction* of an ore? What is necessary for this purpose? How is the gangue separated? How with a fusible metal or ore? How with gold?

2. The separation of the mineralizing ingredients when the ore is pure, is sometimes effected by heat alone; thus the common ores of mercury and lead, both sulphurets, will give up the sulphur in part when heated. In most cases, some material is added to combine with the mineralizing ingredient and carry it off; as when certain iron ores (oxyds of iron) are heated with charcoal, the charcoal takes the oxygen (forming the gas carbonic acid which escapes) and leaves the iron pure.

3. When two or more metals are mixed in the ore, one is sometimes removed by oxydation, or in other words, it is *burnt out*. Thus lead containing silver, is heated in a draft of air; the lead unites with the oxygen of the air and forms an earthy slag, while the silver, which is not thus oxydated, remains untouched. Such a process, carried on in a vessel of bone-ashes, or some material of the kind, which will absorb the oxyd of lead formed, is called *cupellation*. (See beyond under gold.) Much of the iron in the ordinary copper ore (copper pyrites) is removed in the common process of reduction in England by repeated fusions and stirring, while exposed to a draft of air.

4. When there are impurities present, or a mixture of the gangue, which is commonly the case, a material is sought for which will form, when heated, a fusible compound with the gangue and impurities; and this material is called a *flux*. Most iron ores are associated with quartz or clay, quartz being pure silica, and clay containing 75 per cent. of silica. Common limestone readily fuses into a glass with silica, when used in the requisite proportions, and hence it is generally employed as a flux in iron furnaces. A salt of soda or potash would produce the same result, for these are the ingredients which form with silica common glass. The glass formed is more or less frothy, and is called *slag* or *scoria*.

Before reduction, the volatile impurities and any water present, are often removed by a process called *roasting*.

The processes of reducing the ordinary metallic ores in the arts are combinations of the different steps here pointed out. There are other chemical methods for certain cases, which it is unnecessary to allude to in this place.

How is the mineralizing ingredient separated in some cases? How in others? Explain by examples. How in cases of mixture. Explain the process of cupellation. How in still other cases, and explain the use of fluxes by an example. What is said in conclusion of the processes of reduction?

1. 2. 3. CERIUM, YTTRIUM, LANTHANUM.

Cerium and Yttrium are not used in the arts. The species are infusible alone before the blowpipe or only in the thinnest splinters.

YTTROCERITE.

Massive, of a violet-blue color, somewhat resembling a purple fluor spar; sometimes reddish-brown. Opaque. Luster glistening. H=4—5. Gr=3·4—3·5.

Composition: fluoric acid 25·1, lime 47·6, oxyd of cerium, 18·2, and yttria 9·1. Infusible alone before the blowpipe.

Obs. From Finbo and Broddbo, near Fahlun in Sweden, with albite and topaz in quartz. Also from Massachusetts, probably in Worcester county, and from Amity, Orange county, N. Y.

Flucerine and *Basic Flucerine.* These two fluorids of cerium have a bright yellow or yellowish-red color. Infusible alone in the blowpipe flame. They are from Sweden.

Parisite. Occurs in bipyramidal dodecahedrons, (fig. 65, page 39,) of a reddish-brown or brownish-yellow color and vitreous fracture. Cleavage easy parallel to the base. Gr=4·35. Infusible alone. *Composition:* carbonic acid 23·5, protoxyds of cerium, lanthanum and didymium 59·4, lime 3·2, fluorid of calcium 11·5, water 2·4. From New Grenada.

Lanthanite. Trimetric. In thin minute tables or scales of whitish or yellowish color. H=2·5—3. *Composition:* carbonate of lanthanum 77·48, water 24·09. From Bastnäs, Sweden, and Saucon valley in Lehigh Co., Pa.

MONAZITE.

Monoclinic. In modified oblique rhombic prisms; M : M=93° 10′, $\bar{e}$ on $\bar{a}$=140° 40′, M : $\bar{e}$=136° 35′. Perfect and brilliant basal cleavage. Observed only in small imbedded crystals.

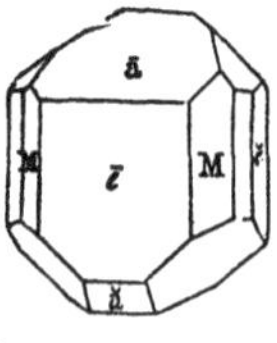

Color brown, brownish-red; subtransparent to nearly opaque. Luster vitreous inclining to resinous. Brittle. H=5. Gr=4·8—5·1.

Composition: oxyd of cerium 26·0, oxyd of lanthanum 23·4, thoria 17·95, phosphoric acid 28·5, with

---

What is said of the blowpipe action of ores of cerium and yttrium? What is the appearance and composition of yttrocerite? What is monazite?

oxyd of tin 2·1, protoxyd of manganese 1.9, lime 1·7. Infusible or nearly so. Decomposed by muriatic acid, evolving chlorine.

Dif. The brilliant easy transverse cleavage distinguishes monazite from sphene.

Obs. Occurs near Slatoust, Russia. In the United States it is found in small brown crystals, disseminated through a mica slate at Norwich, Conn.; also at Chester, Conn., and Yorktown, Westchester county, N. Y.

Cryptolite. A phosphate of the oxyd of cerium in minute prisms (apparently six-sided,) found with the apatite of Arendal, Norway Color pale wine yellow. Gr=4·6.

ALLANITE.

Monoclinic. In oblique rhombic prisms like epidote. Cleavage only in traces. Also massive and in acicular aggregations, the needles sometimes a foot long.

Color pitch-brown, brownish-black, streak greenish or brownish-gray, luster pitchy and submetallic. Opaque or nearly so. Brittle. H=5·5—6. Gr=3·3—4·2.

Varieties and *Composition. Allanite, cerine,* and *orthite* are names of different varieties of this species. The last occurs in acicular crystals as well as massive. They consist of silica and alumina, with oxyds of iron, cerium, lanthanum, and lime. They fuse before the blowpipe to a black glassy globule or pearl.

Dif. Allanite differs from garnet, some varieties of which it resembles, in its inferior hardness, and colored streak. Gadolinite fuses with more difficulty and glows on charcoal, besides gelatinizing in nitric acid.

Obs. Allanite was first brought from Greenland. It occurs in Norway, Sweden, and the Ural.

In the United States it has been found in large crystals in Allen's vein, Haddam Conn.; at Bolton, Athol, and South Royalston, Mass.; at Monroe, Orange county, N. Y.

Pyrorthite. This appears to be an impure orthite, containing some carbon, in consequence of which it burns when heated. Hence the name from the Greek *pur, fire,* and *orthite.* It comes from near Fahlun, Sweden. *Mosandrite* is related to allanite.

Cerite. A hydrated silicate of cerium. Color between clove-brown and cherry-red. Luster adamantine. Crystals hexagonal. From Bastnäs, Sweden.

How is it distinguished from sphene? What is the appearance and composition of allanite? what are its varieties?

Bodenite is a cerium ore, resembling orthite. From Boden in Saxony.

PYROCHLORE.

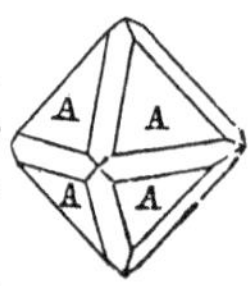

In small octahedrons, with a cleavage parallel to the faces of the octahedron sometimes distinct.

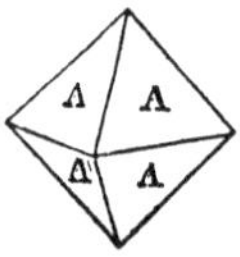

Color yellow to brown. Subtransparent to opake. Luster vitreous inclining to resinous. H=5. Gr=3·8—4·3.

Composition: essentially columbic acid, with oxyds of cerium, thorium, and lime. Titanic acid sometimes replaces part of the columbic acid. Fuses with very great difficulty before the blowpipe.

The *microlite* of Prof. Shepard appears to be pyrochlore.

Dif. The color, difficult fusibility and colored streak distinguish this species from others crystallizing in octahedrons. It is much softer than spinel.

Obs. Occurs in syenite in Norway, and also in Siberia. In the United States it is found in minute octahedrons at the Chesterfield albite vein, Mass.

The following species contain *yttrium* or *cerium* as a characteristic ingredient :—

Xenotime is a phosphate of yttria, having a yellowish-brown color pale brown streak, opaque, and resinous in luster. Crystals square prisms, with perfect lateral cleavage. H=4—5. Gr=4·6. Infusible alone before the blowpipe ; insoluble in acids. From Lindesnaes, Norway.

Gadolinite has a black or greenish-black color, resinous or subvitreous luster, greenish-gray streak. Crystalline form an oblique rhombic prism, with no distinct cleavage. H=6·5—7. Gr.=4·1—4·4. Consists mainly of silica, yttria, glucina, and protoxyd of iron, with also the recently discovered oxyd of lanthanum. From Fahlun and Ytterby, Sweden ; also from Norway and Greenland.

Fergusonite is a columbate of yttria, crystallizing in secondaries to a square prism. Color brownish-black ; luster dull, but brilliantly vitreous on a surface of fracture. Infusible before the blowpipe but loses its color. From Cape Farewell, Greenland.

Yttro-tantalite is a tantalate of yttria containing half as much yttria as the preceding. There are three varieties, the black, the yellow, and the brown or dark-colored. They are infusible. From Ytterby, Sweden, and at Broddbo and Finbo, near Fahlun.

Euxenite is a columbate of yttria with some titanic acid and oxyd of uranium. Massive. Color brownish-black. Streak powder reddish-brown. Infusible. From Norway.

What is the appearance and composition of pyrochlore ?

Tschefkinite. Resembles gadolinite. Color velvet-black. Luster vitreous. Streak dark brown. H=5—5·5. Gr=4·5—4·6. It is a silico-titanate of cerium. Gelatinizes readily on heating in muriatic acid. From the Ilmen Mountains, in Siberia.

Polymignite is principally a titanate of zirconia, yttria, iron and cerium. It has a black color, a brilliant submetallic luster within, a dark brown streak, and a conchoidal fracture. Generally in slender striated crystals, secondaries to a rectangular prism. H=6·5. Gr=4·7—4·9. From Norway. Also, as observed by Prof. C. U. Shepard, from Beverly, Mass.

Polycrase is near polymignite. Massive, and in thin linear crystals, of bright luster. Color black. Streak grayish-brown. H=5·5. Gr =5·1. With orthite in Norway.

Samarskite is a columbate of uranium, yttria and iron. Velvet-black. H=5·5—6. Gr=5·4—5·7. From the Urals, also from North Carolina.

Æschynite. In crystals, black to brownish yellow; luster resinous to submetallic; streak gray to yellowish brown or black. H=5—6. Gr=4·9—5·1. A titanate of zirconia and cerium. From Miask in the Urals, in feldspar with mica and zircon.

Rutherfordite. Blackish brown, with a vitreo-resinous fracture and no cleavage; powder yellowish-brown. From the gold mines of Rutherford Co., N. C., along with rutile, brookite, zircon and monazite. It contains 58·5 per cent. of titanic acid with 10 per cent. of lime, and perhaps cerium and yttrium.

4. TITANIUM.

Titanium occurs in nature combined with oxygen, forming titanic acid or oxyd, and also in combinations with different bases. It has not been met with native.

The ores are infusible alone before the blowpipe, or nearly so. Their specific gravity is between 3·0 and 4·5. With salt of phosphorus, in the inner flame on charcoal, a globule is obtained with some difficulty, which is violet blue when cold.

In the species called silico-titanates, that is, containing silica and titanic acid, the titanic acid is a base. Titanium and iron, and allied elements, are isomorphous, and analogous oxyds replace one another. See author's System of Min., 4th edition.

How does titanium occur? What is said of its ores?

RUTILE.

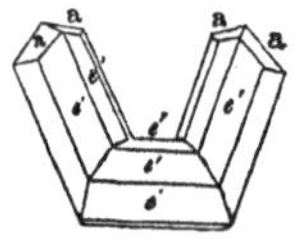

Dimetric. In prisms of eight, twelve, or more sides, with pyramidal terminations, and often bent as in the figure; a : a = 123° 8′. Crystals often acicular, and penetrating quartz. Sometimes massive. Cleavage lateral, somewhat distinct.

Color reddish-brown to nearly red; streak very pale-brown. Luster submetallic-adamantine. Transparent to opaque. Brittle. H=6—6·5. Gr=4·15—4·25.

Composition: titanium 61, oxygen 39. Sometimes contains iron, and has nearly a black color; this variety is called *nigrine.* Unaltered alone before the blowpipe. Forms a hyacinth-red bead with borax.

Dif. The peculiar subadamantine luster of rutile, and brownish-red color, much lighter red in splinters, are striking characters. It differs from tourmaline, idocrase, and augite, by being unaltered when heated alone before the blowpipe; and from tin ore, in not affording tin with soda; from sphene in its crystals.

Obs. Occurs imbedded in granite, gneiss, mica slate, syenite, and in granular limestone. Sometimes associated with specular iron, as at the Grisons. Yrieix in France, Castile, Brazil, and Arendal in Norway, are some of the foreign localities.

In the United States, it occurs in crystals in Maine, at Warren; in New Hampshire, at Lyme and Hanover; in Massachusetts, at Barre, Windsor, Shelburne, Leyden, Conway; in Connecticut, at Monroe and Huntington; in New York, near Edenville, Warwick, Amity, at Kingsbridge, and in Essex county at Gouverneur; in the District of Columbia, at Georgetown; in North Carolina, in Buncombe county; in the gold district of Georgia.

Uses. The specimens of limpid quartz, penetrated by long acicular crystals, are often very elegant when polished. A remarkable specimen of this kind was obtained at Hanover, N. H., and less handsome ones are not uncommon. Polished stones of this kind are called flèches d' amour (love's arrows) by the French.

Describe rutile. Of what does it consist? How is it distinguished from other minerals? What are its uses?

This ore is employed in painting on porcelain, and quite largely for giving the requisite shade of color and enamel appearance to artificial teeth.

Anatase. Brookite. These species have the same composition as rutile. Anatase occurs in slender nearly transparent octahedrons, of a brown color. A : A=97° 56′. H=5·5—6. Gr=3·8—3·9. From Dauphiny, the Tyrol, and Brazil. Said to accompany native titanium in slags from the iron furnaces of Orange county, N. Y.

Brookite is met with in thin hair-brown crystals, attached by one edge. H=5·5—6. The crystals are secondaries to a rhombic prism. From Dauphiny, and Snowdon in Wales. Said to occur at the Phenixville tunnel on the Reading railroad, Pa. *Arkansite* is Brookite.

SPHENE.

Monoclinic. In very oblique rhombic prisms; the lateral faces having angles either of 76° 1′, 113° 28′ (*r* : *r*)

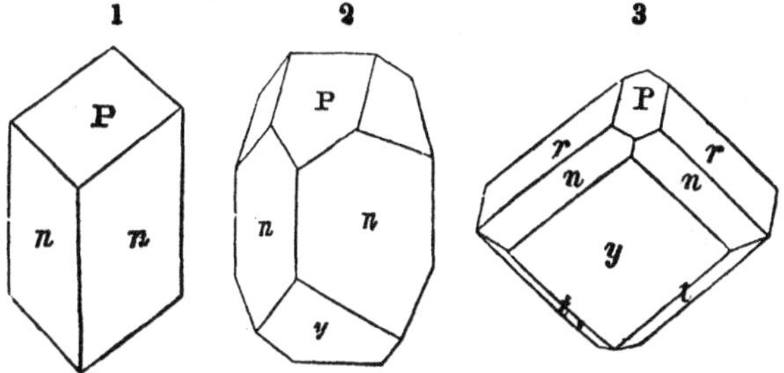

136° 4′ (*n* : *n*), or 133° 48′. The crystals are usually thin with sharp edges. Cleavage in one direction sometimes perfect. Occasionally massive.

Color grayish-brown, gray, brown or black; sometimes yellow or green; streak uncolored. Luster adamantine to resinous. Transparent to opaque. H=5—5.5. Gr=3·2—3·6.

Composition: silica 30·5, titanic acid 41·3, lime 28·2. Before the blowpipe, the yellow varieties are unaltered in color, and others become yellow; on charcoal, they fuse on the edges with a slight intumescence to a dark glass.

The dark varieties of this species were formerly called *titanite*, and the lighter *sphene*. The name *sphene* alludes to the wedge-shaped crystals, and is from the Greek *sphen*, a wedge.

What is said of the crystals of sphene? What are the color, luster, and hardness? the composition?

Dif. The crystals, in general, by their thin wedge shape, readily distinguish this species when crystallized; but some crystals are very complex. From garnet, tourmaline, and idocrase, this species is distinguished by its infusibility before the blowpipe.

Obs. Sphene occurs mostly in disseminated crystals in granite, gneiss, mica slate, syenite, or granular limestone. It is usually associated with pyroxene and scapolite, and often with graphite. It has been found in volcanic rocks. The crystals are commonly $\frac{1}{4}$ to $\frac{1}{2}$ an inch long; but are sometimes 1 to 2 inches.

Foreign localities are Arendal in Norway; at St. Gothard and Mount Blanc; in Argyleshire and Galloway in Great Britain.

In the United States, it is met with in good crystals in New York, at Rogers' Rock on Lake George, with graphite and pyroxene, at Gouverneur, near Natural Bridge in Lewis county, (the variety called *lederite*,) in Orange county in Monroe, Edenville, Warwick, and Amity, near Peekskill in Westchester county, and near West Farms. In Massachusets, at Lee, Bolton, and Pelham. In Connecticut, at Trumbull. In Maine, at Thomaston. In New Jersey, at Franklin. In Pennsylvania, near Attleboro', Bucks county. In Delaware, at Dixon's quarry, 7 miles from Wilmington. In Maryland, 25 miles from Baltimore, on the Gunpowder.

Greenovite is a sphene containing manganese.

Perofskite. This is a titanate of lime. It occurs in minute modified cubes, grayish to iron-black in color. Gr=4·017. H=5·5. From the Urals.

Pyrrhite. In minute regular octahedrons, of a yellowish color. Transparent; vitreous. H=6. From near Mursinsk, Siberia; also from the Western Islands, as first detected by Mr. J. E. Teschemacher of Boston. Supposed to contain titanic acid.

Keilhauite, or *yttro-titanite.* Related to sphene. Brownish-black, with a grayish-brown powder. Gr=3·69. H=6·5. Fuses easily. Contains silica 30·0, titanic acid 29·0, yttria 9·6, lime 18·9, peroxyd of iron 6·4, alumina 6·1. From Arendal, Norway.

Warwickite. It occurs in prismatic crystals, of a brownish to an iron-gray color, often tarnished bluish or copper-red. Luster metallic pearly to imperfectly vitreous or resinous. H=5—6. Gr=3—3·3. Infusible alone before the blowpipe. From magnesian limestone, with ilmenite and spinel, at Amity, Orange county, N. Y.

What are distinctive characteristics of the species sphene? In what rocks does it occur?

The analysis of warwickite, by Smith and Brush, has shown that it contains 20 per cent. of boracic acid, and therefore is a borotitanate.

Schorlomite. Black, and often irised tarnished. Streak grayish-black. H—7—7·5. Gr—3·80. Fuses readily on charcoal. Easily decomposed by the acids, and gelatinizes. Near gadolinite. From the Ozark Mountains, Arkansas.

Besides the ores here described, titanium is an essential constituent also of *ilmenite*, (titanic iron); also in the zirconia and yttria ores *æschynite*, *œrstedite*, and *polymignite*, and in some other rare species; sometimes in *pyrochlore*.

The metal titanium has seldom been obtained in the metallic state, and is not used in the arts. The uses of the oxyd have been mentioned.

5. TIN.

Tin has been reported as occurring native. There are two ores, the oxyd and a sulphuret. It also occurs in some ores of columbium. The specific gravity of the sulphuret is between 4·3 and 4·4; that of the oxyd, between 6·5 and 7·1. With carbonate of soda on charcoal, a globule of tin is obtained. When the tin is in minute quantities in a mineral, it is well to add also some borax, and by this means, especially if any iron present be first removed, or if it be only in small quantaties, even a ½ per cent. of tin may be detected.

Native tin is found in gray metallic grains in the gold washings of the Ural. The crystals of pure tin are either tesseral (cubic), or dimetric, this metal being dimorphous.

TIN PYRITES.—*Sulphuret of Tin.*

In cubes and massive. Color steel-gray or yellowish Streak black. Brittle. H=4. Gr=4·3—4·6.

Composition: sulphur 30, tin 27, copper 30, iron 13.

Obs. This rare ore has been found only in Cornwall, where it is often called *bell-metal ore*, from its frequent bronze appearance.

How does tin occur in the mineral kingdom? How is it detected by the blowpipe? What is the appearance and composition of tin pyrites?

TIN ORE.—*Oxyd of Tin.*

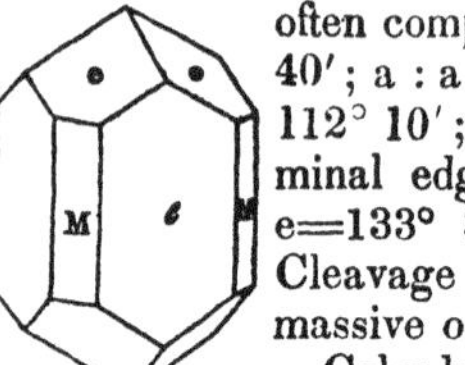

Dimetric. In modified square prisms and octahedrons; often compound : e : e=121° 40′; a : a (over the summit) 112° 10′; a : a (over a terminal edge) 133° 31′; M : e=133° 34′; M : *e*=135°. Cleavage indistinct. Also massive or in grains.

Color brown or black, with a high adamantine luster when in crystals. Streak pale gray to brownish. Nearly transparent to opaque. H=6—7. Gr=6·5—7·1.

Composition: when pure, tin 78·38, oxygen 21·62; often contains a little oxyd of iron, and sometimes oxyd of columbium. Before the blowpipe alone, infusible; with soda, affords a globule of tin.

Stream tin is the gravel-like ore found in debris in low grounds. *Wood tin* occurs in botryoidal and reniform shapes with a concentric and radiated structure; and *toad's-eye tin* is the same on a small scale.

Dif. Tin ore has some resemblance to a dark garnet, to black zinc blende, and to some varieties of tourmaline. It is distinguished by its infusibility, and its yielding tin before the blowpipe on charcoal with soda. It differs from blende also in its superior hardness, and in giving no fumes on charcoal before the blowpipe.

Obs. Tin ore occurs in veins in the crystalline rocks granite, gneiss, and mica slate, associated often with wolfram, copper and iron pyrites, topaz, tourmaline, mica or talc, and albite. Cornwall is one of its most productive localities. It is also worked in Saxony, at Altenberg, Geyer, Ehrenfriedersdorf and Zinnwald; in Austria, at Schlackenwald and other places; in Malacca, Pegu, China, and especially the Island of Banca in the East Indies. It has also been found in Galicia, Spain; at Dalecarlia in Sweden; in Russia; in Mexico, Brazil, and Chili; in the *United States*, at Chesterfield and Goshen, Mass., in some of the Virginia gold mines,

---

What is the crystallization of tin ore? Mention its other physical characters? What is its composition and blowpipe reactions? What is stream tin? wood tin, and toad's eye? How is tin ore distinguished from garnet, blende, and tourmaline?

and in Lyme and Jackson, N. H. At the last mentioned place, where this ore was discovered by Dr. C. T. Jackson, there are sufficient indications to warrant exploration.

GENERAL REMARKS ON TIN AND TIN ORES.

The principal tin mines now worked, are those of Cornwall, Banca and Malacca, Saxony, and Austria.

The Cornwall mines are supposed to have been worked long before the Christian era. Herodotus, 450 years before Christ, is believed to allude to the tin islands of Britain under the cabalistic name *Cassiterides*, derived from the Greek *kassiteros*, signifying *tin.** The Phœnicians are allowed to have traded with Cornubia, (as Cornwall was called, it is supposed from the horn shape of this western extremity of England.) The Greeks residing at Marseilles were the next to visit Cornwall, or the isles adjacent, to purchase tin ; and after them came the Romans, whose merchants were long foiled in their attempts to discover the tin market of their predecessors.

Camden says: " It is plain that the ancient Britons dealt in tin mines from the testimony of Diodorus Siculus, who lived in the reign of Augustus and Timaus, the historian in Pliny, who tells us that the Britons fetched tin out of the Isle of Icta, (the Isle of Wight,) in their little wicker boats covered with leather. The import of the passage in Diodorus, is that the Britons who lived in those parts dug tin out of a rocky sort of ground, and carried it in carts at low water to certain neighboring islands ; and that from thence the merchants first transported it to Gaul, and afterwards on horseback in thirty days to the springs of Eridanus, or the city of Narbona, as to a common mart. Æthicus too, another ancient writer, intimates the same thing, and adds that he had himself given directions to the workmen." In the opinion of the learned author of the Britannica here quoted, and others who have followed him, the Saxons seem not to have meddled with the mines, or according to tradition, to have employed the Saracens ; for the inhabitants of Cornwall to this day call a mine that is given over working *Attal-Sarasin*, that is, the leavings of the Saracens.†

The Cornwall veins, or *lodes*, mostly run east and west, with a dip—*hade*, in the provincial dialect—varying from north to south ; yet they are very irregular, sometimes crossing each other, and sometimes a promising vein abruptly narrows or disappears ; or again they spread out into a kind of bed or *floor*. The veins are considered worth working when but three inches wide. The gangue is mostly quartz, with some chlo-

---

Where are the principal tin mines? What is said of the Cornwall veins?

---

* This term and the *stannum* of the Romans, or *plumbum candidum*, are supposed to include the white compounds of lead and other metals ; and it has even been doubted whether the metal tin was ordinarily included.

† Manuf. in Metals ; London, 1834, iii. 2.

rite. Much of the tin is also obtained from loose stones, (called ***shodes,***) and courses of such stones or tin debris are called ***streams,*** whence the name ***stream tin.***

The ore taken from the mines is first pounded or stamped in a stamping mill, and then washed by running water, which carries off to a great extent the lighter impurities and leaves the heavy ore behind, with still some of the gangue. It is next roasted in a reverberatory furnace, to expel any arsenic or sulphur derived from the presence of other ores, and then again washed. After being thus purified as far as possible, the ore is usually mixed with pit-coal and a little lime, and strongly heated in either a reverberatory furnace or what is called a *blowing* furnace. A state of fusion is kept up for about eight hours. The metal is then drawn off into iron vessels. As it contains still some slag or earthy matters, it is remelted at a lower temperature, which does not fuse the impurities, and kept agitated for a while by wet charcoal or carbonized wood; it is then skimmed and run into blocks, weighing from 275 to 325 pounds each. The tin thus made from the ore derived from the mines, is called ***block tin,*** and is less pure than that from the stream ore; the latter was formerly called ***grain tin,*** though now this is a general term applied to the purest kinds of tin in commerce.

In an assay of tin ore, after pulverizing, washing, roasting, and weighing, the ore should be mixed with lampblack or charcoal, and heated quickly in a covered crucible to a white heat. On removing the crucible from the fire, a button of tin will be found in it. If the ore is not pure, carbonate of soda or borax may be added to the lampblack. The result is good if the tin obtained is malleable and not brittle. The tin may be farther purified by fusing it in a ladle, and pouring it into another vessel whenever the cooling has hardened the alloys, or just before the tin itself begins to harden; it will flow out, leaving the impurities behind.

The best tin ores afford 65 to 70 per cent. of tin in the large way.

The annual production of tin in different countries, is as follows:

Great Britain,	140,000	cwt.
Banca and Malacca,	100,000	"
Saxony,	3,500	"
Austria,	380	"
Sweden,	750	"

Tin is used in castings, and also for coating other metals, especially iron and copper. Copper vessels thus coated were in use among the Romans, thongh not common. Pliny says that the tinned articles could scarcely be distinguished from silver, and his use of the words *incoquere* and *incoctilia,* seems to imply, as a writer states, that the process was the same as for the iron wares of the present day, by *immersing* the *vessels in melted tin.* The sheets of iron for tinning are cleaned with acid, heated, and then cold-rolled; again subjected to dilute acid, and afterwards scoured with sand in pure water: then two or three hundred

---

What are the steps in the process of reduction? Describe the mode of assaying tin ore. What is the yield of Great Britain in tin? What the whole amount from the tin mines of the world? How is iron tinned?

sheets in a vertical position are immersed, first in a vat of grease, and then in a cast iron bath containing about 5 cwt. of melted tin; they remain in the tin for an hour and a half, and are then taken out. As there is now two or three times too much tin on the plates, they are made to undergo a process called *washing*, in a vessel of melted *grain* tin, by which the excess of tin is removed; after which they are cleaned and rubbed in bins of dry bran until they receive the characteristic silver polish.

When tin plate slightly heated is sponged over quickly by an acid, (nitro-muriatic,) the crystalline character of the tin is brought out, and the ware so treated is called *moiré metallique*. The plate before subjecting it to the acid should be well washed with alkali; and after the action it should be immediately washed in clean water and dried.

Tin is also used extensively as tinfoil, the sheets of which are about 1000th of an inch thick; also with quicksilver it is used to cover glass in the manufacture of mirrors. It is alloyed with copper in various proportions, constituting thus 7 to 10 per cent. of bronze; 20 per cent. of the ancient bronze for weapons; 20 per cent. of the metal for cymbals and the Chinese gong; 20 to 30 per cent. of bell metal; and 30 to 40 per cent. of speculum metal.

The oxyd of tin, as obtained by chemical processes, is employed on account of its hardness for forming a paste for sharpening fine cutting instruments. The *chlorid* of tin is an important agent in the precipitation of many colors as lakes, and in fixing and changing colors in dyeing and calico printing. The bisulphuret of tin has a golden luster, and was termed *aurum musivum*, or *mosaic gold*, by the alchemists. It is much used for ornamental painting, for paper hangings and other purposes, under the name of bronze powder.

Pins are tinned by boiling them for a few minutes in a solution of 1 part of cream tartar, 2 of alum, 2 of common salt, in 10 or 12 of water, to which some tin filings or finely granulated tin are added.

Tin medals or castings, are bronzed by being washed over with a solution of 1 part of protosulphate of iron, 1 of sulphate of copper, in 20 of water; this gives a gray tint; they are then brushed over with a solution of 4 parts of verdigris in 11 of distilled vinegar, and then polished with a soft brush and colcothar.

6. MOLYBDENUM.

Molybdenum occurs in nature as a sulphuret, and sparingly as an oxyd. Also as molybdic acid, in *molybdate of lead*.

1. MOLYBDENITE.—*Sulphuret of Molybdenum.*

In hexagonal crystals, plates, or masses, thin foliated like graphite, and resembling that mineral. Color pure lead-gray; streak the same, slightly greenish. Thin laminæ very flexible; not elastic. H=1—1·5. Gr=4·5—4·75.

In what other way is tin used? What alloys are made with it? What are the characters of molybdenite?

Composition: molybdenum 59·0, sulphur 41·0. Infusible before the blowpipe, but when heated on charcoal, sulphur fumes are given off, which are deposited on the coal. Dissolves in nitric acid, excepting a gray residue.

Dif. Resembles graphite, but differs in its paler color and streak, and also in giving fumes of sulphur when heated, as well as by its solubility in nitric acid.

Obs. Occurs in granite, gneiss, mica slate, and allied rocks; also in granular limestone. It is found at Numedahl in Sweden, Arendal in Norway, in Saxony, Bohemia, at Caldbeck Fell in Cumberland, and in the Cornish mines.

In the United States, it occurs in Maine at Blue Hill Bay, Camdage farm, Brunswick, and Bowdoinham; in New Hampshire at Westmoreland, Landaff, and Franconia; in Massachusetts at Shutesbury and Brimfield; in Connecticut at Haddam and Saybrook; in New York, near Warwick; in New Jersey, near the Franklin furnace.

Molydic ocher. An earthy yellow or whitish oxyd of molybdenum, (or rather molybdic acid,) occurring only as an incrustation. Occurs at Westmoreland, N. H.

For *molybdate of lead*, see under *Lead.*

7. TUNGSTEN.

Tungsten is found in combination with iron, lead, and lime, constituting wolfram, (p. 244,) tungstate of lead, (p. 283,) and tungstate of lime. It also occurs sparingly in some ores of columbium, as in certain varieties of the minerals pyrochlore, columbite, and yttro-columbite. It is met with in very small quantities as an ocher, or as *tungstic acid*, forming a yellow powder on other tungsten ores.

Lane's mine, Monroe, Conn., the adjoining town of Huntington, and Camdage farm, Blue Hill Bay, Me., are the only American localities of tungsten ores yet discovered. Lane's mine affords wolfram and the calcareous tungsten, and also the tungstic ocher. These ores are frequent associates of tin ore.

No use in the arts has been made of this metal or its com-

What is its composition? How does it differ from graphite? What are the principal ores of tungsten? Has any use been made of them in the arts?

pounds. Tungstic acid is a fine yellow, even brighter than chrome yellow; but it turns green on exposure to the sun's rays.

The metal tungsten was so called from the Swedish word *tung*, meaning heavy, the calcareous tungsten being peculiarly heavy for an earthy looking mineral. It has also been called *scheelium*, in honor of the chemist Scheele.

Tungstate of lime. In square octahedrons; A : A=100° 8′ and 130° 20′. Cleavage octahedral, perfect. Color yellowish-white, or brownish. Brittle. H=4—4·5. Gr=6·075. *Composition*, tungstic acid 7·8, lime 19·06. Infusible alone, or only on the thinnest edges. Found with wolfram at Lane's mine, Munroe, Conn.

8. VANADIUM.

Vanadium is a rare metal. It is found in nature as vanadic acid in the vanadate of lead (p. 285), and vanadate of copper (p. 302), and also combined with lime. The last mentioned has a brick-red color, a foliated structure, and a bright shining luster.

9. TELLURIUM.

Tellurium occurs native, and also in combination with gold, silver, lead, and bismuth.

The metal is distinguished from arsenic and selenium by giving no odor before the blowpipe; from antimony and bismuth by affording fumes in a glass tube below the temperature of fusing the glass; and when heated on charcoal, the oxyd covers the coal with a brownish-yellow oxyd, like bismuth; but the inner flame directed on this oxyd is tinged bright green, while bismuth gives no color. This last test distinguishes also the *ores* of tellurium.

Native tellurium occurs in six-sided prisms, of a tin-white color, and also massive. Brittle. H—2—2·5. Gr—6·1—6·3. *Composition*, pure tellurium with a little gold. From Transylvania.

Telluric Ochre. Occurs with native tellurium in Transylvania, in small whitish or yellowish masses, radiated in structure, and also as an earthy coating. Supposed to be tellurous acid.

In what minerals is vanadium found? How does tellurium occur in nature? How is this metal distinguished from arsenic and selenium?

10. BISMUTH.

Bismuth occurs native, and also in combination with sulphur, tellurium, oxygen, carbonic acid and silica. The ores fuse easily before the blowpipe, and an oxyd is produced which stains the charcoal brownish or yellow, without rising in fumes.* Specific gravity of the ores, between 4·3 and 9·5.

NATIVE BISMUTH.

Rhombohedral. Cleavage rhombohedral, perfect. In rhombohedrons, near cubes in form, R : R=87° 40′; generally massive, with distinct cleavage: sometimes granular.

Color and streak silver white, with a slight tinge of red. Subject to tarnish. Brittle when cold, but somewhat malleable when heated. H=2—2·5. Gr=9·7—9·8. Fuses at a temperature of 476° F.

Composition: pure bismuth, with sometimes a trace of arsenic. Evaporates before the blowpipe, and leaves a yellow coating on charcoal.

Obs. Bismuth is abundant with the ores of silver and cobalt of Saxony and Bohemia, and occurs also in Cornwall and Cumberland, England. At Schneeberg, it forms arborescent delineations in brown jasper.

In the United States, it has been found at Lane's mine, Monroe, where it occurs with tungsten, galena and pyrites, but is not abundant; also at Brewer's mine, in Chesterfield district, South Carolina.

There are other ores of bismuth, but none of them are common.

Sulphuret of bismuth. Massive and in acicular crystals, of a lead-gray color. H=2—2·5. Gr=6·55. Contains bismuth 81, sulphur

---

What are the color and physical characters generally of native bismuth? What is its temperature of fusion? With what ores is it usually associated?

---

* Tellurium produces a similar stain on charcoal, but on directing the inner flame on the coating, it colors the flame strongly green, while with bismuth no color is obtained. Antimony gives *white fumes*, producing a white coating on charcoal, and the flame directed on it is colored greenish-blue.

18·7. Fuses in the flame of a candle. From Cumberland, Cornwall, Johanngeorgenstadt, and Sweden.

Acicular bismuth. A sulphuret of bismuth, lead and copper, containing a trace of gold. In acicular crystals of a dark lead-gray color, with a pale copper-red tarnish. Gr=6·1. Fuses easily, emitting fumes of sulphur. From Siberia. A *cupreous* bismuth, of a pale lead-gray color, contains 34·7 per cent. of copper.

Tetradymite. Consists of tellurium and bismuth. It has a foliated structure, a pale steel-gray color, and soils like molybdenite. Gr=7·5. From Schemnitz and Retzbanya, Brazil, Virginia and North Carolina.

Bismutite. In acicular crystals and massive. Color greenish or yellowish. H=4—4·5. Gr=6·8—7·7. It is a *carbonate of bismuth.* From Cornwall; also South Carolina. *Bismuth ocher* is an impure oxyd, occurring massive and earthy; color greenish, yellowish, or grayish-white. From Saxony, Bohemia and Siberia.

Bismuth blende is a *silicate of bismuth.* Color dark hair-brown, or yellow. H=3·5—4·5. Gr=5·9—6·0. In dodecahedrons and massive. From Saxony.

GENERAL REMARKS ON BISMUTH AND ITS ORES.

The first notice of the metal bismuth is in the writings of Agricola, in 1529. It is known in the arts under the name of *tin glass*, from the French name *etain de glace.* It is obtained for the arts from the native bismuth alone, and much the greater part of the metal comes from Schneeberg in Saxony. The American mine at Monroe, Conn., has been but little explored, and has afforded only a few small specimens. The metal is obtained by heating the powdered ore in a furnace, when the bismuth melts, and separating from the gangue, is drawn off into cast iron moulds.

Bismuth is employed in the manufacture of the best type metal, to give a sharp, clear face to the letter. Equal parts of tin, bismuth and mercury form the *mosaic gold* used for various ornamental purposes. *Plumber's solder*, used for soldering pewter wares and other purposes, consists of 1 part of bismuth, 5 of lead, and 3 of tin. Bismuth is one of the constituents of *fusible metal*, of which spoons are made, as toys, that will melt on putting them into a cup of hot tea; this fusible alloy consists of 8 parts of bismuth, 5 of lead, and 3 of tin; or better of 10½ parts of bismuth, 5 parts of lead, and 3 of tin. It may be rendered more fusible still by adding mercury. An alloy of tin and bismuth in equal parts melts at 280° F. But with less bismuth, tin is increased in hardness.

The *magestens* of bismuth, a white hydrated oxyd precipitated by adding water to a solution of the nitrate, is used as a cosmetic. It contains a little nitric acid. *Pearl powder* is a similar preparation made in the same way from a nitrate containing some chlorid of bismuth. These powders blacken when exposed to an offensive atmosphere.

What is said of Bismuth and its ores?

11. ANTIMONY.

The metal antimony is occasionally found native. It is usually combined with sulphur, or sulphur and lead. It is also found in combination with arsenic, oxygen, and lime; also with nickel, silver, and copper.

It rises easily in white fumes before the blowpipe without odor, and in one or both of these particulars, it is distinguished from other vaporizable metals. The ores fuse very easily, and all evaporate, some giving off fumes of sulphur. Specific gravity below 7.

NATIVE ANTIMONY.

Rhombohedral. Usually massive, with a distinct lamellar structure. Color and streak tin-white. Brittle. H=3—3·5. Gr=6·6—6·75.

Composition: pure antimony, often with a little silver or iron. Fuses easily and passes off in white fumes.

Obs. Occurs in veins of silver and other ores in Dauphiny, Bohemia, Sweden, the Hartz, and Mexico.

GRAY ANTIMONY.—*Sulphuret of Antimony.*

Trimetric. In right rhombic prisms, with striated lateral faces. M : M=90° 45′. Cleavage in the direction of the shorter diagonal, highly perfect. M : e=145° 29′ e : e=109° 16′. Commonly divergent, columnar or fibrous. Sometimes massive granular.

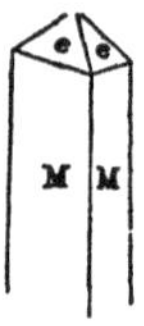

Color and streak lead-gray; liable to tarnish. Luster shining. Brittle; but thin laminæ, a little flexible. H=2. Gr=4·5—4·62.

e e M M

Composition: antimony 73, sulphur 27. Fuses readily in the flame of a candle. On charcoal it is absorbed, giving off white fumes and a sulphur odor.

Dif. Distinguished by its extreme fusibility and its vaporizing before the blowpipe.

Obs. Gray antimony occurs in veins with ores of silver, lead, zinc, or iron, and is often associated with heavy spar

---

How does antimony occur in nature? What are its blowpipe characters? What are the characters of native antimony? What is the crystallization and appearance of gray antimony? What is its composition? How is it distinguished? How does this ore occur?

or quartz. Its most celebrated localities are at Schemnitz, Kremnitz, and Felsobanya, in Hungary. It also occurs in the Hartz, Auvergne, Cornwall, Spain.

In the United States, it has been found sparingly at Carmel, Me., Lyme, N. H., and at "Soldier's Delight," Md.

Uses. This ore affords nearly all the antimony of commerce.

SULPHURETS OF ANTIMONY AND LEAD.

There are several sulphurets of antimony and lead, all of which fuse very easily, giving off white fumes, with a sulphur odor, and covering the charcoal with yellowish oxyd of lead. The color and streak are between lead-gray and dark steel-gray.

Jamesonite. Occurs in right rhombic crystals, and also fibrous or columnar. M : M=101° 20′. Streak and color steel-gray. H=2—2·5. Gr=5·5—5·8. Contains antimony 36 per cent., lead 44, and sulphur 20. From Cornwall, Siberia, and Hungary.

Feather ore. In fine capillary crystallizations, like a cobweb, or plumose. Color dark lead-gray. Contains antimony 31, lead 50, sulphur 19. From the Eastern Hartz.

Boulangerite. In plumose masses. Color bluish lead-gray. H=2·5. Gr=5·97. Contains antimony 24·1, lead 58·0, sulphur 18. From Molières in France; also from Lapland and Russia.

Plagionite. In oblique rhombic crystals. M : M=120° 49′. Color blackish lead-gray. Brittle. H=2·5. Gr=5·4. Contains antimony 38, lead 41, sulphur 21. From Wolfsberg in the Hartz.

Zinkenite. In hexagonal prisms; also fibrous and massive. Color steel-gray. H=3—3·5. Gr=5·3. Contains antimony 44, lead 35, sulphur 22. From Wolfsberg in the Hartz.

Geocronite, Kilbrickenite. Massive, with an imperfect cleavage, and also granular. Color light gray. H=2—2·5. Gr=6·4—6·6. Contains antimony 16·7, (which is sometimes partly replaced by arsenic,) lead 67, sulphur 16·5. From Gallicia, Kilbricken in Ireland, and Sala in Sweden.

Kobellite. Radiated like gray antimony. Gr=6·3. Contains 33 per cent. of sulphuret of bismuth, along with 46 of sulphuret of lead, and 13 of sulphuret of antimony. From Hvena in Sweden.

Steinmannite. In cubes with cubic cleavage, and massive. H=2·5. Gr=6·83. Color lead-gray. Affords before the blowpipe fumes of sulphur and antimony, and a globule of lead containing silver.

Besides these, there are also—

Berthierite, (called also *haidingerite,*) which resembles gray antimony, but contains 27 per cent. of sulphuret of iron with sulphuret of antimony. Another species contains 15 per cent. of sulphuret of iron. From Chazelles in Auvergne.

Arsenical antimony. Granular, massive; color tin-white or reddish-gray. H=2—4. Gr=6·2. *Composition,* antimony 36·4, arsenic 63·6. From Allemont and Bohemia.

Are there other ores of antimony? What is their general constitution?

WHITE ANTIMONY.

In white, grayish, or reddish rectangular crystals, with perfect cleavage, affording a rhombic prism of 136° 58′. Also in tabular masses, and columnar and granular. H = 2·5—3. Gr = 5·57. Luster adamantine to pearly. From Bohemia, Saxony, Hungary, Dauphiny. It is an oxyd of antimony containing 84·3 per cent. of antimony.

The *antimonic* and *antimonous acids* have been observed in a white pulverulent form. *Stiblite* is the name of a compound of oxyd of antimony and an antimony acid, (an antimonate of antimony.)

Red antimony is a compound of oxyd and sulphuret of antimony. Occurs usually in tufts of capillary crystals, or in flakes. Color cherry-red; streak brownish-red. Luster adamantine. H=1—1·5. Gr=4·4—4·6. From Hungary, Dauphiny, Saxony, and the Hartz.

Romeine is an *antimonate of lime*. It occurs in Piedmont in groups of minute square octahedral crystals, of a hyacinth or honey-yellow color. Scratches glass.

Antimonate of lead. A rare mineral consisting of antimonic acid 31·7, oxyd of lead 61·8, water 6·5. Amorphous, compact. Color yellow; also grayish, green, or black. Luster resinous. Gr=4.6—4·76. From Nertschinsk, Russia.

Senarmontite is the same compound as white antimony in octahedrons. Gr=5·2—5·3. From Algiers.

GENERAL REMARKS ON ANTIMONY AND ITS ORES.

The antimony of commerce is obtained from the sulphuret of antimony. This ore is worked at Schemnitz and Kremnitz in Lower Hungary, where it is associated with ores of silver, copper, lead, zinc, and manganese, and some gold. This region affords 6000 quintals of antimony annually. It has also been brought in considerable quantities from Borneo to Boston and then reduced. Several mines have been opened and abandoned in Auvergne and Dauphiny, but they are not now worked. There are also mines in France and Great Britain.

To obtain the crude antimony of the shops, the ore is placed in crucibles having a hole at bottom, and these are inserted in other vessels: heat is applied above, and the ore melts from its gangue and flows into the vessel below, where it becomes solid. It is not altered in composition. It is reduced by carefully roasting the crude antimony in a reverberatory furnace, and thus obtaining a gray oxyd. This oxyd is then mixed with a tenth of its weight of crude tartar, placed in large melting pots, and heated in a wind furnace. The metal antimony (called *regulus* of antimony) is thus obtained pure, excepting generally some little iron. By melting it again with one-fourth its weight of the oxyd of antimony, the impurities separate and form a slag above, leaving the metal beneath. It is a silver-white, brittle metal, coarsely crystalline in texture. It fuses at about 800° F.

What ore affords the antimony of commerce? Where is it mostly obtained? How is crude antimony obtained, and how reduced?

The sulphuret may be reduced also by heating it with iron filings; the iron takes the sulphur and liberates the antimony.

Antimony forms an important part of *type metal*. The proportions vary in different establishments; they have been stated at 1 of antimony to 4 to 12 of lead. A little tin is sometimes used, and also bismuth for the best type. The alloy is specially fitted for this purpose because it expands a little on cooling, filling well the mould and making a sharp, clear letter. The *Britannia metal*, which has superseded the use of pewter, consists of 100 parts of the best block tin, with 8 parts of the metal antimony, and either 2½ parts of each copper and brass, or 2 parts of copper and bismuth. A soft solder is used in the manufacture of Britannia ware, consisting of fine tin alloyed with about 30 per cent. of lead. Antimony with tin, forms the metal on which music is engraved.

The *glass of antimony*, which is much used for making pharmaceutical preparations, is a mixture of the sulphuret and oxyd of antimony, usually 85 of the latter to 15 of the former; it is formed by partially reducing the sulphuret to an oxyd by roasting, and then raising the heat till the whole melts.

Antimony in the condition of tartrate of antimony and potassa, is the *tartar emetic* of the apothecary.

12. ARSENIC.

The metal arsenic occurs native, and united with oxygen or sulphur. It also occurs in combinations with various metals, as iron, cobalt, nickel, silver, copper, manganese, and antimony; also as an acid in combination with the oxyds of iron, cobalt, nickel, copper, lead, and with lime. Its ores are distinguished readily by giving off an odor like garlic when heated on charcoal before the blowpipe. Its compounds with the metals and bases have already been described.

NATIVE ARSENIC.

Rhombohedral. R ; R=185° 41′. Cleavage basal, imperfect. Also massive, columnar, or granular.

Color and streak tin-white, but usually dark grayish from tarnish. Brittle, H=3·5. Gr=5·65—5·95.

Volatilizes very readily before fusing, with the odor of garlic; also burns with a pale bluish flame when heated just below redness.

Obs. Occurs with silver and lead ores. It is found in considerable quantities at the silver mines of Freiberg and

---

How is crude antimony reduced? For what is antimony used? What is Britannia metal? How does arsenic occur in the mineral kingdom? How is it distinguished? Describe native arsenic. With what is it found?

Schneeberg; also in Bohemia, the Hartz, at Kapnik in Upper Hungary, in Siberia in large masses, and elsewhere.

In the United States, it has been observed at Haverhill, N. H., in mica slate, and also at Jackson in the same state.

The name *arsenic* is derived from the Greek *arsenikon*, or *arrenikon*, masculine, a term applied to *orpiment*, a sulphuret of arsenic, on account of its potent properties.

WHITE ARSENIC.—*Arsenous Acid.*

In minute capillary crystals, and botryoidal or stalactitic. Color white. Soluble; taste astringent, sweetish. H=1·5—Gr=3·7. *Composition*, arsenic 75·8, oxygen 24·2.

This is the same compound with the common *arsenic* of the shops. It is found but sparingly native, accompanying ores of silver, lead and arsenic in the Hartz, Bohemia, and elsewhere.

Uses. It is a well known poison.

Pharmacolite, is an *arsenate of lime,* occurring in white or grayish crystals. H=2—2·5; Gr=2·6—2·8.

Haidingerite. Haidingerite is another arsenate of lime.

SULPHURETS OF ARSENIC.

There are two sulphurets of arsenic.

Orpiment or the *yellow sulphuret of arsenic.* In foliated masses, and sometimes in prismatic crystals, with a perfect diagonal cleavage. Color and streak fine yellow. Luster brilliant pearly, or metallic pearly on the face of cleavage. Subtransparent to translucent: sectile. H=1·5—2. Gr=3·4—3·5. *Composition*, sulphur 39·0, arsenic 61·0. Wholly evaporates before the blowpipe with an alliaceous odor, and on charcoal burns with a blue flame. From Hungary, Koordistan in Turkey in Asia, China, and South America. Occurs at Edenville, N. Y., as a yellow powder, resulting from the decomposition of arsenical iron.

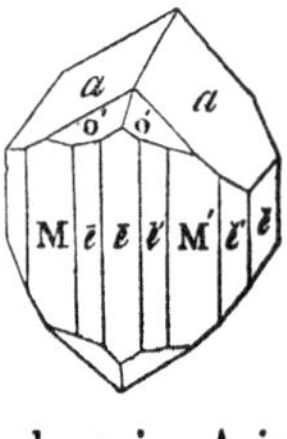

Realgar, or *Red sulphuret of arsenic.* In oblique prisms, and also massive: cleavage much less perfect than in orpiment. Color fine clear red, aurora red to orange. Luster resinous. Transparent to translucent. H=1·5—2. Gr=

What is white arsenic? What are the characters of orpiment? what of realgar?

3·35—3·65. *Composition*, sulphur 30, arsenic 70. Like the preceding before the blowpipe. From Hungary, Bohemia, Saxony, the Hartz, Switzerland, and Koordistan in Asiatic Turkey. It has been observed in the lavas of Vesuvius.

GENERAL REMARKS ON ARSENIC AND ITS ORES.

Arsenic is most used in the state of arsenous acid, called also white arsenic. This substance is prepared principally at Joachimstahl in Bohemia, and in Hungary, and is obtained from arsenical cobalt and iron. These ores are roasted in reverberatory furnaces, (the cobalt ores for the cobalt they contain,) and the vapors (which are white arsenic) are condensed in a long horizontal chimney; after undergoing a second sublimation, usually with a little potash, it is ready for commerce. The manufacture is very destructive to life, and those engaged in it seldom live over 30 or 35 years.

White arsenic, besides its use as a poison, is employed as a flux for glass, and also to give a peculiar milky or porcelain-like hue to glass ware. When too much is added, the glass becomes unsafe for domestic use.

The sulphurets afford valuable pigments. Orpiment is the basis of the pigment called *king's yellow.* The *ammoniacal solution* of orpiment is recommended for dyeing. It affords a yellow which is permanent, but is injured by soap. Realgar is used in the preparation of the pyrotechnical compound called *white Indian fire,* which consists of 24 parts of saltpeter, 7 of sulphur, and 2 of realgar, finely powdered and well mixed. It burns with a white flame and great brilliancy.

The sulphurets are obtained for commerce by distilling arsenical pyrites and iron pyrites, (sulphuret of iron,) or from white arsenic and rough brimstone; the product is realgar or orpiment according to the proportions employed.

A combination of the arsenous acid with oxyd of copper, obtained by mixing arsenite of potash and sulphate of copper, produces a fine green pigment called Scheele's green.

Arsenic is mixed in a small quantity (less than 1 per cent.) with lead, in the manufacture of shot, as it renders the metal more ready to break up into minute drops when caused to fall through a sieve from a height, as in the shot tower, and the grains assume a more spherical form on the descent, besides being less malleable than if of pure lead. In shot towers, the melted lead falls usually about 150 feet into a vessel of water at the bottom of the tower. They are afterwards sifted in sieves of different degrees of fineness, from No. 1, the finest, to No. 12, and thus the several sizes of shot are separated and assorted. There are still some imperfect shot among them; and to separate them the shot are made by a shake to roll from trays a little inclined into a bin; those that are imperfect roll sluggishly and are behind in the movement, and are thus separated to be melted over again.

How do orpiment and realgar differ in composition? From what ores is arsenic obtained? How is white arsenic prepared? For what is arsenic used? How are shot made?

13. URANIUM.

The uranium ores have a specific gravity not above 7, and a hardness below 6. The ores are either of some shade of light green or yellow, or they are dark brown or black and dull, or submetallic without a metallic luster when powdered. They are not reduced when heated with carbonate of soda; and the brown or black species fuse with difficulty on the edges or not at all.

PITCHBLENDE.—*Oxyd of Uranium.*

Massive and botryoidal. Color grayish, brownish, or velvet-black. Luster submetallic or dull. Streak powder black. Opaque. H=5·5. Gr=6·47.

Composition: 79 to 87 per cent. of protoxyd of uranium with silica, lead, iron, and some other impurities. Infusible alone before the blowpipe, but forms a gray scoria with borax. Dissolves slowly in nitric acid, when powdered.

Obs. Occurs in veins with ores of lead and silver in Saxony, Bohemia and Hungary; also in the tin mines of Cornwall, near Redruth. In the United States, at Middletown and Haddam, Conn.

Uranic ochre is a light yellow pulverulent mineral, becoming orange yellow when gently heated. It is believed to be peroxyd of uranium, sometimes combined with carbonic acid. Accompanies pitchblende in Cornwall and in Bohemia. It occurs sparingly in a yellow powder with columbite and uranite at the feldspar quarry, near Middletown, Conn.

Uses. The oxyds of uranium are used in painting upon porcelain, yielding a fine orange in the enameling fire, and a black color in that in which the porcelain is baked.

Coracite (Le Conte). An ore resembling pitchblende, and probably that species. From the north shore of Lake Superior, in a vein 2 inches wide, near the junction of trap and syenite.

Eliasite. A similar ore, containing 10½ per cent. of water.

URANITE.

Dimetric. In short square prisms, thinly foliated parallel

What is said of the ores of uranium? Describe pitchblende. What is its composition? What are the uses of the oxyds?

to the base, almost like mica; laminæ brittle and not flexible.

Color bright clear yellow and green; streak a little paler. Luster of laminæ pearly. Transparent to subtranslucent. H=2—2·5. Gr=3—3·6.

Composition. There are two ores here included, the yellow one containing phosphoric acid 16, oxyd of uranium 63, and lime 6, with water 15; the other of a green color, (sometimes called *chalcolite*,) containing oxyd of copper in place of lime. They fuse before the blowpipe to a blackish mass, and the green variety colors the flame green.

Dif. The micaceous structure connected with the light color is a striking character. The folia of mica are not brittle like those of uranite.

Obs. Occurs with uranium, silver and tin ores. It is found at St. Symphorien, near Autun, and also near Limoges, and in the Saxon and Bohemian mines. Cornwall affords splendid crystallizations of the green variety.

Found sparingly at Middletown, Conn., and Chesterfield, Mass., of a yellow color.

Samarskite (formerly named *uranotantalite* and *yttro-ilmenite*) is a compound of oxyd of uranium with columbic and tungstic acids, from Miask in the Ural. It is of a dark brown color and submetallic luster. H=5·5. Gr=5·4—5·7. Also occurs in North Carolina.

Johannite or *uranvitriol* is a sulphate of uranium. It has a fine emerald-green color, and a bitter taste. From Bohemia.

14. IRON.

Iron occurs native or alloyed with nickel in meteoric iron. Its most abundant ores are the oxyds and sulphurets. It is also found combined with other metals, and with silica and carbonic and other acids. Its ores are widely disseminated. They are the ordinary coloring ingredients of soils and many rocks, tinging them red, yellow, dull green, brown and black.

The ores have a specific gravity below 8, and the ordinary workable ores seldom exceed 5. Many of them are infusible before the blowpipe, and a great part become attractable by the magnet after heating, when not so before. When undisguised by other metals, they afford with borax, in the

---

What is the color and structure of uranite? its composition? How is it distinguished from other species? What is said of the modes of occurence of iron? What characters of its ores are mentioned?

inner flame, a bottle-green glass. By their difficult fusibility, the species with a metallic luster are distinguished from ores of silver and copper, and also more decidedly from these and other ores by blowpipe rëaction and reduction.

NATIVE IRON.

Monometric. In regular octahedrons; cleavage parallel to the faces of the octahedron. Usually massive, with a more or less fine granular structure.

Color and streak iron-gray. Fracture hackly. Malleable and ductile. H=4·5. Gr=7·3—7·8. Acts strongly on the magnet.

Obs. Native iron, as it occurs in meteorites, is usually alloyed with nickel and other metals. Whether terrestrial native iron has been observed, is a question of some doubt. A mass from Canaan, Conn., reported as of this character, has been shown by Dr. A. A. Hayes to be artificial, beyond doubt. Steinbach and Eibenstock in Saxony, and the mine of Hackenberg have been mentioned as foreign localities. Another occurs in Western Africa.

Meteoric iron occurs in nearly all meteorites, and almost wholly constitutes a large part of those that have been discovered. A mass weighing 1635 pounds is now in the cabinet of Yale College; it came from Texas. It contains 90 to 92 per cent. of iron, and 8 to 10 per cent. of nickel, the alloy not being uniform throughout. Meteoric iron often has a very broad crystalline structure, long lines and triangular figures being developed by putting nitric acid on a polished surface. The coarseness of this structure differs in different meteorites, and serves to distinguish specimens not identical in origin. The Texas iron is remarkable for the large size of the crystallization.

The most remarkable masses of meteoric iron occur in the district of Chaco-Gualamba in South America, where there is one whose weight is estimated at 30,000 pounds. The large Pallas meteorite weighed originally 1600 pounds; it contains imbedded crystals of chrysolite.

Besides nickel, which sometimes amounts nearly to 20 per cent., meteoric iron often contains a small per centage of

---

What is the crystallization of iron? its hardness, gravity, and other character? How does it occur native? What is said of meteoric iron?

cobalt, tin, copper and manganese; and frequently nodules of magnetic iron pyrites are imbedded in the mass. Chlorine has been detected in some specimens by Dr. C. T. Jackson.

Of still greater interest is the occurrence of a phosphuret of nickel (called *Schreibersite*) in most iron meteorites. It is in steel-gray masses, grains or folia, imbedded in the iron, and consists, according to Dr. J. L. Smith, of phosphorus 13·9, iron 57·2, nickel 25 8, cobalt 0·3, copper *a trace*, silica 1·6, alumina 1·6, lime *a trace*, chlorine 0·1. Its special interest arises from the fact that no phosphuret occurs among terrestrial minerals, and could not occur in any planet where oxygen is an abundant constituent, as on the earth. This mineral therefore, as stated by Dr. Smith, confirms the testimony from the native iron, that these meteoric bodies in space are in general without an atmosphere like ours, although not wholly destitute of oxygen, since there are several siliceous minerals present in many of them, as chrysolite, augite, feldspar, &c.

Meteoric iron is perfectly malleable, and may be worked like manufactured iron. The nickel diminishes much its tendency to rust.

IRON PYRITES.—*Bisulphuret of Iron.*

Monometric. Usually in cubes (fig. 1) simple or modified,

1

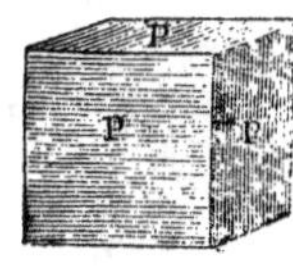

2

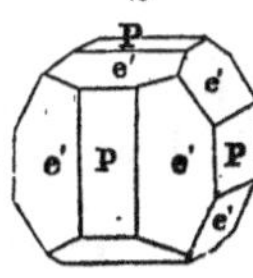

3

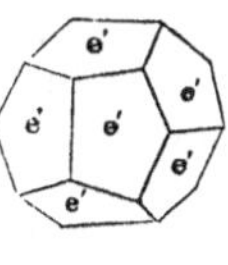

4

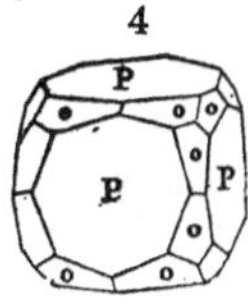

(2, 4,) or in pentagonal dodecahedrons (3); also in octahedrons. Faces of cubes often striated as in figure 1. Occurs also in imitative shapes, and massive.

Color bronze-yellow; streak brownish-black. Luster of crystals often splendent metallic. Brittle. H=6—6·5. Gr=4·8—5·1. Strikes fire with steel.

Composition: iron 46·7, sulphur 53·3. Before the blowpipe gives off sulphur, and ultimately affords a globule attractable by the magnet.

What is the crystallization of iron pyrites? its color and other characters? its composition?

Pyrites sometimes contains a minute quantity of gold, and is then called *auriferous pyrites*.

Dif. Distinguished from copper pyrites in being too hard to be cut by a knife, and also in its paler color. The ores of silver, at all approaching pyrites, instead of having its pale bronze-yellow color, are steel-gray or nearly black; and besides, they are easily cut with a knife and quite fusible. Gold is sectile and malleable; and besides, it does not give off a sulphur odor before the blowpipe, like pyrites.

Obs. Iron pyrites is one of the most common ores on the globe. It occurs in rocks of all ages. Cornwall, Elba, Piedmont, Sweden, Brazil, and Peru, have afforded magnificent crystals. Alston Moor, Derbyshire, Kongsberg in Norway, are well known localities. It has also been observed in the Vesuvian lavas.

In the United States, the localities are numerous. Fine crystals have been met with at Rossie, N. Y.; also in New York state at Scoharie, at Johnsburg and Chester, Warren county; at Champion and near Oxbow, in Jefferson county; at Warwick and Deerpark, Orange county. In Vermont, crystals occur at Shoreham; in Massachusetts, at Heath, Barre, and Boxborough; in Maine, at Corinna, Peru, Waterville and Farmington; in Connecticut, at Monroe, Orange, Milford and Stafford; in Pennsylvania, at Little Britain, Lancaster county. Massive pyrites occurs in Connecticut at Colchester, Ashford, Tolland, Stafford, and Union; in Massachusetts, at Hawley and Hubbardston; in Maine, at Bingham, Brooksville, and Jewell's Island; in New Hampshire, at Unity; in Vermont, at Strafford, where there is a vein in mica slate four rods wide, and also abundantly at Woodbury, and other places; in New York, in Franklin, Putnam and Orange counties, and elsewhere; in Maryland, abundant and worked at Cape Sable.

Uses. This species is of the highest importance in the arts, although not affording good iron on account of the difficulty of separating entirely the sulphur. It affords the greater part of the sulphate of iron (green vitriol or copperas) and sulphuric acid (oil of vitriol) of commerce, and also a considerable portion of the sulphur and alum. The py-

How is iron pyrites distinguished from copper pyrites? from silver ores? from gold? What is said of the occurrence of pyrites? Why does not this ore afford good iron? What are its uses? How is vitriol obtained from it?

rites is sometimes heated in clay retorts, by which about 17 per cent. of sulphur is distilled over and collected. The ore is then thrown out into heaps, exposed to the atmosphere, when a change ensues, by which the remaining sulphur and iron become sulphuric acid and oxyd of iron, and form sulphate of iron or copperas.* The material is lixivated, and partially evaporated, preparatory to its being run off into vats or troughs to crystallize. In other instances, the ore is coarsely broken up and piled in heaps and moistened. Fuel is sometimes used to commence the process, which afterwards the heat generated continues. Decomposition takes place as before, with the same result. At Strafford, Vermont, about 1000 tons of copperas have been produced annually, valued at 2 cents a pound, or $40,000. The quantity manufactured might easily be much increased. The pyrites of Cape Sable, Maryland, also affords large quantities of copperas. The lixivated liquid is often employed in Germany for the production of sulphuric acid; at a red heat, the acid passes off, leaving behind a red oxyd of iron, which is called *colcothar*. Cabinet specimens of pyrites, especially granular or amorphous masses, often undergo a spontaneous change to copperas, particularly when the atmosphere is moist.

The name *pyrites* is from the Greek *pur*, fire, because, as Pliny states, "there was much fire in it," alluding to its striking fire with steel. This ore is the *mundic* of miners.

White iron pyrites. This ore has the same composition as common iron pyrites, but crystallizes in secondaries to a right rhombic prism; M : M=106° 36′. The color is a little paler than that of common pyrites, and it is more liable to decomposition; hardness the same; specific gravity 4·6—4·85. *Radiated pyrites, hepatic pyrites, cockscomb pyrites,* (alluding to its crested shapes,) and *spear pyrites* are names of some of its varieties. It occurs in crystals at Warwick and Phillipstown, N. Y. Massive varieties are met with at Cummington, Mass.; Monroe, Trumbull, and East Haddam, Conn.; and at Haverhill, N. H.

PYRRHOTINE.—MAGNETIC PYRITES.—*Sulphuret of Iron.*

Hexagonal. Occurs occasionally in hexagonal prisms, which are often tabular; generally massive.

Color between bronze-yellow and copper-red; streak dark

How is sulphuric acid obtaine? and what is colcothar? What is the origin of the name pyrites? What is the crystallization and appearance of magnetic pyrites?

---

* This change consists in the union of oxygen with the sulphur and iron.

grayish-black. Brittle. H=3·5—4·5. Gr=4·4—4·65 Slightly attracted by the magnet. Liable to speedy tarnish.

Composition: sulphur 39·5 iron 60·5. Before the blowpipe on charcoal in the outer flame it is converted into a globule of red oxyd of iron. In the inner flame it fuses and glows, and affords a black globule which is magnetic, and has a yellowish color on a surface of fracture.

Dif. Its inferior hardness and shade of color, and its magnetic quality distinguish it from common iron pyrites; and its paleness of color from copper pyrites. It differs from the cobalt and nickel ores in affording a magnetic globule before the blowpipe.

Obs. Crystallized specimens have been found at Kongsberg in Norway, and at Andreasberg in the Hartz. The massive variety is found in Cornwall, Saxony, Siberia, and the Hartz; also at Vesuvius and in meteoric stones.

In the United States, it is met with at Trumbull and Monroe, New Fairfield, and Litchfield, Conn.; at Strafford and Shrewsbury, Vt.; at Corinth, New Hampshire; and in many parts of Massachusetts and New York. This ore at Litchfield is quite abundant.

Uses. Same as for common pyrites.

MISPICKEL.—*Arsenical Iron Pyrites.*

Trimetric. In rhombic prisms, with cleavage parallel to the faces M; M : M=111° 40′ to 112°. Crystals sometimes elongated horizontally, producing a rhombic prism of 100° nearly, with M and M the end planes. Occurs also massive.

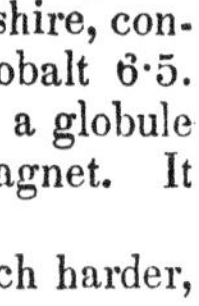

M M

Color silver-white; streak dark grayish-black. Luster shining. Brittle. H=5·5—6. Gr=6·3.

Composition: iron 34·4, arsenic 46·0, sulphur 19·6. A cobaltic variety contains 4 to 9 per cent. of cobalt in place of part of the iron. The *Danaite* of New Hampshire, consists of iron 32·9, arsenic 41·4, sulphur 17·8, cobalt 6·5. Affords arsenical fumes before the blowpipe, and a globule of sulphuret of iron which is attracted by the magnet. It gives fire with a steel and emits a garlic odor.

Dif. Resembles arsenical cobalt; but is much harder,

What is the constitution of magnetic pyrites? How is it distinguished from common iron pyrites? how from copper pyrites? from cobalt and nickel ores. For what is it used? What is the form and appearance of mispickel?

it giving fire with steel; it differs also in yielding a magnetic globule before the blowpipe and in not affording the reaction of cobalt with the fluxes.

Obs. Mispickel is found mostly in primitive regions, and is commonly associated with ores of silver, lead, iron, or copper. It is abundant at Freiberg, Munzig, and elsewhere in Europe, and also in Cornwall, England.

It occurs in crystals in New Hampshire, at Franconia, Jackson, and Haverhill; in Maine, at Blue Hill, Corinna, Newfield, and Thomaston; in Vermont, at Waterbury; in Massachusetts, massive at Worcester and Sterling; in Connecticut, at Chatham, Derby, and Monroe; in New Jersey, at Franklin; in New York, in Lewis, Essex county, and near Edenville and elsewhere in Orange county; in Kent, Putnam county.

Leucopyrite. This is the name of an arsenical iron, containining no sulphur, or but few per cent. It resembles the preceding in color and in its crystals; M : M=122° 26′. It has less hardness and higher specific gravity. H=5—5·5. Gr=7·2—7·4. Contains iron 32·4, arsenic 65·9, with some sulphur. From Styria, Silesia, and Carinthia. A crystal weighing two or three ounces has been found in Bedford county, Penn.; and in Randolph county, N. C., a mass was found weighing two pounds.

MAGNETITE.—*Octahedral Iron Ore.*

Monometric. Often in octahedrons and dodecahedrons, Cleavage octahedral; sometimes distinct. Also granularly massive

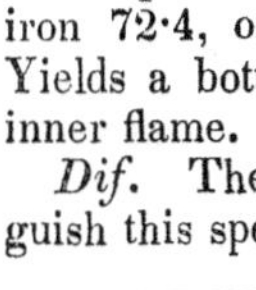

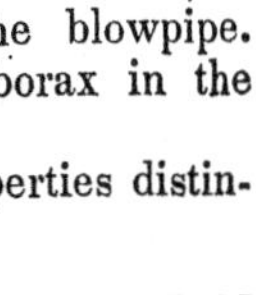

Color iron-black. Streak black. Brittle. H=5·5—6·5. Gr=5·0—5·1. Strongly attracted by the magnet, and sometimes having polarity.

Composition: peroxyd of iron 69, protoxyd of iron 31; or iron 72·4, oxygen 27·6. Infusible before the blowpipe. Yields a bottle-green glass when fused with borax in the inner flame.

Dif. The black streak and magnetic properties distinguish this species from the following.

What are the constituents of mispickel? What is the effect before the blowpipe? How does it differ from arsenical cobalt? What is the crystallization of magnetic iron? its other physical characters? its composition? What is the action of magnetic iron before the blowpipe? How is it distinguished from specular iron?

Obs. Magnetic iron ore occurs in extensive beds, and also in disseminated crystals. It is met with in granite, gneiss, mica slate, clay slate, syenite, hornblende, and chlorite slate ; and also sometimes in limestone.

The beds at Arendal, and nearly all the Swedish iron ore, consist of massive magnetic iron. At Dannemora and the Taberg in Southern Sweden, and also in Lapland at Kurunavara and Gelivara, there are mountains composed of it.

In the United States, extensive beds occur in Warren, Essex, and Clinton counties, N. Y. ; also in Orange, Putnam, Saratoga, and Herkimer counties ; at Mount Desert and Marshall's Island, Maine ; in Somerset, Vermont ; in Bernardstown and Hawley, Massachusetts ; at Franconia, Lisbon, and Winchester, New Hampshire. The mountainous districts of New Jersey and Pennsylvania afford this ore, and also the eastern side of Willis mountain in Buckingham county, Virginia. Crystals occur in New Hampshire, at Franconia in epidote ; also at Swanzey, (near Keene,) Unity, and Jackson ; in Vermont, at Marlboro', Bridgewater and Troy, in chlorite slate ; in Connecticut, at Haddam ; in Maine, at Raymond, Davis's Hill, in an epidotic rock ; in New York, at Warwick, Orange county, and also at O'Neil mine ; in New Jersey, at Hamburgh, near the Franklin furnace ; in Maryland, at Deer Creek ; in Pennsylvania, at Morgantown, Berks county ; also in the south part of Chester county.

Masses of this ore in a state of magnetic polarity, constitute what is called *lodestone* or *native magnets.* They are met with in many beds of the ore. Siberia and the Hartz have afforded fine specimens ; also the island of Elba. They also occur at Marshall's Island, Maine ; also near Providence, Rhode Island. The lodestone is called *magnes* by Pliny, from the name of the country, Magnesia, (a province of ancient Lydia,) where it was found ; and it hence gave the terms *magnet* and *magnetism* to science.

Uses. No ore of iron is more generally diffused than the magnetic ore, and none is superior for the manufacture of iron. The ore after pounding may be separated from impurities by means of a magnet ; and machines are in use in northern New York and elsewhere, for cleaning the ore on a large scale for furnaces.

How does magnetic iron occur? What are its uses? What is said of lodestone?

SPECULAR IRON ORE.—HEMATITE.

Rhombohedral. In complex modifications of a rhombohe-

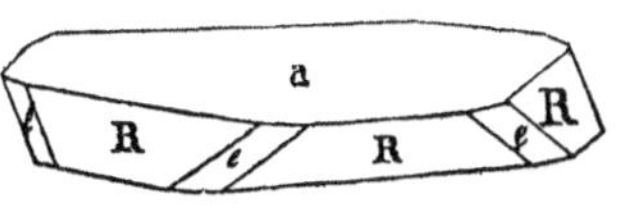

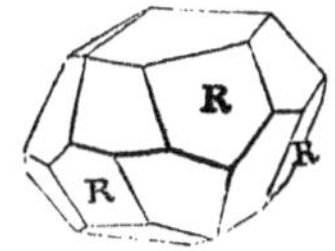

dron of 85° 58′; crystals occasionally thin tabular. Cleavage usually indistinct. Often massive granular; sometimes lamellar or micaceous. Also pulverulent and earthy.

Color dark steel-gray or iron-black, and often when crystallized having a highly splendent luster; streak-powder cherry-red or reddish-brown. The metallic varieties pass into an earthy ore of a red color, having none of the external characters of the crystals, but perfectly corresponding to them when they are pulverized, the powder they yield being of a deep red color, and earthy or without luster. Gr=4.5—5·3. Hardness of crystals 5·5—6·5. Sometimes slightly attracted by the magnet.

Varieties and *Composition*.

Specular iron. Specimens having a perfectly metallic luster.

Micaceous iron. Specular iron, with a foliated structure.

Red hematite. Submetallic, or unmetallic, and of a brownish-red color.

Red ocher. Soft and earthy, and often containing clay.

Red chalk. More firm and compact than red ocher, and of a fine texture.

Jaspery clay iron. A hard impure ore, containing clay, and having a brownish-red jaspery look and compactness.

Clay iron stone. The same as the last, the color and appearance less like jasper.

This is one variety of what is called "clay iron stone." Much of it belongs to the following species, and a large part also is spathic iron, as is the case with that of the English coal measures.

Lenticular argillaceous ore. A red ore, consisting of small flattened grains, something like an oolite.

Oligiste iron, *iron-glance*, and *rhombohedral iron ore*, are other names of the species specular iron.

What is the crystallization of specular iron? What are its physical characters? Describe the varieties.

Composition of the pure ore: iron 70, and oxygen 30. The varieties without a perfect metallic luster often contain more or less clay or sand. Before the blowpipe alone infusible; with borax in the inner flame gives a green glass, and a yellow glass in the outer flame.

Dif. This ore is distinguished from magnetic iron ore by its red powder; and from any silver or copper ores by its hardness and infusibility. The word *hematite*, from the Greek *haima*, blood, alludes to the color of the powder.

Obs. This ore occurs in both crystalline and stratified rocks, and is of all ages. The more extensive beds of pure ore abound in the primary rocks; while the argillaceous varieties occur in stratified rocks, being often abundant in coal regions and other strata. Crystallized specimens occur also in some lavas.

Splendid crystallizations of this ore come from Elba, whose beds were known to the Romans; also from St. Gothard; Arendal, Norway; Langbanshyttan, Sweden; Lorraine and Dauphiny. Etna and Vesuvius afford handsome specimens.

In the United States, this is an abundant ore. The two iron mountains of Missouri, situated 90 miles south of St. Louis, consist mainly of this ore, piled "in masses of all sizes from a pigeon's egg to a middle size church." One of them is 300 feet high, and the other, the "Pilot knob," is 700 feet. Both the massive and micaceous varieties occur there together with red ochreous ore. Large beds of specular iron have been explored in St. Lawrence and Jefferson counties, N. Y.; Plymouth, Bartlett and elsewhere in New Hampshire; Woodstock and Aroostock, Maine, and Liberty, Maryland, are other localities; also the Blue Ridge, in the western part of Orange county, Va. The micaceous variety occurs at Hawley, Mass., Piermont, N. H., and in Stafford county, Va. Lenticular argillaceous ore is abundant in Oneida, Herkimer, Madison, and Wayne counties, N. Y., constituting one or two beds 12 to 20 inches thick in a compact sandstone; it contains 50 per cent. of oxyd of iron, with about 25 of carbonate of lime, and more or less magnesia and clay. The coal region of Pennsylvania affords abundantly the clay iron ores, but they are mostly the argillaceous carbonate of iron or limonite.

What is the composition of specular iron? What are its distinguishing characters? What is its mode of occurrence? What is said of the iron mountains of Missouri?

Uses. Valuable as an iron ore, though less easily worked when pure and metallic than the magnetic and hematitic ores. Pulverized red hematite is used for polishing metals. *Red chalk* is a well known material for red pencils.

LIMONITE.—BROWN IRON ORE.

Usually massive, and often with a smooth botryoidal or stalactitic surface, having a compact fibrous structure within. Also earthy.

Color dark brown to ocher-yellow; streak yellowish-brown to dull yellow. Luster sometimes submetallic; often dull and earthy; on a surface of fracture frequently silky. H=5—5·5. Gr=3·6—4.

Varieties and *Composition.* The following are the principal varieties:

Brown hematite. The botryoidal, stalactitic and associated compact ore.

Brown ocher, Yellow ocher. Earthy ochreous varieties, of a brown or yellow color.

Brown and *yellow clay iron stone.* Impure ore, hard and compact, of a brown or yellow color.

Bog iron ore. A loose earthy ore of a brownish-black color, occurring in low grounds.

Composition when pure: peroxyd of iron 85·6, (*seven-tenths* of which is pure iron,) and water 14·4; or it is a *hydrous peroxyd of iron*, containing when pure about two-thirds its weight of pure iron. Before the blowpipe, blackens and becomes magnetic. Gives with borax in the inner flame a green glass.

Dif. This is a much softer ore than either of the two preceding, and is peculiar in its frequent stalactitic forms, and in its affording water when heated in a glass tube.

Obs. Occurs connected with rocks of all ages, but appears, as shown by the stalactitic and other forms, to have resulted in all cases from the decomposition of other iron ores, probably the sulphuret.

This is an abundant ore in the United States. The following are a few of its localities. Extensive beds exist at Salisbury and Kent, Conn., in mica slate; also in the neigh-

---

What is said of the uses of specular iron? What is the appearance of brown iron ore? its composition? Describe its varieties. What are distinguishing characters? How does this ore occur?

boring towns of Beekman, Fishkill, Dover, and Amenia, N. Y.; also in a similar situation north, at Richmond and Lenox, Mass.; also at Bennington, Monkton, Pittsford, Putney, and Ripton, Vermont. Large beds are found in Pennsylvania, the Carolinas, near the Missouri iron mountains, and also in Tennessee, Iowa and Wisconsin.

Uses. This is one of the most valuable ores of iron. It is also pulverised and used for polishing metallic buttons and other articles. As yellow ocher, it is a common material for paint.

Göthite, Lepidokrokite. These are names given to crystals of a hydrous peroxyd of iron, differing in composition from brown iron ore by containing half as much water. The crystals are of a brown color, and blood-red by transmitted light when subtransparent. Streak brownish-yellow to ocher-yellow. H=5. Gr=4·0—4·2. Occurs with hematite at Eiserfeld in Nassau; at Clifton in Cornwall; in Siberia and elsewhere.

FRANKLINITE.

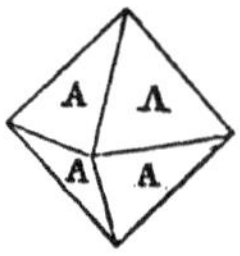

Monometric. In octahedral and dodecahedral crystals, and also coarse granular massive. Color iron-black; streak dark reddish-brown. Brittle. H=5·5—6·5. Gr=4·85—5·1; acts slightly on the magnet.

A A A A

Composition: peroxyd of iron 66, sesquoxyd of manganese 16, oxyd of zinc 17. Alone infusible. At a high temperature zinc is driven off, and is deposited on the charcoal; with borax on a platinum wire, in the outer flame, it gives the violet color due to manganese; and in the inner flame on charcoal, the green color due to iron.

Dif. Resembles magnetic iron, but the exterior color is a more decided black. The streak is not black, and the blowpipe reactions are different.

Obs. This is an abundant ore at Sterling and Hamburgh, in New Jersey, near the Franklin furnace; at the former place, the crystals are sometimes 4 inches in diameter. It is said to occur also in the mines of Altenberg, near Aix-la-Chapelle.

Uses. The attempts to work this ore for zinc have not been successful.

What is said of the uses of brown iron ore? What is the appearance of franklinite? What is its composition? How is it distinguished from magnetic iron ore?

ILMENITE.—*Titanic iron.*

In crystallization near specular iron. R : R=85° 59′. Often in thin plates or seams in quartz; also in grains. Crystals sometimes very large and tabular.

Color iron-black; streak metallic. Luster metallic or submetallic. H=5—6. Gr=4·5—5; acts slightly on the magnetic needle.

Composition: oxyd of iron, with a variable proportion of titanic acid or oxyd of titanium. Infusible alone before the blowpipe.

Crichtonite, ilmenite, menaccanite, hystatite, and *iserine*, are names of some of the varieties of this species. The hystatite variety includes the *washingtonite* of Professor Shepard. *Octahedral* and *cubic* crystals of this mineral have been found with titaniferous sand, which are supposed to be pseudomorphous.

Dif. Near specular iron, but differs in the less luster of its crystals, and its metallic streak.

Obs. Crystals an inch or so in diameter occur in Warwick, Amity, and Monroe, Orange county, N. Y.; also near Edenville and Greenwood furnace; also at South Royalston and Goshen, Mass.; at Washington, South Britain, and Litchfied, Conn.; at Westerly, Rhode Island.

Uses. Of no value in the arts.

CHROMIC IRON.—*Chromate of Iron.*

Monometric. In octahedral crystals, without distinct cleavage. Usually massive, and breaking with a rough unpolished surface.

Color iron-black and brownish-black; streak dark brown. Luster submetallic; often faint. H=5·5. Gr=4·3—4·5. In small fragments attractable by the magnet.

Composition: green oxyd of chromium 60·0, protoxyd of iron 20·1, alumina 11·8, magnesia 7·5. The alumina and magnesia are variable. Infusible alone before the blowpipe. Fuses slowly with borax to a beautiful green globule.

Dif. The little luster of this ore on a surface of fracture is peculiar; also its fine green glass with borax, which distinguishes it from ores of iron and other metals.

Describe titanic iron. Of what does it consist? How does it differ from specular iron? What is the appearance of chromic iron? its composition? How is it distinguished from other ores?

Obs. Occurs usually in serpentine rocks, in imbedded masses or veins. Some of the foreign localities are the Gulsen mountains in Styria; the Shetland Islands; the department of Var in France; Silesia, Bohemia, etc.

In the United States, it is abundant in Maryland in the Bare Hills near Baltimore, and also in Montgomery county, at Cooptown in Harford county, and in the north part of Cecil county; occurs also in Townsend and Westfield, Vermont, and at Chester and Blandford, Mass. It is also found at Hoboken, N. Y., and at Milford and West Haven, Conn.; in Pennsylvania in Little Britain, Lancaster county, and West Branford, Chester county, and on the Wisahicon, 11 miles from Philadelphia.

Uses. The compounds of chrome are extensively used as pigments. These compounds are obtained either from chromic iron or the native chromate of lead, (see under lead.) The chromate of lead and copper (vauquelinite) is too rare to be employed for this purpose. The chromate of potash is readily formed by mixing equal parts of nitre and the powdered chromic iron and exposing the mixture in a crucible to a strong heat for some hours. The soluble part is then washed out, and the process is repeated with the insoluble portion (digesting it first in muriatic acid to remove the free oxyd of iron and alumina) till all the ore is decomposed. The colored liquid obtained from the washings is carefully saturated with nitric acid, and concentrated by evaporation till crystals of nitre cease to be deposited. Being then set aside for a week or two, it gradually deposits abundant crystals of the yellow chromate of potash. Chromate of lead, called also *chrome yellow*, is the most common chrome paint used. It is made by adding to the liquid obtained as above stated, before its crystallization, a solution of acetate of lead (sugar of lead) till it is saturated. The yellow precipitate washed out and dried, is the chrome yellow of commerce. It is used as a yellow pigment both in oil and water colors, calico printing, dyeing, and porcelain painting. This material is largely manufactured at Baltimore, Md. The native nitrate of soda of Peru, has been suggested as a substitute for nitre in the above process.

Another mode of this manufacture recently proposed, con-

Where does chromic iron occur? What are its uses? How is the ore treated? What is chrome yellow, and how is it made?

sists in making a chromate of lime from the chromic iron. It is as follows: 1. Pulverize very finely chalk and chromic iron, and mix the sifted material well by means of a revolving barrel. 2. Calcine for nine or ten hours at a bright red heat in a reverberatory furnace, when, if complete, the whole has a yellowish-green color, and dissolves entirely in muriatic acid. 3. The porous mass after being crushed under a mill, to be mixed with hot water and kept agitated, adding a little sulphuric acid till it slightly reddens blue litmus paper. 4. Triturated chalk should then be added, and the oxyd of iron is thus removed. 5. After being left quiet for a while, the clear supernatant liquid is to be drawn off: it contains bichromate, with a little sulphate of lime. The chromate of potash may then be made from it by adding carbonate of potash; the chromate of lead, by adding acetate of lead; chromate of zinc, by adding chlorid of zinc.

The *bichromate of potash* has a fine red color, and is much used by calico printers. It is made from the chromate by adding nitric or acetic acid to its solution, (enough to give it a sour taste,) and setting it aside to crystallize. The *green oxyd of chromium* gives the fine green color to glass of borax in blowpipe experiments with chromic iron; and it is used to produce this tint in porcelain and enamel painting. It is the coloring ingredient of the emerald, and the emerald-colored chrysoberyl of the Urals; and occurs in some varieties of diallage and serpentine. It has been found native. *Chromic acid* is said to be the coloring matter of the red sapphire or ruby. With oxyd of tin, it affords a pink color, which is used in porcelain painting.

COLUMBITE.

Trimetric. In rectangular prisms, more or less modified. Also massive. Disseminated in the gangue. Cleavage parallel to the lateral faces of the prism, somewhat distinct.

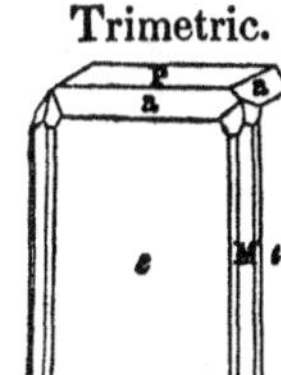

Color iron-black, brownish-black; often with a characteristic iridescence on a surface of fracture; streak dark brown, slightly reddish. Luster submetallic, shining. Opaque.

Describe another mode of treating chromic iron? What is the coloring ingredient of the emerald? what of the red sapphire? What are the color, luster and form of columbite?

Brittle. H=5—6. Gr=5·3—6·4. American 5·3—5·71. Bavarian 5·7—6·4.

Composition of an American specimen: columbic acid 79·6, protoxyd of iron 16·4, protoxyd of manganese 4·4, oxyd of tin 0·5, oxyds of copper and lead 0·1. The Bavarian columbite, besides having a higher specific gravity than the American, has also a black streak. Damour has proposed for it the name *Baierine;* but the differences, as far as yet known, are not important.

Infusible alone before the blowpipe. With borax in a fine powder fuses quite slowly, but perfectly, to a dark green glass, which indicates only the presence of iron.

Dif. Its dark color, submetallic luster, and a slight iridescence, together with its breaking readily into angular fragments, will generally distinguish this species from the ores it resembles.

Obs. Occurs in granite at Bodenmais in Bavaria, and also in Bohemia. In the United States, it is found in the same rocks, feldspathic or albitic, at Middletown and Haddam, Conn.; at Chesterfield and Beverly, Mass., and at Acworth, N. H. A crystal was found at Middletown, which originally weighed 14 pounds avoirdupois; and a part of it, 6 inches in length and breadth, weighing 6 lbs. 12 oz., is now in the collections of the Wesleyan University of that place.

This mineral was first made known from American specimens, by Mr. Hatchett, an English chemist, and the new metal it was found to contain was named by him *columbium.*

Tantalite or *Ferrotantalite.* This is an allied mineral, often called, from its locality at Kimito in Finland, *kimito-tantalite.* It is a *neutral* tantalate of iron. H=5—6. Gr=7·2—8·0. A variety from *Broddbo* contains 8 per cent. of oxyd of tin, with 6 of tungstic acid. Sp. gr. =6·5.

Note.—The metal named *Columbium* by Hatchett, is the same that has since been called *Niobium;* and the *Tantalum* of the Swedish ores is a different metal. For other ores of columbium and tantalum, see pages 208, 209.

WOLFRAM.—*Tungstate of Iron and Manganese.*

Trimetric. In modified rhombic or rectangular prisms; sometimes pseudomorphous in octahedrons imitative of tungstate of lime. Also massive. Color dark grayish-black;

Of what does columbite consist? How does it differ from other ores? Describe wolfram.

streak dark reddish brown. Luster submetallic, shining, or dull. H=5—5·5. Gr=7·1—7·9.

Composition: tungstic acid 75·89, protoxyd of iron 19·24, protoxyd of manganese 4·97. Fuses with difficulty. Gives a green bead with borax, and a deep red globule with salt of phosphorus.

Found often with tin ores. Occurs in Cornwall, and at Zinnwald and elsewhere in Europe. In the United States, it is found at Monroe and Trumbull, Conn.; on Camdage farm near Blue Hill, Me.; near Mine la Motte, Missouri; in the gold regions of North Carolina.

SILICATES OF IRON.

There are several compounds of silica and oxyd of iron, none of which are of special interest in an economical point of view.

Hedenbergite is a variety of augite, consisting essentially of these ingredients, (see page 151.)

Iron chrysolite differs from ordinary chrysolite in containing oxyd of iron in place of magnesia.

Isopyre is a black glassy amorphous mineral, found in granite. H=6—6·5. Gr=2·9—3. Consists of silica 47·1, alumina 13·9, peroxyd of iron 20·1, lime 15·4, oxyd of copper 1·9.

Lievrite, (called also *yenite* and *ilvaite.*) Occurs in rhombic prisms, often with the sides much striated or fluted; color black or brownish black. Luster submetallic. Streak black, greenish or brownish. H=5·5—6. Gr=3·8—4·1. Contains about 50 to 55 per cent. of oxyd of iron with 14 of lime and 29 of silca. Fuses to a black globule. From the island of Elba in large crystallizations; also from Norway, Siberia, Silesia. At Cumberland, Rhode Island, yenite occurs in slender black or brownish-black crystals, in quartz.

The following are hydrous species, giving off water when heated in a tube before the blowpipe.

Nontronite and *pinguite,* are earthy almost like clay, of a yellowish or greenish color.

Chloropal is a harder species, (H=3—4,) of a greenish-yellow or pistachio-green color. *Grengesite, thuringite, knebelite,* and *kirwanite,* are other allied species.

Green earth. Includes different compounds of a green earthy appearance. The *green earth* occupying cavities in amygdaloid is near *chlorite.* It is a silicate of the peroxyd of iron with some potash, magnesia and water; often with other ingredients. The *green grains* of the *green sand* of New Jersey, consist of silica 51·5, alumina 6·4, protoxyd of iron 24·3, potash 9·96, water 7·7.

Hisingerite, cronstedtite, anthosiderite, polyhydrite, sideroschisolite, chamoisite, stilpnomelane, and *xylite,* are names of dark brown or black species.

Of what does wolfram consist? With what ores is it usually associated? What is said of the compounds of oxyd of iron with silica?

Crocidolite has a fibrous structure much resembling asbestus, and has been called *blue asbestus.* Color lavender-blue or leek-green. H=4. Gr=3·2—3·3. From Southern Africa.

Pyrosmalite occurs in hexagonal prisms with a perfect basal cleavage, and pearly surface. Color pale liver-brown, grayish, or greenish. H=4—4·5. Gr=3·8. Contains 14 per cent. of chlorid of iron, and gives off fumes of muriatic acid before the blowpipe.

Iron-zeolite. A hydrous silicate of the oxyds of iron and manganese, forming incrustations at a mine near Freyberg.

COPPERAS.—*Sulphate of Iron, or Green Vitriol.*

Monoclinic. In acute oblique rhombic prisms. M : M=82° 21′; P : M=80° 37′. Cleavage parallel to P, perfect. Generally pulverulent or massive.

Color greenish to white. Luster vitreous. Subtransparent to translucent. Taste astringent, sweetish, and metallic. Brittle. H=2. Gr=1·83.

Composition: oxyd of iron 25·42, sulphuric acid 29·01, water 45·57. Becomes magnetic before the blowpipe. Yields a green glass with blowpipe; and a black color with a tincture of nut galls. On exposure, becomes covered with a yellowish powder, which is a persalt of iron.

Obs. This species is a result of the decomposition of pyrites, which readily affords it if moistened while exposed to the atmosphere, as stated under *pyrites.* The old mine of Rammelsberg in the Hartz, near Goslar, is its most noted locality; but it occurs wherever pyrites is found.

Copperas is much used by dyers and tanners, on account of its giving a black color with tannic acid, an ingredient in nutgalls and many kinds of bark. It for the same reason forms the basis of ordinary *ink,* which is essentially an infusion of nutgalls and copperas. It is also employed in the manufacture of *Prussian blue.* With prussiate of potash, any soluble persalt of iron, even in minute quantity, gives a fine blue color to the solution, (due to the formation of Prussian blue,) and this is a common test of the presence of iron.

About 1800 tons of copperas are used in the United States annually. The *colcothar of vitriol* is the browish-red oxyd of iron, obtained from copperas by calcination and other processes. It is much used as a polishing powder.

Coquimbite, or *white copperas,* and *yellow copperas,* are names of two sulphates of the peroxyd of iron. *Pittizite, fibro-ferrite,* are allied

---

What is the appearance and taste of copperas? its composition? What is its origin in nature? For what is it used?

compounds. *Apatelite* is still another, peculiar in containing but 4 per cent. of water.

Voltaite is a double sulphate of iron, alumina, potash and water, crystallizing like alum in octahedrons. From the Solfatara, near Naples.

SPATHIC IRON.—*Carbonate of Iron.—Chalybite.*

Hexagonal. In rhombohedrous and six-sided prisms, easily cleavable parallel to a rhombohedron of 107°. Faces often curved. Usually massive, with a foliated structure, somewhat curving. Sometimes in globular concretions or implanted globules.

Color light grayish to brown; often dark brownish-red, or nearly black on exposure. Streak uncolored. Luster pearly to vitreous. Translucent to nearly opaque. H=3—4·5. Gr=3·7—3·85.

Composition, when pure: protoxyd of iron 62·07, carbonic acid 37·93. Often contains some oxyd of manganese or magnesia, replacing part of the oxyd of iron. Before the blowpipe it blackens and becomes magnetic; but alone it is infusible. Colors borax green. Dissolves in nitric acid, but scarcely effervesces unless pulverized.

The ordinary crystallized or foliated variety is called *spathic* or *sparry* iron, because the mineral has the aspect of a spar. The globular concretions found in some amygdaloids or lavas, have been called *spherosiderite*. An argillaceous variety, occurring in nodular forms, is often called *clay iron stone*, and is abundant in the English coal measures.

Dif. This mineral is foliated like calc spar and dolomite; but it has a much higher specific gravity. It readily becomes magnetic before the blowpipe.

Obs. Spathic iron occurs in rock of various ages, and often accompanies metallic ores. The largest beds are found in gneiss and graywacke, and also in the coal formation. In Styria and Carinthia, it is very abundant in gneiss, and in the Hartz it occurs in graywacke. Cornwall, Alstonmoor and Devonshire, are English localities.

A vein of considerable extent occurs at Roxbury, near New Milford, Conn., in quartz, traversing gneiss; at Plymouth, Vt., and Sterling, Mass., it is also abundant. It oc-

---

Describe spathic iron. What is its constitution? What are its chemical characters? How does it differ from calc spar? What are its varieties? How does it occur?

curs also at Monroe, Conn.; in New York state, in Antwerp, Jefferson county, and in Hermon, St. Lawrence county. The argillaceous carbonate in nodules and beds, is very abundant in the coal regions of Pennsylvania.

Uses. This ore is employed extensively for the manufacture of iron and steel.

Thomaite is a carbonate of iron occurring in rhombic prisms. Gr=3·1. From the Siebengebirge mines. *Junkerite* has proved to be common spathic iron.

Mesitine spar, (Breunnerite.) A carbonate of iron and manganese, occurring in yellowish rhombohedrons of 107° 14′. H=4. Gr=3·3—3·6. This includes much of what is called *rhomb spar,* or *brown spar,* which becomes rusty on exposure.

Oligon spar. A carbonate of iron and manganese. Angle of rhombohedron 107° 3′. Color yellow or reddish-brown. Gr=3.75.

VIVIANITE.

Monoclinic. In modified oblique prisms, with cleavage in one direction highly perfect. Also radiated, reniform, and globular, or as coatings.

Color deep blue to green. Crystals usually green at right angles with the vertical axis, and blue parallel to it. Streak bluish. Luster pearly to vitreous. Transparent to translucent; opaque on exposure. Thin laminæ flexible. H=1·5—2. Gr=2·66.

Composition: protoxyd of iron 42·4, phosphoric acid 28·7, water 28·9. Loses its color before the blowpipe and becomes opaque; and if pulverized, fuses to a scoria, which is magnetic. Affords water in a glass tube, and dissolves in nitric acid.

Dif. The deep blue color connected with the softness, are decisive characteristics. The blowpipe affords a confirmatory test.

Obs. Found with iron, copper and tin ores, and sometimes in clay, or with bog iron ore. St. Agnes in Cornwall, Bodenmais, and the gold mines of Vöröspatak in Transylvania, afford fine crystallizations. In the United States, good crystals have been found at Imleytown, N. J. At Allentown, Monmouth county, and Mullica Hill, Gloucester county, N. J., are other localities. It often fills the interior of certain fossils. Occurs also at Harlem, N. Y., in Somerset and

---

For what is spathic iron used? What is the color and structure of vivianite? Of what does it consist?

Worcester counties, Md., and with bog ore in Stafford county, Va.

The *blue iron earth* is an earthy variety, containing about 30 per cent. of phosphoric acid. The mineral from Mullica Hill has been called *mullicite*.

Anglarite, from Anglar, France, is a similar mineral, with less phosphoric acid.

Triphyline occurs in cleavable masses, of a greenish-gray or bluish color. H=5. Gr=3·6. It is an anhydrous phosphate of the protoxyds of iron, and manganese, with some lithia. From Bodenmais in Bavaria, and Norwich, Mass.

Triplite. Another phosphate of iron and manganese, of brown or blackish-brown color. From Limoges, in France.

Green iron stone, (*kraurite*,) *alluaudite*, *melanchlor*, and *beraunite*, are names of phosphates of the peroxyd of iron. Color of the first two, dull leek-green; structure fibrous. Luster silky. Color of the third, black; of the fourth, hyacinth-red, becoming darker on exposure.

Cacoxene. This is a handsome species, occurring in radiated silky tufts of a yellow or yellowish-brown color. H=3—4. Gr=3·38. It is a phosphate of alumina and iron. It differs from wavellite, which it resembles in its more yellow color and iron reactions. It also resembles carpholite, but has a deeper color. It occurs on brown iron ore in Bohemia.

Carphosiderite is another yellow phosphate of iron from Greenland. It occurs in reniform masses.

ARSENATES OF IRON.

Cube ore. Occurs in cubes of dark green to brown and red colors Luster adamantine, not very distinct. Streak greenish or brownish. H=2·5. Gr=3. It is a hydrous arsenate of the peroxyd of iron, containing 38 per cent. of arsenic acid. From the Cornwall mines; also from France and Saxony.

Scorodite. Crystallizes in rhombic prisms, modified. M : M=120° 10′. Color pale leek-green or liver brown. Streak uncolored. Luster vitreous to subadamantine. Subtransparent to nearly opaque. H=3·5—4. Gr=3·1—3·3. Scorodite is a hydrous arsenate of the peroxyds of iron, containing 50 per cent. of arsenic acid. From Saxony, Carinthia, Cornwall, and Brazil.

It occurs in minute crystals near Edenville, N. Y., with arsenical pyrites. The name of this species is from the Greek *skorodon*, garlic, alluding to the odor before the blowpipe.

Iron sinter is a yellowish or brownish hydrous arsenate of the peroxyd of iron, containing but 30 per cent. of arsenic acid. *Arseno-siderite* is another fibrous arsenate, containing 34 per cent. of arsenic acid.

Symplesite is a blue or green mineral, supposed to be an arsenate of the protoxyd of iron. Its crystals are right rhomboidal, with a perfect cleavage. H=2·5. Gr=2.96. From Voigtland.

Oxalate of iron. This is a soft, yellow, earthy mineral of rare occurrence. It blackens instantly in the flame of a candle. Occurs in Bohemia; it is supposed to have resulted from the decomposition of succulent plants.

GENERAL REMARKS ON IRON AND ITS ORES.

The metal iron has been known from the most remote historical period, but was little used until the last centuries before the Christian era. Bronze, an alloy of copper and tin, was the almost universal substitute, for cutting instruments as well as weapons of war, among the ancient Egyptians and earlier Greeks; and even among the Romans (as proved by the relics from Pompeii) and also throughout Europe, it continued long to be extensively employed for these purposes.

The *Chalybes*, bordering on the Black Sea, were workers in iron and steel at an early period; and near the year 500 B. C., this metal was introduced from that region into Greece, so as to become common for weapons of war. From this source we have the expression *chalybeate* applied to certain substances or waters containing iron.

The iron mines of Spain have also been known from a remote epoch, and it is supposed that they have been worked " at least ever since the times of the later Jewish kings; first by the Tyrians, next by the Carthagenians, then by the Romans, and lastly by the natives of the country." These mines are mostly contained in the present provinces of New Castile and Aragon. Elba was another region of ancient works, " inexhaustible in its iron," as Pliny states, who enters somewhat fully into the modes of manufacture. The mines are said to have yielded iron since the time of Alexander of Macedon. The ore beds of Styria in Lower Austria, were also a source of iron to the Romans.

Iron ores. The ores from which the iron of commerce is obtained, are the spathic iron or carbonate, magnetic iron, specular iron, brown iron ore or hematite, and bog iron ore. In England, the principal ore used is an argillaceous carbonate of iron, called often clay iron stone, found in nodules and layers in the coal measures. It consists of carbonate of iron, with some clay, and externally has an earthy, stony look, with little indication of the iron it contains except in its weight. It yields from 20 to 35 per cent. of cast iron. The coal basin of South Wales, and the counties of Stafford, Salop, York, and Derby, yield by far the greater part of the English iron. Brown hematite is also extensively worked. In Sweden and Norway, at the famous works of Dannemora and Arendal, the ore is the magnetic iron ore, and is nearly free from impurities as it is quarried out. It yields 50 to 60 per cent. of iron. The same ore is worked in Russia, where it abounds in the Urals. The Elba ore is the specular iron. In Germany, Styria, and Carinthia, extensive beds of the spathic iron are worked. The bog ore is largely reduced in Prussia.

In the United States, all these different ores are worked. The localities are already mentioned. The magnetic ore is reduced in New England, New York, northern New Jersey, and sparingly in Pennsyl-

---

What was the usual substitute for iron among the ancients? What is said of the Chalybes? What of the working of the Spanish mines? What of the Elba mines? What are the common ores of iron? What is said of the most common in England? in Sweden and Norway? at Elba, Styria, and Carinthia? What ores abound in the United States?

vania and other states. The brown hematite is largely worked along Western New England and Eastern New York, in Pennsylvania, and many states south and west. The earthy argillaceous carbonate like that of England, and the hydrate, are found with the coal deposits, and are a source of much iron.

The several kinds of ore differ somewhat in the quality of the iron they afford; but the greatest part of the supposed difference, if we except the bog ore, depends on the mode of working, and the use of proper fluxes in the right proportion. The bog ore (a bog formation) often contains phosphorus from animal decomposition, and generally yields a brittle product, though from its fusibility good for some kinds of casting.

Mode of Assay. In the assay of ores in the dry way, for economical purposes, somewhat different means are used for the different ores. As in the reduction in the large way, the object is to separate the iron from the oxygen with which it is united, and from the impurities clay, lime, or quartz, if such be present.

With the pure oxyds, or the carbonate in a pure state, a simple mixture of the pulverized ore and charcoal strongly heated in a crucible, will effect a reduction. But it is found better to add carbonate of lime or burnt lime, with clay, or glass, or borax, which fuse into a slag, and besides aiding the reduction, protect the reduced iron from combustion. For ***specular iron***, with 10 parts of the ore finely pulverized, mix as much chalk or limestone, 6 to 8 parts of bottle glass, and sixteenth or a twentieth of the whole by weight of charcoal. For a ***magnetic iron ore***, mix with 10 parts of the ore 12 of glass, and as much chalk, with one part of charcoal; or, say 3 parts of each burnt lime and burnt clay, and 2½ of charcoal. For a ***brown hematite***, 10 parts of burnt lime, as many of burnt clay, and 3 of charcoal. These proportions, taken from Mushet, are not given as invariably necessary, but simply to guide the experimenter. The fitness of the proportions is to be determined from the result. If the slag is clear and nearly colorless, the reduction is perfect. If dark colored, it contains unreduced oxyd, and too much glass or clay may have been added; if opaque or porcellanous, too much lime has been used. In the case of an ***argillaceous*** ore, the proportions of lime and glass should be determined from the proportions of lime and clay in the ore.

The prepared ore with the fluxes, well mixed, is placed in a crucible lined with moistened and well compacted charcoal dust; the crucible is filled with charcoal, and closed with a luted lid of fire clay. The heat should be very slowly raised, not using the bellows for three quarters of an hour, and finally sustained for a quarter of an hour at a white heat, and then the crucible may be removed and the button of cast iron, after cooling, taken out.

Reduction of ores. In the reduction of iron ores, the simplest and oldest process consists in heating the pounded ore with charcoal in an open forge, (see beyond, page 237.) By the improved process, the ore is heated in a blast furnace along with charcoal, coke, or mineral coal,

---

What is said of the iron from different ores? Describe the general mode of assaying iron ores? What is the usual mode of reduction? Describe the blast furnace.

and also a certain proportion of some flux, usually limestone. The lime forms a glass with the silicious impurities of the ore, while the carbon (first becoming carbonic oxyd) takes the oxygen which is in combination with the metal. A small proportion of the carbon also enters into the metal after it is reduced, giving it the fusibility it has as *cast iron.*

Before describing the process, a brief description may be given of a blast furnace.* The following figure (excluding the structure on the right, to be afterwards explained,) represents the essential features of a furnace, in an exterior side view.

1

It is essentially a broad truncated four-sided pyramid of brick and stone, containing within a cavity where the ore is heated and reduced.

---

* I am indebted to Mr. S. S. Haldeman for the following figures and their descriptions. They are 1-20th of an inch to a foot. The furnace was built for anthracite, as is explained beyond. It is a model of the fine works near Columbia, Pa., owned by the Messrs. Haldeman.

The annexed figure 2, exhibits the interior laid open. The main structure is called the *stack*. Of the interior cavity, the lower part, H, *h*, is the hearth, H is four-sided; B B, the *boshes*,* having nearly the shape of a funnel, except that it is square below; above *b*, is the proper *furnace*, usually about 30 feet high; below the crucible, lies the hearth, commonly of refractory grit rock. The furnace is circular, and is lined with fire brick (*l*); next to this, is a layer of dry sand (*r*,) and then one of brick (*r'*,) constituting the inner part of the stack. The layer of sand allows the interior to expand by heat, without cracking the exterior; and moreover, the whole, *l*, *r*, *r'*, may be removed for repairs without injuring the exterior work.

2

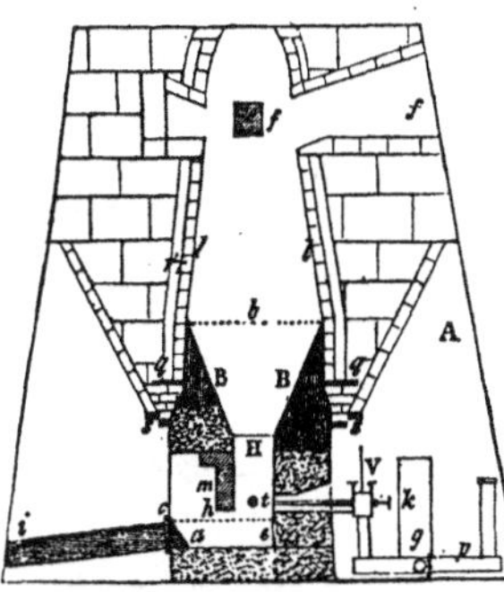

At *t*, is one of the *twiers*, (or tuyeres,) the tubes by which the blast of air is driven into the furnace. At *m*, is a partial partition of fire brick, called the *tymp*, separating the back and front of the hearth, but not extending to the bottom or hearth-stone. The *hearth-stone* is made of a refractory grit rock.

In each side of the four-sided stack, at bottom, there is a door-like or arched opening, (A, figs. 1, 2,) which extends in to the stonework that encloses the hearth. Three of these opening are called the *twier-arches*, and the other is the *front* or *working* arch; the twiers enter by the twier-arches to the interior, and at *t*, (fig. 1,) is shown the place of entrance of one. The view in figure 1, gives a front view of a twier arch; and in figure 2, at A, there is a side view, with the twier in place. To prevent the melted metal, which often rises above the twiers, from flowing into the blast pipe, in case of the blast being accidentally checked, there is at V (fig. 2) a valve, which is raised by the blast and closes when it stops; and at *k*, a place for inserting a rod to remove any slag that may cling to the twier.

3

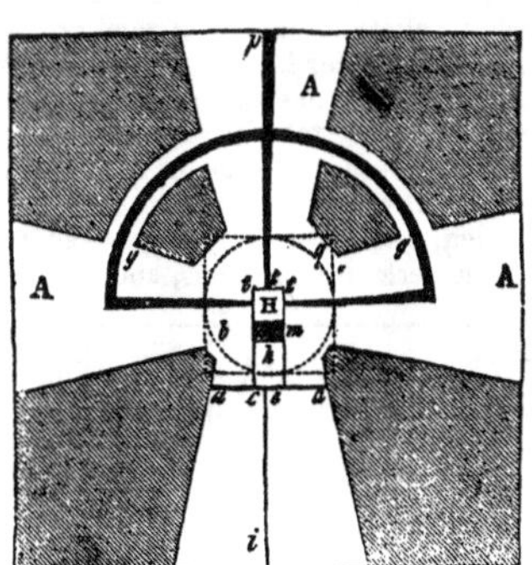

Figure 3, is a horizontal section, at bottom; A, A, A, are the twier arches, separated by the masonry of the stack; H, *h*, the position of the hearth or crucible; *m* is the *tymp* between H and *h*; *t*, *t*, *t*, are the

* This word is from the German word *böschung*, a slope. H.

twiers, the three blast tubes of which connect with a common tube that extends round, by the passage *g g*, (figs. 1, 3,) in the form of a semicircle, and receives the blast through the tube *p*. The dotted circle within corresponds to the inner outline of the fire brick lining of the widest part of the furnace.

The melted iron runs into the lower part of the hearth, and is covered by the cinder. It is prevented from running out by the *damstone c*, (figs. 2, 3) ; and farther to hinder the metal from being forced out by the blast, clay is rammed beneath the tymp around the twiers and upon the surface at *h*, where it is retained by heavy iron plates. These plates are raised every few hours to allow the cinder to run off, which passes out over the *damstone*, along the *dust-plate, c i*, (figs. 2, 3.) The metal is drawn off every twelve hours at the lower level *a*, through an aperture at the bottom of the damstone.

Great economy in making iron has of late been secured by heating the blast to three to six hundred Fahrenheit. The cooling effect of the vast volumes of air thrown into the furnace is avoided ;* and this is absolutely necessary when anthracite coal is used, as is the case in many works of recent construction. In the view above given, *f, f*, (fig. 2,) represent two (out of three) passages in the upper part of the furnace, by which the waste flame is led off, first to heat boilers at W, W, (fig. 1,) and then to a hot-oven chamber, *o*. In the last there is a great number of iron pipes, arranged in series ; the blast by the action of the engine, is thrown through all the pipes in succession, and after being thus heated, flows on to *p*, (fig. 3,) whence it passses to the twiers, (*t, t, t*.) When the engine is separated from the furnace, the oven is usually placed upon the front side (instead of back) of the top, and the flame passes in by a single aperture. The works here figured are situated upon a side hill. It is important that the blast should not be too great, as it wastes the metal by oxydation ; and at the same time it should be sufficiently copious to supply the requisite qantity of oxygen.

The first step in the process of reduction, consists in roasting the ore to drive off any volatile ingredients, and open its texture. This is effected by piling the ore in heaps, made of alternate layers of coal or coke and ore, covering up the heap loosely with earth and firing it. The carbonic acid, if it contains any, the moisture, and any sulphur present, are thus expelled, and the ore is in a looser state for reduction. The furnace is filled with coal and slowly heated up—ten or twelve days being required for this, to avoid the effect of too sudden heat on the furnace. The charge, next to be added, consists of coal, the roasted ore, and limestone, (if this be the flux,) in certain proportions, and it is car-

---

What is said of the hot blast? Describe the method of heating the engine, and air of the blast. Mention the several steps in the process of reduction.

---

* The weight of air thrown into a Glasgow furnace in 24 hours, has been estimated at 6192 cwt., or 6292 cubic feet per minute, while the whole weight of coke, ore and limestone added in the same time, was only 666½ cwt. In ordinary cases, the weight of the air is at least four times as much as that of the charges.

ried to the top of the furnace, often by a railway, and thrown in at intervals of an half hour or so, as the coal sinks, so that the furnace is kept full. The charge at the top of the furnace is two days or more in descending to where it comes within the direct action of the blast. The fusion of the ore finally takes place a short distance above the twiers, and its reduction is completed at the same time by the burning coal and flux; in a few hours the hearth fills with metal and slag, and as it accumulates, the fused iron displaces the slag which is continually running over and conveyed off by the workmen: the metal being let out below by removing a luting of clay, is run into moulds of sand, to form pigs—oblong masses of about 180 pounds each. The slag in this process serves to protect the metal from combustion as it is reduced. Its color and condition indicate the success of the reduction. If of a dark color and heavy, it shows that all the ore is not reduced, and much metal lost; probably owing to too little coal or too rapid working. If dark vitreous, with streaks of green, there is some oxyd of iron carried off by the silica, which may probably be remedied by adding more lime to take up the silica. If light colored, all is going on well.*

The proportion of fluxes depends on the ore and its condition, and no general rule can be given. With the argillaceous carbonate of iron of Staffordshire, limestone alone is used, 10 to 12 per cent. being employed for 45 per cent. of ore, and 45 of coke. Even this addition is unnecessary when the ore is associated with much lime. For the ordinary argillaceous ores, the weight of limestone used is about *one-fourth* the weight of the ore, or from one-third to one-sixth. When there is no silica in the ore, it is added in nearly equal proportions with the lime and other earthy ingredients present. Previous assays must determine what is required for each variety of ore. The brown hematite is easily reduced, and requires much coal with a slow process, or only a white iron is produced; 8 to 12 per cent. of limestone is added to a charge as a flux.

Good metal is strong of a dark gray color, with a granular texture, and runs fluid when melted; while the bad metal is light colored and brittle, and runs thick and sluggish. There are numbers 1, 2, 3, 4, in market, including the two kinds just described and two intermediate grades. Number 1 is best fitted for castings, as it contains the most carbon and is more fusible than the others. Cast iron sometimes contains a trace of silicium without injury, and according to Berzelius, the best Swedish iron contains after it is made into wrought iron 1-20 per cent. of silicium. Sulphur and phosphorus are highly deleterious, except when a fusible metal is desired with the strength comparatively unessential.

Wrought or malleable iron. As cast iron owes its fusibility principally to the carbon present, the change of *cast* to *wrought* iron, called

---

What is said of the slag? On what does the proportion of fluxes depend?

---

* The slag from Merthyr Tydvil, in South Wales, afforded Berthier on analysis, silica 40·4, lime 38·4, magnesia 5·2, alumina 11·2, protoxyd of iron 3·8, and a trace of sulphur.

refining, must consist in the removal of this carbon and any remaining impurities. This is done by *burning* it out, and for this purpose the poorer kinds of cast iron answer as well as the best. Formerly the metal was melted three or four times, and then hammered with a large forging hammer to remove the scoria. In the next improvement, the metal while in fusion was stirred for a while to effect the more complete combustion of the carbon; and in this way it gradually lost its fusibility and became stiff enough for forging. This process is called puddling. The metal passes first through one fusion as preparatory. It is next placed on plates in a furnace of the reverberatory kind, the metal being loosely piled in the middle of the horizontal furnace; 3½ cwt. is an ordinary charge. The flame plays over it, and in half an hour it begins to melt. The workmen now stir it about, occasionally dashing in a scoopful of water. The metal gives off freely bubbles of gas, which burn with a blue flame, (carbonic oxyd); in about twenty minutes the whole falls to pieces like a coarse gravel, and a lurid flame appears over it. The whole is still kept in motion and well heated, and soon it begins to unite again, when it is separated into several lumps of the size of three or four bricks. These masses as they assume a clotty consistency (sometimes called "coming into nature,") are drawn from the furnace and *dolleyed* or stamped into cakes with hammers. The plates are thrown while hot into water, which renders them brittle; they are then broken into pieces, again placed together in the furnace, heated to a welding heat, and finally forged under a ponderous hammer, moved by machinery, into short thick bars called *blooms*. 100 parts of cast iron yield about 63 of blooms. Some of the steps in this process are often neglected in making the ordinary iron.

It has been found that full 24 per cent. of the gas escaping from an iron furnace is carbonic oxyd, and in the boshes this is the only gas. This gas has been used as fuel in the refining of the iron, and by this means the whole expense of fuel for refining is saved. (See the Amer. Jour. Sci., vols. i. and ii., 2d ser., where the theory of the blast furnace is well explained.)

The iron produced is said to be *cold short* if it is brittle when cold, and this has been attributed to the presence of silicium. It is termed *red short* when it becomes brittle on heating.

Cast iron is also changed to *malleable iron* by covering castings with powdered hematite or other oxyd of iron, and exposing to heat below fusion. The carbon is removed by the oxygen of the oxyd. The scales of oxyd thrown off in the forging of iron are much used. This process was first introduced in 1804, and is one of great importance in the arts.

Malleable iron is also obtained directly from the ore by a single fusion in what is called a Catalan forge. It has a rectangular crucible or basin below the fire, about 18 inches by 21 in width and 17 inches deep. The twier enters about 9½ inches above the bottom and receives the blast from a water-blowing machine; and it admits of a change of position so as to give a change of direction to the blast as is required in the

---

Describe the manufacture of wrought from cast iron. How is the gas used in heating? What are *cold short* and *red short* iron? What other mode is there of rendering cast iron malleable? Describe a mode of obtaining malleable iron direct from the ore.

different stages of the process. The ore after a previous roasting in a kiln, is pounded up and sifted; the coarser part is piled up in the forge on the side opposite the blast, and charcoal fills up the rest of the space. After the heat is well up, the finer siftings are thrown at intervals upon the charcoal fire. The basin below, which has been previously lined with two or three coats of pounded charcoal, or loam and charcoal, receives the iron as it is reduced and runs down. The slag is occasionally removed from the surface of the basin through holes opened for the purpose. The iron, when sufficiently accumulated, is taken out in a pasty state and at once forged. The process usually lasts five or six hours. A lump or bloom of malleable iron is thus produced in three or four hours. This cheap and simple process has long been used in Catalonia, and it is hence called the method of the Catalan forge. By a slow operation, and but a small quantity of siftings, worked with an upraised twier, the proportion of steel obtained by the process is increased. This mode of reduction is adapted only for the purer and more fusible ores; and moreover it requires a large consumption of fuel and is attended by a considerable loss. The argillaceous ore of the coal region would yield only an iron glass in a Catalan forge.

By another mode of reduction, the iron ore coarsely powdered is mixed with coal in certain proportions, or a material containing the requisite amount of carbon, and the charge is heated in a reverberatory furnace till reduction has taken place. The carbon carries off the oxygen of the ore, and if the proper proportions have been employed, it leaves a mass of malleable iron behind.

Steel. Wrought iron is changed to steel by a process called *cementation.* The best iron is heated with charcoal; a portion of carbon is thus absorbed, and the iron at the same time acquires a *blistered* surface, and becomes fine grained and fusible. When the *blistered steel* is drawn down into smaller bars and beaten, it forms *tilted steel;* and this broken up, heated, welded, and again drawn out into bars, forms *shear steel. Cast steel* is prepared by fusing blistered steel with a flux and casting it into ingots, and then by gentle heating and careful hammering or rolling, giving it the form of bars.

Steel is also formed direct from certain ores of iron, more particularly when oxyd of manganese is associated with them, and especially from the spathic iron, which often contains a portion of carbonate of manganese. The oxygen of the manganese is said to remove part of the carbon from the cast iron, and thus reduce it to the state of steel. There are 1 or 2 per cent. of manganese in the metal thus obtained. The product is of inferior quality as steel, but is largely manufactured in Germany. The *wootz* of India is a steel obtained from a black ore of iron, in a furnace even simpler than the Catalan forge. It is said to contain a minute proportion of silicium and aluminium.

The amount of iron manufactured in the United States in 1853, (over a third in Pennsylvania,) was 1,000,000 tuns; in Great Britain, in 1852, 2,700,000 tuns; in France, in 1849, 514,000; in Russia, in 1845, 400,000; in Sweden, in 1846, 145,000; other parts of Europe, (Austria, Belgium, Germany,) 700,000 tons.

How is steel made? Describe the kinds of steel. How is steel made direct from ores of iron?

15. MANGANESE.

The ores of manganese have a specific gravity below 5·2. They afford a violet-blue color with borax or salt of phosphorus, in the outer flame of the blowpipe; and on heating the oxyd with muriatic acid, fumes of chlorine are given out which are derived from the acid.

RHODONITE.—MANGANESE SPAR.

Monoclinic? In oblique rhombic prisms, isomorphous with pyroxene; usually large massive, the cleavage often indistinct. Possibly triclinic, and the same as *Fowlerite.*

Color reddish, usually deep flesh-red; also brownish, greenish, or yellowish, when impure; streak uncolored. Luster vitreous. Transparent to opaque. Becomes black on exposure. H=5·5—6·5. Gr=3·4—3·7.

Composition: oxyd of manganese 52·6, silica 39·6, oxyd of iron 4·6, lime and magnesia 1·5, water 2·7. The impure varieties, *Bustamite*, *Photizite*, and *Allagite*, contain variable proportions of carbonate of iron, lime, or manganese, beside alumina. Becomes dark brown when heated, and fuses with borax in the outer flame, giving a hyacinth red globule.

Dif. Resembles somewhat a flesh-red feldspar, but differs in greater specific gravity, in blackening on long exposure, and in the glass with borax.

Obs. Occurs in Sweden, the Hartz, Siberia, and elsewhere. In the United States it is found in masses, at Plainfield, and Cummington, Mass.; also abundantly at Hinsdale, and on Stony Mountain, near Winchester, N. H.; at Blue Hill Bay, Me. The black exterior is a more or less pure hydrated oxyd of manganese.

Uses. Dr. Jackson has suggested the use of this ore for making a violet-colored glass, and also for a colored glazing on stone ware. The finely pulverized mineral, spread on stone ware as a paste, will afford a permanent glazing, which will have a black color if it be of considerable thickness, and of a deep violet-blue if quite thin. It may be used along with the usual salt glazing.

What is said of the ores of manganese? What is the appearance of manganese spar? its composition and blowpipe characters? How is it distinguished from feldspar? For what may it be used?

It receives a high polish, and is sometimes employed for inlaid work.

Tephroitc. A silicate of manganese, occurring massive and cleavable. Color ash-gray. H=5·5. Gr=4. From Franklin, N. J. *Composition:* silica 29·8, protoxyd of manganese 70·2. Fuses easily to a black scoria. *Knebelite* is a related species.

Fowlerite (Paisbergite). Probably same as Rhodonite. Form *triclinic.* In crystals at Franklin, N. J., and Paisberg, Sweden.

PYROLUSITE—*Binoxyd of Manganese.*

Trimetric. In small rectangular prisms, more or less modified. M : M=93° 40′; M : ē=136° 50′. Sometimes fibrous and radiated or divergent. Often massive and in reniform coatings.

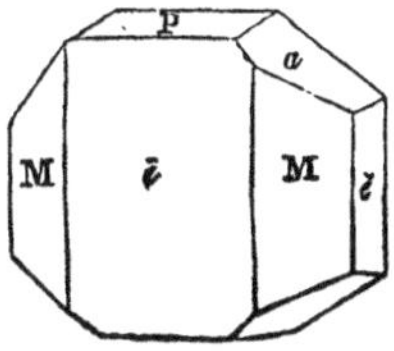

Color iron-black; streak black, unmetallic. H=2—25. Gr=4·8—5·0. *Composition:* essentially the binoxyd of manganese, consisting of oxygen 37, and manganese 63. With borax it gives an amethystine globule. It yields no water in a matrass.

Dif. Differs from psilomelane by its inferior hardness, and from ores of iron by the violet glass with borax.

Obs. This ore is extensively worked in Thuringia, Moravia, and Prussia. It is common in Devonshire, Somersetshire, and Aberdeenshire, in England. In the United States it is associated with the following species in Vermont, at Bennington, Brandon, Monkton, Chittenden, and Irasburg; it occurs also in Maine, at Conway, and Plainfield, in Massachusetts; at Salisbury, and Kent, in Conn., on hematite.

The name *pyrolusite* is from the Greek *pur*, fire, and *luo*, to wash, and alludes to its property of discharging the brown and green tints of glass, for which it is extensively used.

Uses. Besides the use just alluded to, this ore is extensively employed for bleaching, and for affording the gas oxygen to the chemist.

PSILOMELANE.

Massive and botryoidal. Color black or greenish-black. Streak reddish or brownish-black, shining. H=5—6. Gr=4—4·4.

Describe pyrolusite. What is its constitution? What are its uses? Describe psilomelane? How does it differ from pyrolusite.

Composition: essentially binoxyd of manganese with one per cent. of water, and also some baryta or potassa. The compound is somewhat varying in its constitution. Before the blowpipe like pyrolusite, except that it affords water.

Obs. This is an abundant ore, and is associated usually with the pyrolusite. Prof. Silliman, jr., has lately detected oxyd of cobalt mixed with this ore. It occurs at the different localities mentioned under pyrolusite, and the two are often in alternating layers; it has been considered only an impure variety of the pyrolusite. The name is from the Greek *psilos*, smooth or naked, and *melas*, black.

Uses. Same as with pyrolusite.

Heteroclin and *marceline* are similar ores, containing 10 to 16 per cent. of silica.

WAD.—*Bog manganese.*

Massive, reniform or earthy; also in coatings and dendritic delineations.

Color and streak black or brownish-black. Luster dull, earthy. H=1. Gr=3·7. Soils.

Composition. Consists of peroxyd of manganese, in varying proportions, from 30 to 70 per cent. along with peroxyd of iron, 20 to 25 per cent. of water, and often several per cent. of oxyd of cobalt or copper. It is a hydrated peroxyd, mechanically mixed with other oxyds, organic acids and other impurities, and like bog iron ore, is formed in low places from the decomposition of minerals containing manganese. Gives off much water when heated, and affords a violet glass with borax.

Obs. Wad is abundant in Columbia and Dutchess counties, N. Y., at Austerlitz, Canaan Center, and elsewhere; also at Blue Hill Bay, Dover, and other places in Maine; at Nelson, Gilmanton, and Grafton, N. H.; and in many other parts of the country.

Uses. May be employed like the preceding in bleaching, but is too impure to afford good oxygen. It may also be used for umber paint.

TRIPLITE.—*Ferruginous Phosphate of Manganese.*

Massive, with cleavage in three directions. Color blackish-brown. Streak yellowish-gray. Luster resinous; nearly or quite opaque. H=5—5·5. Gr=3·4—3·8.

What is wad? its composition? its origin? For what may it be used? What is triplite?

Composition: protoxyd of manganese 33·2, protoxyd of iron 33·6, phosphoric acid 33·2, with some phosphate of lime. Fuses easily to a black scoria, before the blowpipe; dissolves in nitric acid, and gives a violet glass with borax.

Obs. From Limoges in France. Rather abundant at Washington, Conn., and sparingly found at Sterling, Mass.

Heterosite is another *phosphate of the oxyds of manganese and iron,* of a greenish-gray or bluish color. Contains 41·77 per cent. of phosphoric acid. *Huraulite* is a *hydrous phosphate* of the same oxyds, containing 18 per cent. of water and 38 of phosphoric acid. Occurs in transparent, oblique, reddish-yellow crystals. Both heterosite and hureaulite are regarded as either altered *triphyline* or *triplite.*

Hausmannite. A sesquioxyd of manganese containing 72·1 per cent. of manganese, when pure. Brownish-black and submetallic, occurring massive and in square octahedrons; H=5—5·5. Gr=4·7. From Thuringia and Alsatia.

Braunite. A protoxyd of manganese, containing 69 per cent. of manganese when pure. Color and streak dark brownish-black, and luster submetallic. Occurs in square octahedrons; H=6—6·5. Gr=4·8. From Piedmont and Thuringia.

Manganite. A hydrous sesquioxyd of manganese. Occurs massive and in rhombic prisms. Color steel-black to iron-black. H=4—4·5. Gr=4·3—4·4. From the Hartz, Bohemia, Saxony, and Aberdeenshire.

Peloconite is an ore of manganese and iron, of a bluish-black color, and liver brown streak, with a weak vitreous luster. From Chili.

Manganblende, or *Alabandine.* A sulphuret of manganese, of an iron-black color, green streak, submetallic luster. H=3·5—4. Gr=3·9—4·0. Crystals, cubes and regular octahedrons. From the gold mines of Nagyag, in Transylvania.

Hauerite is a sulphuret, containing twice the proportion of sulphur in the last. Color reddish-brown and brownish-black, resembling zinc blende. H=4. Gr=3·46. From Hungary.

There is also an *arseniuret* of manganese, of a grayish-white color, and metallic luster, which gives off alliaceous fumes. G=5·55. From Saxony.

Diallogite. A carbonate of manganese. Color rose-red to brownish; streak uncolored. Luster vitreous, inclining to pearly. Translucent to subtranslucent. Crystals rhombohedral. H=3·5. Gr=3·59. Infusible alone. From Saxony, Transylvania, and the Hartz. Also from Washington, Conn., with triplite.

GENERAL REMARKS ON THE ORES OF MANGANESE.

Manganese is never used in the arts in the pure state; but as an oxyd it is largely employed in bleaching. The importance of the ore for this purpose, depends on the oxygen it contains, and the facility with which

---

On what does the value of manganese ores depend in the art of bleaching?

this gas is given up. As the ores are often impure, it is important to ascertain their value in this respect. This is most readily done by heating gently the pulverised ore with muriatic acid, and ascertaining the amount of chlorine given off. The chlorine may be made to pass into milk of lime, to form a chlorid, and the value of the chlorid then tested according to the usual modes. The amount of chlorine derived from a given quantity of muriatic acid depends not only on the amount of oxygen in the ore, but also on the presence or absence of baryta and such other earths as may combine with this acid. The binoxyd of manganese when pure, affords 18 parts by weight of chlorine, to 22 parts of the oxyd; or 23½ cubic inches of gas from 22 grains of the oxyd. The best ore should give about three-fourths its weight of chlorine, or about 7000 cubic inches to the pound avoirdupois.

The chlorine for bleaching is used commonly in combination with lime. To make the chlorid of lime, the chlorine is generally obtained either through the action of muriatic acid on the ore, (3 to 4 parts by weight of the former, to 1½ of the latter,) or more commonly by mixing 1 part of the ore with 1¼ parts of common salt, 2 or 2¼ parts of concentrated sulphuric acid, and as much water. As the chlorine passes off, it is conveyed into chambers containing slaked lime, by which it is absorbed.

Manganese is also employed to give a violet color to glass. The sulphate and the chlorid of manganese are used in calico printing. The sulphate gives a chocolate or bronze color.

The best beds of manganese ores in the United States, which have been opened, are at Brandon, Chittenden, and Irasburg, Vt.

16. CHROMIUM.

The ores of chromium are the chromates of lead and chromic iron, which are described under Lead and Iron. There is also a native *chromic ochre*, supposed to consist of silica chromic acid, alumina, and iron. *Wolchonskoite* is an allied mineral. *Miloschine* or *Serbian* is considered a chromiferous clay.

17. NICKEL.

The ores of nickel, excepting one or two, have a metallic luster, and pale color; their specific gravity is between 3 and 8, and hardness mostly between 5 and 6, (in one, about 3.) They resemble some cobalt ores, but do not like them give a deep blue color with borax.

How is manganese used? For what other purpose is manganese used? What is said of the ores of chromium? What is said of the ores of nickel?

COPPER NICKEL.—*Arsenical Nickel.*

Hexagonal. Usually massive. Color pale copper-red; streak pale brownish-red. Luster metallic. Brittle. H=5—5·5. Gr=7·3—7·7.

Composition: nickel 44, and arsenic 56; sometimes part of the arsenic is replaced by antimony. Gives off arsenical (alliaceous) fumes before the blowpipe, and fuses to a pale globule, which darkens on exposure. Assumes a green coating in nitric acid, and is dissolved in aqua-regia.

Dif. Distinguished from iron and cobalt pyrites by its pale reddish shade of color; also from the former by its arsenical fumes, and from the latter by not giving a blue color with borax. None of the ores of silver with a metallic luster have a pale color, excepting native silver itself.

Obs. Accompanies cobalt, silver, and copper ores in the mines of Saxony, and other parts of Europe; also sparingly in Cornwall.

It is found at Chatham, Conn., in gneiss, associated with white nickel or chloanthite.

CLOANTHITE.—*White Nickel.*

Monometric. In cubes. Color tin-white. Streak grayish-black. H=5·5—6. Gr=6·4—6·7.

Composition: nickel 28·40, arsenic 70·34, (from Kamsdorf.) Often contains cobalt, and graduates into smaltine.

It also sometimes contains iron, and this variety is the *safflorite* of Haidinger, or *chathamite* of Shepard. The ore from Chatham, Conn., affords 10 to 12 per cent. of nickel, 1 to 3 of cobalt, and 12 to 18 of iron.

Found usually with smaltine at its various localities.

Nickel glance is another arsenical ore, occurring in cubes and massive. Color silver-white to steel-gray. Contains 28 to 30 per cent. of nickel with arsenic and sulphur. H=5·5. Gr=6·1. From Helsingland, in Sweden, and also in the Hartz. Also at Schladming, in Austria, containing 38 per cent. of nickel, and having the specific gravity 6·6—6·9. This ore has been called *Gersdorffite.*

Nickel Stibine. An antimonial sulphuret, called sometimes *Nickeliferous antimony ore,* containing 25 to 28 per cent. of nickel. Color steel-gray, inclining to silver-white. In cubical crystals and also massive. H=5—5·5. Gr=6·45. From the Duchy of Nassau.

What is the crystallization and appearance of copper nickel? of what does it consist? How is it distinguished from iron and cobalt pyrites? how from silver ores? Where does it occur?

Antimonial nickel. Contains 29 per cent. of nickel and no sulphur. It has a pale copper-red color, inclining to violet. H=5·5—6. Gr=7·5. Crystals hexagonal. From the Andreasberg mountains.

Nickel pyrites or *capillary pyrites.* A brass-yellow sulphuret of nickel, occurring usually in delicate capillary forms; also in rhombohedral crystals. Gr=5·28. Contains 64·3 per cent. of nickel. From Bohemia, Saxony and Cornwall. Also occurs in needles at Antwerp, N. Y., and in Lancaster Co., Penn. The mineral has been named *Millerite.*

A *sulphuret of iron and nickel,* of a light bronze-yellow, has been reported from southern Norway. It contains 22 per cent. of nickel. Gr=4·6.

Grünauite. Still another sulphuret, (called *bismuth nickel,*) contains 10 to 14 per cent. of bismuth, with 22 to 40·7 of nickel. Color light steel-gray to silver-white; often tarnished yellowish. H=4·5. Gr=5·13. From the district of Altenkirchen, Prussia.

Nickel green. An arsenate of nickel, containing 37·6 per cent. of oxyd of nickel. Color fine apple-green. Occurs with other nickel ores in Dauphiny, Prussia, and elsewhere. It is found with copper nickel at Chatham, Conn.

EMERALD NICKEL.

Incrusting, minute globular or stalactitic. Color bright emerald green. Luster vitreous. Transparent or nearly so. H=3—3·25. Gr=2·5—2·7.

It is a carbonate of nickel, containing 28·6 per cent. of water. Infusible before the blowpipe alone, but loses its color.

Obs. Occurs with chromic iron and carbonate of magnesia, on serpentine, in Lancaster county, Pennsylvania.

An earthy oxyd of nickel and sulphuret occurs with black cobalt, at Mine la Motte, Missouri.

Pimelite is a clay colored by green oxyd of nickel. Klaproth found 15·6 per cent. in one specimen. Quartz is sometimes colored by nickel. *Chyroprase* is a chalcedony thus colored.

GENERAL REMARKS ON NICKEL AND ITS ORES.

The nickel of commerce is obtained mostly from the copper nickel and chloanthite, or from an artificial product called *speiss,* (an impure arseniuret,) derived from roasting ores of cobalt with which arseniuretted nickel ores are mixed. The ores are nowhere very abundant, and the most productive are those of Saxony and Germany.

Nickel also occurs in meteoric iron, forming an alloy with the iron, which is characteristic of most meteorites. The proportion sometimes amounts to 20 per cent. The great Texas meteorite, now in the Yale College collections, contains 8·8 to 9·7 per cent. of this metal.

Nickel is obtained in the pure state from the speiss, by the following

---

Describe the green hydrate of nickel. What is pimelite? What ores afford the nickel of commerce. Where else is it found?

process, proposed by Wöhler: 1 part of the ore is fused with 3 of pearlash and 3 of sulphur. The arsenic forms a soluble compound with the sulphur and potash, and the nickel an insoluble sulphuret. This is well washed with water and dissolved in nitric acid; and the solution, after any lead, copper, or bismuth, that may be present, have been precipitated by a current of sulphuretted hydrogen, is precipitated by caustic or carbonated potash or soda. The washed precipitate is now acted on by an excess of oxalic acid, which forms with the peroxyd of iron, that is generally present, a soluble, and with the oxyd of nickel an insoluble, oxalate, which of course includes any cobalt that the ore may have contained. The oxalate is now dissolved in an excess of ammonia, and the solution exposed to the air. As the ammonia escapes, the nickel is deposited as an insoluble double oxalate, while the cobalt remains dissolved as a soluble double oxalate of the metallic oxyd with ammonia. The nickel salt, being ignited, leaves an oxyd which may be reduced by heating with charcoal; or it may be dissolved in acid and again converted into oxalate, which this time is free from cobalt and appears as an apple-green powder. The oxalate of nickel, being well washed, dried and ignited in a closed crucible, with an aperture for the escape of gas, leaves metallic nickel, which, if the heat be very intense, is fused to a button. Its color is between that of silver and tin.

As nickel does not rust or oxydize, (except when heated,) it is superior to steel, for the manufacture of many philosophical instruments.

An alloy of copper, nickel, and zinc, has been much used for various purposes, under the name of *German silver*, or *argentane*. Good German silver consists of copper 8 parts, nickel 3, zinc 3½. An inferior article is made of copper 8, nickel 2, zinc 3½. Below the proportion of nickel last stated, the alloy approaches pale brass and tarnishes readily, while the better kind has the appearance of silver, and retains well its polish. It is, however, easily distinguished from silver by a somewhat greasy feel.

But "German silver" is not a very recent discovery. In the reign of William III, an act was passed making it felony to ***blanch copper*** in imitation of silver, or mix it with silver for sale. "***White copper***" has long been used in Saxony for various small articles; the alloy employed is stated to consist of copper 88·00, nickel 8·75, sulphur with a little antimony 0·75, silex, clay and iron, 1·75. A similar alloy is well known in China, and is smuggled into various parts of the East Indies, where it is called ***packfong***. It has been sometimes identified with the Chinese ***tutenague***. M. Meurer analyzed the white copper of China, and found it to consist of copper 65·24, zinc 19·52, nickel 13, silver 2·5, with a trace of cobalt and iron. Dr. Fyfe obtained copper 40·4, nickel 31·6, zinc 25.4, and iron 2·6. It has the color of silver, and is remarkably sonorous. It is worth in China about one-fourth its weight of silver, and is not allowed to be carried out of the empire.

Nickel alloyed with iron, as in meteoric iron, renders it less liable to rust; but with steel the tendency to rust is increased.

Articles are now plated with nickel, by galvanic precipitation from the sulphate.

How is nickel obtained from the ore? For what is nickel used? What is German silver? What is the Chinese packfong?

18. COBALT.

Cobalt has not been found native. The ores of cobalt having a metallic luster, vary in specific gravity from 6·2 to 7·2; and the color is nearly tin-white or pale steel-gray, inclining to copper-red. The ores without a metallic luster have a clear red or reddish color, and specific gravity of nearly 3. The ores are remarkable for giving a deep blue color to glass of borax, even when the proportion of cobalt is small.

SMALTINE.—*Tin-white Cobalt.*

Monometric. Occurs in octahedrons, cubes, and dodecahedrons, more or less modified. (See figs. 1, 2, 3, page 25, and 32, 37, page 36.) Cleavage octahedral, somewhat distinct. Also reticulated; often massive.

Color tin-white, sometimes inclining to steel-gray. Streak grayish-black. Fracture granular and uneven. H=5·3—Gr=6·4—7·2.

Composition: essentially cobalt and arsenic; the cobalt varies from 18 to 23·5 per cent. and the arsenic from 69 to 79 per cent. A variety contains 9 to 14 per cent. of cobalt and is called *radiated white cobalt;* another variety contains iron. See further, *Chloanthite.*

Gives off arsenical fumes in a candle. Colors borax and other fluxes blue, and affords a pink solution with nitric acid.

Dif. The arsenical cobalts are at once distinguished from mispickel or white iron pyrites, by the blue color they give with borax; and also by their crystals and specific gravity.

Obs. Usually in veins with ores of cobalt, silver, and copper. Occurs in Saxony, especially at Schneeberg; also in Bohemia, Hessia, and Cornwall.

In the United States it is found in gneiss with copper nickel, at Chatham, Conn.

Cobaltine. This is another arsenical ore of cobalt, containing sulphur as well as arsenic. Color silver-white, inclining to red. Contains 33 to 37 per cent. of cobalt. Forms of crystals, figures 42, 46, page 37. From Sweden, Norway, Siberia, and Cornwall. The most

---

What is said of the ores of cobalt? Describe tin-white cobalt? What is its composition? its blowpipe characters? How is it distinguished from mispickel and white iron pyrites?

productive mines are those of Wehna, in Sweden, which were first opened in 1809.

Cobalt pyrites is a *sulphuret of cobalt,* of a pale reddish or steel-gray color. H=5·5. Gr=6·3—6·4. Crystals cubic. From Sweden, and also Prussia; also Mine La Motte, Missouri. Named *Linnæite.*

Another sulphuret of cobalt, with a less proportion of sulphur than in the last, has been observed in Hindostan. Color steel-gray, a little yellowish. Named *Syepoorite.*

EARTHY COBALT.—*Black oxyd of Cobalt.*

Earthy, massive. Color black or blue-black. Soluble in muriatic acid, with an evolution of fumes of chlorine.

Obs. Occurs in an earthy state mixed with oxyd of manganese, and in Missouri has been mistaken for black oxyd of copper. It is quite abundant at Mine La Motte, Missouri, and also near Silver Bluff, South Carolina. The analyses vary in the proportion of oxyd of cobalt associated with the manganese, as the compound is a mere mixture. Sulphuret of cobalt occurs with the oxyd. The Carolina ores afforded Dr. J. L. Smith, oxyd of cobalt 24, oxyd of manganese 76. The ore from Missouri, as analyzed by Prof. Silliman, Jr., afforded 40 per cent. of oxyd of cobalt, with oxyds of nickel, manganese, iron and copper. It has also been detected with hematite, in Chester Ridge, Pa.

This ore has been found abroad in France, Germany, Austria, and England, but much of it contains very little oxyd of cobalt.

Uses. The ore of Missouri is exported to England in large quantities, and there purified and made into smalt, for the arts.

ERYTHRINE.—COBALT BLOOM.—*Arsenate of Cobalt.*

Monoclinic. In oblique crystals having a highly perfect cleavage and foliated structure like mica. Laminæ flexible in one direction. Also as an incrustation, and in reniform shapes, sometimes stellate.

Color peach and crimson red, rarely grayish or greenish; streak a little paler, the powder dry lavender blue. Luster of laminæ pearly; earthy varieties without luster. Transparent to subtranslucent. H=1·5—2. Gr=2·95.

Composition: oxyd of cobalt 37·6, arsenic acid 38·4, wa-

What is said of the black oxyd of cobalt? What is the appearance and structure of cobalt bloom? of what does it consist?

ter 24·0. Gives arsenical fumes when heated, and fuses; yields a blue glass with borax.

The earthy ore is sometimes called *peach blossom ore*, from its color; and also *red cobalt ochre.*

Dif. Resembles red antimony, but that species wholly volatilizes before the blowpipe. From red copper ore it differs in giving a blue glass with borax; moreover the color of the copper ore is more sombre.

Obs. Occurs with ores of lead and silver, and other cobalt ores. Schneeberg, in Saxony, Saalfield in Thuringia, and Riechelsdorf, in Hessia, are noted European localities. It is found also in Dauphiny, Cornwall, and Cumberland. Occurs in the U. States, at Mine La Motte, Missouri.

Uses. Valuable as an ore of cobalt, when abundant.

Roselite. A rose-red mineral, related to, if not identical with, cobalt bloom.

Arsenite of cobalt is a compound of arsenous acid and oxyd of cobalt, and results from the decomposition of other cobalt ores.

Sulphate of cobalt, or *Cobalt vitriol.* It has a flesh or rose-red tint, and astringent taste. Consists of sulphuric acid, oxyd of cobalt and water.

GENERAL REMARKS ON COBALT AND ITS ORES.

The two arsenical ores of cobalt afford the greater part of the cobalt of commerce. The earthy oxyd is so abundant in the United States, that it promises to be a profitable source of this metal. Cobalt is never employed in the arts in a metallic state, as its alloys are brittle and unimportant. It is chiefly used for painting porcelain and pottery, and is required for this purpose in the state of an oxyd, or the silicated oxyd called smalt and azure.

Cobalt comes from Germany mostly in the silicated condition. The *zaffre* is prepared by calcining the ores of cobalt in a reverberatory furnace; the sulphur and arsenic are thus volatilized, and an impure oxyd remains, which is next mixed and heated with about twice its weight of finely powdered flints.

By another process the ore is pulverized and roasted, to expel the greater part of the arsenic; a sulphate is then formed by heating for an hour with concentrated sulphuric acid. The sulphate is dissolved in water, and a solution of carbonate of potash added to separate the iron; and when the blue color of the cobalt begins to be thrown down, the supernatant liquid is decanted and filtered, and the cobalt is precipitated by means of a solution of silicated potash, (prepared by heating together 10 parts of potash, 15 of finely pulverized quartz, and 1 of charcoal, and afterwards treating the melted mass with boiling water.) The silicate of cobalt thus prepared is said to be superior to that procured

---

How does cobalt bloom differ from red antimony? From what ores is the cobalt of commerce obtained? For what is cobalt used? In what condition is it imported from Germany? What is zaffre?

in any other way, for staining porcelain, or for the manufacture of blue glass.

Smalt and azure, which have a rich blue color, are made by fusing zaffre with glass; or by calcining a mixture of equal parts of roasted cobalt ore, common potash, and ground glass. The zaffre is used for coloring glass, and for painting enamel and pottery ware. The arsenic volatilized in the above process is condensed in chambers; it constitutes the greater part of the arsenic of commerce. The separation of the nickel from ores rich in this metal, is sometimes effected by exposing the moistened ore to the atmosphere. The nickel is unaltered, while the other metals are oxydized.

The annual yield of zaffre or smalt, in Saxony, amounts to 8000 cwt.; in Bohemia, mainly from Schlackenwald, 4000 cwt.; in the Reisengebirge, in Prussia, 600 cwt.; at Kongsberg, in Norway, 4000 cwt.

19. ZINC.

Zinc occurs in combination with sulphur, oxygen, silica, carbonic acid, and sulphuric acid. It is also found in combination with alumina, constituting one variety of the species spinel.

The ores of zinc are infusible, or very nearly so; but they yield on charcoal, with more or less difficulty, white fumes of the oxyd of zinc. Specific gravity below 4·5.

BLENDE.—*Sulphuret of Zinc.*

Monometric. In dodecahedrons, octahedrons, and other allied forms, with a perfect dodecahedral cleavage. Also massive; sometimes fibrous.

1

2

Color wax-yellow, brownish-yellow, to black, sometimes green or red; streak white, to reddish-brown. Luster resinous or waxy, and brilliant on a cleavage face; sometimes submetallic.—Transparent to subtranslucent. Brittle. H=3·5—4. Gr=4·0—4·1. Some specimens become electric with friction, and give off a yellow light when rubbed with a feather.

Composition: zinc 66·72, sulphur 33·28. Contains frequently a portion of sulphuret of iron when dark colored;

What are smalt and azure? How are they used in porcelain painting? What is said of the ores of zinc? What is the crystallization of blende. What are its luster, color, and other physical characters? Of what does it consist?

often also 1 or 2 per cent. of sulphuret of cadmium, especially the red variety. Infusible alone and with borax. Dissolves in nitric acid, emitting sulphuretted hydrogen. Strongly heated on charcoal yields fumes of zinc.

Dif. This ore is characterized by its waxy luster, perfect cleavage, and infusibility. Some dark varieties look a little like tin ore, but their cleavage and inferior hardness distinguish them; and some clear red crystals which resemble garnet are distinguished by the same characters and also by their infusibility.

Obs. Occurs in rocks of all ages, and is associated generally with ores of lead; often also with copper, iron, tin, and silver ores. The lead mines of Missouri and Wisconsin, afford this ore abundantly. Other localities are in Maine, at Lubec, Bingham, Dexter, Parsonsfield; in New Hampshire, at Eaton, Warren, Haverhill, Shelburne; in Vermont, at Thetford; in Massachusetts, at Sterling, Southampton, and Hatfield; in Connecticut, at Brookfield, Berlin, Roxbury, and Monroe; in New York, at the Ancram lead mine, the Wurtzboro lead vein, at Lockport, Root, 2 miles S. E. of Spraker's basin, in Fowler, at Clinton; in Pennsylvania, at the Perkiomen lead mine; in Virginia, at Austin's lead mine, Wythe county; in Tennesse, near Powell's River, and at Haysboro.

This ore is the *Black Jack* of miners.

Uses. Blende is a useful ore of zinc, though more difficult of reduction than calamine. By its decomposition, (like that of pyrites,) it affords sulphate of zinc or white vitriol.

ZINCITE.—RED ZINC ORE.—*Red oxyd of zinc.*

Trimetric. Usually in foliated masses, or in disseminated grains; cleavage eminent, nearly like that of mica, but the laminæ brittle, and not so easily separable.

Color deep or bright red; streak orange-yellow. Luster brilliant, subadamantine. Translucent or subtranslucent. H=4—4·5. Gr=5·4—5·56. Thin scales by transmitted light deep yellow.

Composition: zinc 80·3, oxygen 19·7=100. Infusible

What is the action of zinc blende before the blowpipe? How is it distinguished? How does it occur? What is the appearance of red zinc ore? its composition?

alone, but yields a yellow transparent glass with borax. Dissolves in nitric acid, without effervescence.

Dif. Resembles red stilbite, but distinguished by its infusibility and also by its mineral associations.

Obs. Occurs with Franklinite at Franklin and Sterling, N. J.

Uses. A good ore of zinc when abundant, and easily reduced. It may be readily and economically converted into sulphate of zinc, or white vitriol.

Voltzite. A compound of sulphuret and oxyd of zinc. Occurs in implanted globules of a dirty rose-red color, with a pearly luster on a cleavage surface. From France.

GOSLARITE.—SULPHATE OF ZINC.—*White Vitriol.*

Trimetric. Cleavage perfect in one direction. Crystals rhombic prisms, of 90° 42′.

Color white. Luster vitreous. Easily soluble ; taste astringent metallic, and nauseous. Brittle. H=2—2·5.—Gr=1·9—2·1.

Composition : oxyd of zinc 28·09, sulphuric acid 27·97, water 43·94. Gives off fumes of zinc when heated on charcoal, which cover the coal.

Obs. Results from the decomposition of blende. Occurs in the Hartz, in Hungary, in Sweden, and at Holywell in Wales.

Uses. Sulphate of zinc is extensively employed in medicine and dyeing. For these purposes it is prepared to a large extent from blende, by decomposition like pyrites, though this affords, owing to its impurities, an impure sulphate. It is also obtained by direct combination of zinc with sulphuric acid ; zinc is exposed to the action of dilute sulphuric acid, and the solution obtained is then evaporated for crystallization. The red oxyd of zinc, of New Jersey, may become an abundant source of this salt.

White vitriol, as the term is used in the arts, is one form of sulphate of zinc, made by melting the crystallized sulphate, and agitating till it cools and presents an appearance like loaf sugar.

How does it differ from red stilbite ? For what may it be used ? What is the appearance and taste of white vitriol ? Of what does it consist ? How is it formed ? For what is it used ?

SMITHSONITE.—*Carbonate of Zinc.*

Rhombohedral. R : R=107° 40′. Cleavage rhombohedral, perfect. Massive or incrusting; reniform and stalactitic.

Color impure white, sometimes green or brown; streak uncolored. Luster vitreous or pearly. Subtransparent to translucent. Brittle. H=5. Gr=4·3—4·45.

Composition: oxyd of zinc 64·54, (*four-fifths* of which is pure zinc,) and carbonic acid 35·46. Often contains some cadmium. Infusible alone before the blowpipe, but carbonic acid and oxyd of zinc are finally vaporized. Effervesces in nitric acid. Negatively electric by friction.

Dif. The effervescence with acids distinguishes this mineral from the following species; and the hardness, difficult fusibility, and the zinc fumes before the blowpipe, from the carbonate of lead or other carbonates.

Obs. Occurs commonly with galena or blende, and usually in calcareous rocks. Found in Siberia, Hungary, Silesia; at Bleiberg in Carinthia; near Aix-la-Chapelle in the Lower Rhine, and largely in Derbyshire and elsewhere in England. In the United States, it is abundant at Vallée's Diggings in Missouri, and at other lead "diggings" in Iowa and Wisconsin; also in Claiborne county, Tenn. Sparingly also at Hamburg, near the Franklin furnace, N. J.; at the Perkiomen lead mine, Pa., and at a lead mine in Lancaster county; at Brookfield, Conn.

Zinc bloom is an earthy carbonate of zinc, containing 69 per cent. of oxyd of zinc, and 15 of water. From Bleiberg, Carinthia.

CALAMINE.—*Silicate of zinc.*

Trimetric. In modified rhombic prisms, the opposite extremities with unlike planes. M : M=103° 54′. Cleavage perfect parallel to M. Also massive and incrusting, mammillated or stalactitic.

Color whitish or white, sometimes bluish, greenish, or brownish. Streak uncolored. Transparent to translucent. Luster vitreous or subpearly. Brittle. H=4·5—5. Gr=3·35—3·49. Pyro-electric.

What is the usual appearance of calamine? What is its constitution and the effects before the blowpipe? What effect is produced by friction? What are distinguishing characteristics? How does it occur? What is electric calamine?

Composition: silica 25·1. oxyd of zinc 67·4, water 7·5. Before the blowpipe it slowly intumesces and emits a green phosphorescent light; but alone it is infusible. Forms a clear glass with borax. In heated sulphuric acid it dissolves, and the solution gelatinizes on cooling.

Dif. Differs from carbonate of lime or aragonite by its action with acids; from a salt of lead or any zeolite, by its infusibility; from chalcedony, by its inferior hardness and its gelatinizing with heated sulphuric acid.

Obs. Occurs with calamine. In the United States, it is found at Vallée's Diggings, at the Perkiomen lead mines on the Susquehanna, opposite Selimsgrove, and abundantly at Austin's mines, Wythe county, Va., and Friedersville, Pa.

Uses. Valuable as an ore of zinc.

Willemite is an anhydrous silicate of zinc, of a yellowish or brownish color. H=5—5·5. Gr=4—4·1. Occurs in large grayish hexagonal prisms with rhombohedral terminations, at Stirling Hill, N. J., being the mineral formerly called *Troostite.* Consists of silica 27·15, and oxyd of zinc 72·85. Also obtained at Moresnet in Belgium.

Hopeite is a rare mineral occurring in grayish-white crystals or massive, with calamine, and supposed to be a phosphate of zinc.

Franklinite, an ore of iron, manganese and zinc, is described under Iron, on page 240.

Auricha1cite is a hydrous carbonate of zinc and copper, occurring in drusy incrustations of acicular crystals, having a verdigris green color. From Siberia, Hungary, England, and Lancaster, Pa.

GENERAL REMARKS ON ZINC AND ITS ORES.

The metal zinc (*spelter* of commerce) is supposed to have been unknown in the metallic state to the Greeks and Romans. It has been long worked in China, and was formerly imported in large quantities by the East India Company. The ores from which it is obtained are the carbonate and silicate of zinc, (calamine and electric calamine,) and to some extent the sulphuret, (blende,) and the oxyd. Blende, the *black jack* of English miners, was considered useless until the year 1738, when a mode of reducing it was introduced.

The principal mining regions of zinc in the world are in Upper Silesia at Tarnowitz and elsewhere; in Poland; in Carinthia at Raibel and Bleiberg; in Netherlands at Limberg; at Altenberg, near Aix-la-Chapelle in the Prussian province of the Lower Rhine; in England, in Derbyshire, Alstonmoor, Mendip Hills, etc.; in the Altai in Russia; besides others in China, of which little is known. In the United States, the calamine and electric calamine occur with the lead of the west in large

---

How is electric calamine distinguished from calc spar and chalcedony? From what ores is the metal zinc obtained? What is zinc called in commerce? When was blende first used in England? Where are zinc mines in the United States?

quantities, and till a recent period were considered worthless and thrown aside under the name of "dry bone." In Tennessee, Claiborne county, there are workable mines of the same ores. Calamine is successfully worked at Friedersville, Pennsylvania. The red oxyd of zinc of Franklin, New Jersey, contains 75 per cent. of pure zinc, and the ore is a valuable one. Blende is sufficiently abundant to be worked at the Wurtzboro' lead mine, Sullivan county, N. Y.; at Eaton and Warren in New Hampshire; at Lubec in Maine; and at Austin's mine, Wythe county, Virginia.

The calamine and electric calamine are prepared for reduction by breaking the ore into small fragments, separating the impurities as far as possible, and then calcining in a reverberatory furnace. This furnace differs little from that figured on a following page under Silver, except that the sole is flat. The ore is frequently stirred, and after five or six hours it is taken out; by this process, water and carbonic acid are expelled. The prepared ore is then mixed with about one-seventh by weight of charcoal, and in the English process, is reduced in large crucibles.

Figure 1, represents a vertical section of the furnace, and figure 2,

1

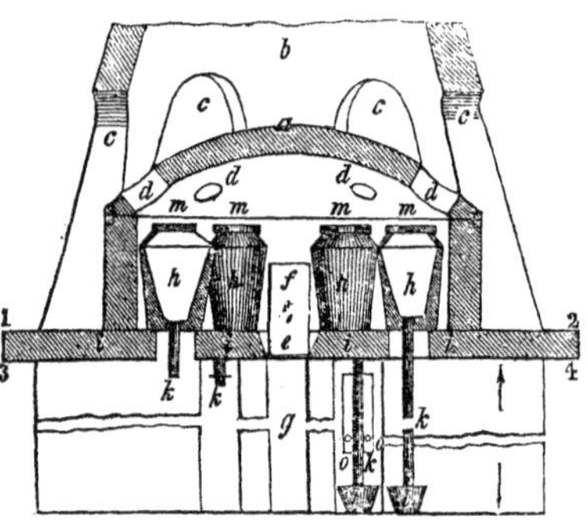

half of a horizontal section across the line 1, 2. The oven has an arched or cupola top, (*a*,) and contains 6 or 8 crucibles or pots, (*h*, *h*, *h*, *h*,)

2

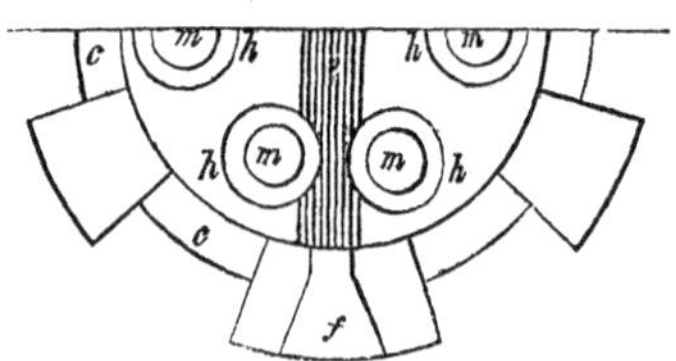

placed upon the sole of the earth, (*i*, *i*, *i*, *i*.) The crucibles have a hole

How is calamine reduced?

at bottom, to which a sheet iron tube (*k*) is adapted, which tube extends down to small vessels of water, or condensers, (*l, l*); and the sole of the hearth is perforated accordingly below each crucible. If one of the tubes becomes clogged with metal, it is cleared by a hot iron bar. In charging, the hole in the bottom of the crucible is stopped by a wooden plug, which afterwards becomes reduced to charcoal by the heat. The pots are charged and cleared out through holes (*d, d, d, d*) in the cupola (*a*.) The covers (of fire-tile, *m*) are placed on whenever a blue flame begins to appear, as this indicates the vaporization of the zinc.

The fire is made on the grate *e*, through the door *f*; *g* is the ash-pit below; *m, m, m, m*, in figure 2, show the position of the pots as seen in a bird's-eye view. The smoke escapes from the oven by the apertures *d*, (fig. 1,) into a conical chimney, (*b*,) by which a strong draught is kept up. In this chimney there are as many doors (*c, c, c, c*) as there are pots; and in the cupola there are the same number of openings for inserting or removing the pots, which are afterwards closed up by brickwork; the pots are many times refilled without removal. The refuse after an operation, is shaken out through the hole in the bottom of each pot, after the tube *k* is removed.

The zinc as it is reduced, rises in vapor and passes down the tubes into the condensers, where it collects in drops or powder with some oxyd; the metal is afterwards melted and cast into bars; and the oxyd which is skimmed off is returned to the crucibles. A charge occupies about three days, and the ore affords from 25 to 40 per cent. of zinc.

In Liege, where the ore from Altenberg is reduced, the ore is heated in horizontal earthen tubes, 3 feet long and 4 to 6 inches in diameter set thickly across a furnace, and around which the heat circulates. From the description given, it is obvious how the process might be varied, and larger combinations of pots or tubes arranged.

The blende is roasted in a reverberatory furnace, 8 or 10 feet square, the ore being placed in the furnace several inches deep, and kept constantly stirred for 10 or 12 hours. The roasted ore is then reduced in crucibles in the same manner as above explained. In England, the roasted blende is mixed with as much calcined calamine and twice the quantity of charcoal.

The annual production of zinc in different countries is as follows:

Great Britain, ...	1,000 tons.
Upper Silesia and Poland, ...	36,000 "
Belgium, ...	16,000 "
Carinthia, ...	1,500 "
United States, ...	5,000 "

Brass is made directly from the ore by heating copper with calcined calamine and charcoal. At Holywell, England, 40 pounds of copper and 60 of calamine yield about 60 pounds of brass. It is also made from copper and roasted blende, but the product is less pure. Dr. Jackson states that he has obtained brass of an inferior quality by heating together in a crucible copper pyrites and blende after roasting them. Brass is commonly made in this country by melting together the metals zinc and copper.

How is blende reduced? How is brass made?

The proportions of zinc in its alloys with copper are given in the remarks on copper. Zinc is a brittle metal, but admits of being rolled into sheets when heated to about 212° F. In sheets it is extensively used for roofing and other purposes, it being of more difficult corrosion, much harder, and also very much lighter than lead. Its combustibility is a strong objection to it as a roofing material.

The *Biddery* ware of the East Indies is made from an alloy of copper 16oz., lead 4oz., and tin 2oz., which is melted together and then mixed with 16oz. of spelter to every 3oz. of alloy.

The white oxyd of zinc is much used for white paint, in place of white lead.

An impure oxyd of zinc called *cadmia*, often collects in large quantities in the flues of iron and other furnaces, derived from ores of zinc mixed with the ores undergoing reduction. A mass weighing 600 pounds was taken from a furnace at Bennington, Vt. It has been observed in the Salisbury iron furnace, and at Ancram in New Jersey, where it was formerly called *ancramite*.

20. CADMIUM.

There is but a single known ore of this rare metal. It is a sulphuret, and is called *greenockite*. It occurs in hexagonal prisms, with pyramidal terminations, of a yellow color, high luster, and nearly transparent. H=3—3·5. Gr=4·8—4·9. From Bishopton, Scotland.

Cadmium is often associated in small quantities with zinc blende and calamine. In a black fibrous blende from Przibram, Löwe found 1·5 to 1·8 per cent.

21. LEAD.

Lead occurs rarely native ; generally in combination with sulphur ; also with arsenic, tellurium, selenium, and various acids.

The ores of lead vary in specific gravity from 5·5—8·2. They are soft, the hardness of the species with metallic luster not exceeding 3, and others not over 4. They are easily fusible before the blowpipe, (excepting plumbo-resinite) ; and with carbonate of soda on charcoal, (and often alone,) malleable lead may be obtained. The lead often passes off in yellow fumes, when the mineral is heated in the outer flame, or it covers the charcoal with a yellow coating.

Where have we the first notice of the metal bismuth ? From what source is it obtained for the arts ? What is it often called in the arts How is the metal obtained ? For what is bismuth used ? How does lead occur in nature ? What is said of the tests ?

NATIVE LEAD.

A rare mineral, occurring in thin laminæ or globules Gr=11·35. Said to have been seen in the lava of Madeira; at Alston in Cumberland with galena; in the county of Kerry, Ireland; and in an argillaceous rock at Carthagena.

GALENA.—*Sulphuret of Lead.*

Monometric. Cleavage cubic, eminent. Occurs under the form of the cube and its secondaries.

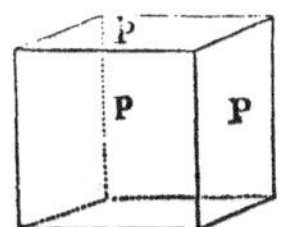

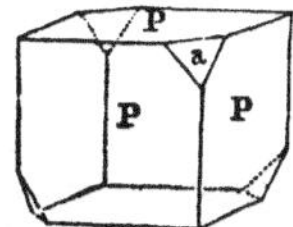

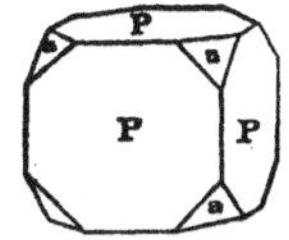

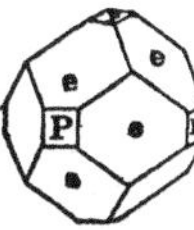

Cleavage cubic, perfect, and very easily obtained. Also coarse or fine granular; rarely fibrous.

Color and streak lead gray. Luster shining metallic. Fragile. H=2·5. Gr=7.5—7·7.

Composition: when pure, lead 86·55, sulphur 13·45. Often contains some sulphuret of silver, and is then called *argentiferous galena*, and at times sulphuret of zinc is present. Before the blowpipe on charcoal, it decrepitates unless heated with caution, and fuses, giving off sulphur, and finally yields a globule of lead.

Dif. Galena resembles some silver and copper ores in color, but its cubical cleavage, or granular structure when massive, will usually distinguish it. Its sulphur fumes obtained before the blowpipe prove it to be a sulphuret; and the lead reaction before the blowpipe show it to be a lead ore.

Obs. Galena occurs in granite, limestone, argillaceous and sandstone rocks, and is often associated with ores of zinc, silver and copper. Quartz, heavy spar, or carbonate of lime, is generally the gangue of the ore; also at times fluor spar. The rich lead mines of Derbyshire and the northern districts of England, occur in mountain limestone; and the same rock contains the valuable deposits of Bleiberg

---

Where has native lead been found? What is the structure of galena? its physical characters? its composition and blowpipe characters? How is it distinguished from silver and copper ores? Where does it occur?

and the neighboring deposits of Carinthia. At Freiberg in Saxony, it occupies veins in gneiss; in the Upper Hartz, and at Przibram in Bohemia, it traverses clay slate; at Sahla, Sweden, it occurs in crystalline limestone; the ore of Lead-hills, England, is in graywacke. There are other valuable beds of galena, in France at Poullaouen and Huelgoet, Brittany, and at Villefort, department of Lozère; in Spain in the granite hills of Linares, in Catalonia, Grenada and elsewhere; in Savoy; in Netherlands at Vedrin, not far from Namur; in Bohemia, southwest of Prague; in Joachimstahl, where the ore is worked principally for its silver; in Siberia in the Daouria mountains in limestone, argentiferous and worked for the silver.

The deposits of this ore in the United States are remarkable for their extent. They abound in what has been called "cliff limestone," in the states of Missouri, Illinois, Iowa, and Wisconsin; argillaceous iron, iron pyrites, calamine, ("dry bone" of the miners,) blende, ("black jack,") carbonate and sulphate of lead, are the most common associated minerals, together often with ores of copper and cobalt. In 1720, the lead mines of Missouri were discovered by Francis Renault and M. La Motte; and the La Motte mine is still known by this name. Afterwards, the country passed into the hands of the Spaniards, and during that period a valuable mine was opened by Mr. Burton, since called *Mine à Burton*. The mines of Missouri are contained in the counties of Washington, Jefferson, and Madison.

The lead region of Wisconsin, according to Dr. D. D. Owen, comprises 62 townships in Wisconsin, 8 in Iowa, and 10 in Illinois, being 87 miles from east to west, and 54 miles from north to south. The ore, as in Missouri, is inexhaustible, and throughout the region, there is scarcely a square mile in which traces of lead may not be found. The principal indications in the eyes of miners, as stated by Mr. Owen, are the following: fragments of calc spar in the soil, unless very abundant, which then indicate that the vein is wholly calcareous or nearly so; the red color of the soil on the surface, arising from the ferruginous clay in which the lead is often imbedded; fragments of lead ("gravel mineral,") along with the crumbling magnesian limestone, and dendritic specks distributed over the rock; also, a depression of the

---

What is said of the extent of the United States mines?

country, or an elevation, in a straight line; or "sinkholes;" or a peculiarity of vegetation in a linear direction. The "diggings" seldom exceed 25 or 30 feet in depth; for the galena is so abundant that a new spot is chosen rather than the expense of deeper mining. From a single spot, not exceeding 50 yards square, 3,000,000 lbs. of ore have been raised; and at the diggings in the west branch of the Peccatonica, not over 12 feet deep, two men can raise 2000 lbs. per day; in one of the townships, two men raised 16,000 lbs. in a day; 500 lbs. is the usual day's labor from the mines of average productiveness.

Galena also occurs in the region of Chocolate river and elsewhere, Lake Superior copper region; at Cave-in-Rock in Illinois, along with fluor; in New York at Rossie, St. Lawrence county, in gneiss, in a vein 3 to 4 feet wide; near Wurtzboro' in Sullivan county, a large vein in millstone grit; at Ancram, Columbia county; Martinsburg, Lewis county, N. Y., and Lowville, are other localities. All these mines have been worked, but they are now abandoned. Dr. Beck says of the Sullivan county and St. Lawrence mines, "in the latter the ore is in small veins with good associates, and is easily reduced; but the situation of the mines is bad. In the former, the ore is in large veins with bad associates, (zinc blende,) and is more difficult of separation and reduction; but the mines are admirably situated, whether we regard the removal of the ore or the facility of transporting produce to them."

In Maine, veins of considerable extent occur at Lubec; also of less interest at Blue Hill Bay, Birmingham and Parsonsfield. In New Hampshire, galena occurs at Eaton, Bath, Tamworth and Haverhill. In Vermont, at Thetford; in Massachusetts, at Southampton, Leverett, and Sterling, but without promise to the miner. In Virginia, in Wythe county, Louisa county, and elsewhere. In North Carolina, at King's mine, Davidson county, where the lead appears to be abundant. In Tennessee, at Brown's creek, and at Haysboro', near Nashville. An argentiferous variety occurs sparingly at Monroe, Conn., which afforded Prof. Silliman 3 per cent. of silver; also at Middletown, Ct.

Uses. The lead of commerce is obtained from this ore. It is often worked also for the silver it contains. It is also employed in glazing common stone ware: for this purpose it is ground up to an impalpable powder and mixed in water with clay; into this liquid the earthen vessel is dipped and then baked.

Cuproplumbite is a galena containing 24·5 per cent. of sulphuret of copper. From Chili.

ARSENURETS, SELENIDS, AND TELLURIDS OF LEAD.

These various ores of lead are distinguished by the fumes before the blowpipe, and by yielding ultimately a globule of lead.

Cobaltic lead ore is an arseniuret of lead, containing a trace of cobalt. From the Hartz. Gives an alliaceous odor (from. the arsenic) before the blowpipe. Gr=8·44.

Dufrenoysite is an arseniuret and sulphuret of lead; in dodecahedrons of a dark steel-gray color. Gr=5·55. From the Dolomite of St. Gothard.

Clausthalite, or *selenid of lead*, has a lead-gray color, and granular fracture. Gr=7·19. Gives a horse-radish odor (that of selenium) before the blowpipe. From the Hartz. There are three *selenids of lead and copper* which give the reaction of all the different constituents before the blowpipe. The sp. gr. of one is 5·6; of the second 7·0; the third 7·4. From the Hartz. There is also a *selenid of lead and mercury* occurring in foliated grains or masses, of a lead-gray to bluish and iron-black color.

Tellurid of lead. This is a tin-white cleavable mineral. Gr=8·16. From the Altai mountains.

Foliated tellurium is a less rare species, remarkable for being foliated like graphite. Color and streak blackish lead-gray. H=1—1·5. Gr=7·085. It contains tellurium 32·2, lead 54·0, gold 9·0, with often silver, copper, and some sulphur. From Transylvania.

MINIUM.—*Oxyd of Lead.*

Pulverulent. Color bright red, mixed with yellow. Gr=4·6. It is a sesquioxyd of lead. Affords globules of lead in the reduction flame of the blowpipe.

Obs. Occurs at various mines, usually associated with galena, and is found abundantly at Austin's mines, Wythe county, Virginia, with white lead ore.

Uses. Minium is the *red lead* of commerce: but for the arts it is artificially prepared. Lead is calcined in a reverberatory furnace, and a yellow oxyd (massicot) is thus formed: the massicot is afterwards heated in the same furnace in iron trays, at a low temperature, by which the lead absorbs more oxygen and becomes *red lead.* A much better material is obtained by the slow calcination of white lead.

Plumbic ocher is another similar ore, of a yellow color; it is a protoxyd of lead. Occurs in Wythe county, Va., also in Mexico.

What is minium? What are its characters?

ANGLESITE.—*Sulphate of Lead.*

Primary form a right rhombic prism, with imperfect lateral cleavage. M : M=103° 38′. Often in slender implanted crystals. Also massive; lamellar or granular.

Color white or slightly gray or green. Luster adamantine; sometimes a little resinous or vitreous. Transparent to nearly opaque. Brittle. H=2·75—3. Gr=6·25—6·3.

Composition: a sulphate of lead, containing about 73 per cent. of oxyd of lead. Fuses before the blowpipe to a slag, and yields lead with carbonate of soda.

Dif. Resembles somewhat some of the zeolite minerals, and also arragonite and some other earthy species; but this and the other ores of lead are at once distinguished by specific gravity, and also by their yielding lead in blowpipe trials. Differs from the carbonate of lead in not dissolving with effervescence in nitric acid.

Obs. Usually associated with galena, and results from its decomposition. Occurs in fine crystals at Leadhills and Wanlockhead, Great Britain, and also at other foreign lead mines. In the United States, it is found at the lead mines of Missouri and Wisconsin; in splendid crystallizations at Phenixville, Pa.; sparingly at the Walton gold mine, Louisa county, Va.; at Southampton, Mass.

Cupreous anglesite. A hydrous azure-blue *sulphate of lead and copper.* It is remarkable for a very perfect cleavage in one direction, and another inclined to the first 102° 45′. Gr=5·3—5·5. From Leadhills and Roughten Gill, England. Very rare.

CERUSITE.—WHITE LEAD ORE.—*Carbonate of Lead.*

Trimetric. In modified right rhombic prisms. M : M=

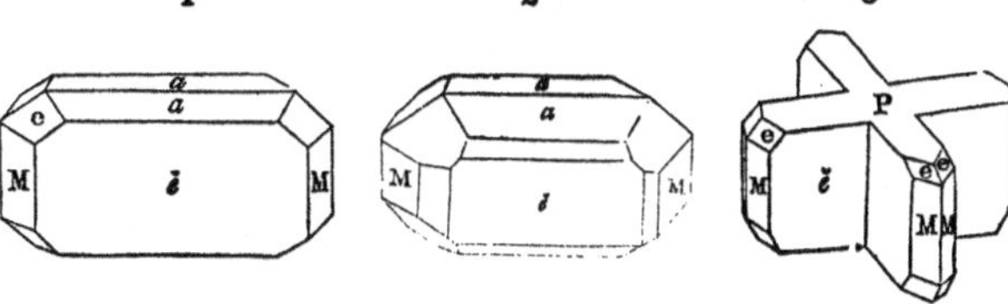

117° 13′. M : ĕ=121° 24; *a* : *a*=140° 15′. Often in

What is the appearance of anglesite? its composition? How is it distinguished from arragonite and the zeolites? What is the crystallization of white lead ore?

compound crystals, either six-sided prisms like aragonite, or wheel-shaped groups of 4 or 6 rays (fig. 3.) Also massive · rarely fibrous.

Color white, grayish, light or dark. Luster adamantine. Brittle. H=3—3·5. Gr=6·46—6·48.

Composition: oxyd of lead 83·46, carbonic acid 16·54 Decrepitates before the blowpipe, fuses, and with care affords a globule of lead. Effervesces in dilute nitric acid.

Dif. Like anglesite, distinguished from most of the species it resembles by its specific gravity and yielding lead when heated. From anglesite it differs in giving lead *alone* before the blowpipe, as well as by its solution and effervescence with nitric acid, and its less glassy lustre.

Obs. Associated usually with galena. Leadhills, Wanlockhead, and Cornwall, have afforded splendid crystallizations; also other lead mines on the continent of Europe.

In the United States, very handsome specimens are obtained at Austin's mines, Wythe county, Virginia, and at King's mine in Davidson's county, North Carolina. At the latter place it constitutes a wide vein, and has been worked for lead. It is associated with native silver and phosphate of lead. Perkiomen and Phenixville, Penn., afford good crystals. It occurs also at "Vallée's Diggings," Jefferson county, Missouri; at Brigham's mine near the Blue Mounds, Wisconsin; at "Deep Diggings" in crystals; and at other places in the West, both massive and in fine crystallizations. Rossie, N. Y., and Southampton, Mass., have afforded this ore.

Uses. When abundant, this ore is wrought for lead. Large quantities occur about the mines of the Mississippi valley. It was formerly buried up in the rubbish as useless, but it has since been collected and smelted. It is an exceedingly rich ore, affording in the pure state 75 per cent. of lead.

Carbonate of lead is the "white lead" of commerce, sc extensively used as a paint. The material for this purpose is, however, artificially made. In most manufacturing establishments, sheets of lead are suspended over a liquid made of vinegar and wine lees, and a gentle heat is applied either

---

What are the color and luster of white lead ore? its composition and blowpipe reaction? How is it distinguished from anglesite? How from minerals not lead ores? What use is made of white lead? How is white lead manufactured?

by stoves or from fermenting bark; the result is that the lead becomes carbonated from the acid fumes that rise from beneath.* The carbonate is then removed by shaking the plates smartly, and after washing and levigation, it is dried for market. According to another good process, (Thenard's,) carbonic acid, either from burning coke, brewers' vats, or some other source, is made to pass through a solution of subacetate of lead, the solution of subacetate being formed by digesting litharge and neutral acetate of lead. In place of this solution, litharge moistened slightly with vinegar, has been proposed. In the processes in the arts more litharge is made than is demanded in trade, and this use of it is considered more economical than its reduction to lead.

Carbonate of lead, mixed with sulphate of barytes, forms what is called *Venice white.*

Carbonate and sulphate of lead. There are two whitish or grayish ores of this composition called *dioxylite* and *leadhillite,* or respectively *sulphato-carbonate* and *sulphato-tricarbonate* of lead. The former contains 71 per cent. of carbonate of lead; the latter 47. Dioxylite has a perfect basal cleavage. Gr=6·2—6·5. Leadhillite cleaves into laminæ that are flexible like gypsum. Gr=6·8—7. From Leadhills.

Caledonite is a compound of the carbonates of lead and copper and sulphate of lead, and is called the *cupreous sulphato-carbonate of lead.* In crystals of a deep verdigris or bluish green color. Gr=6·4. From Leadhills and Red Gill; also from the Missouri mines.

PYROMORPHITE.—*Phosphate of Lead.*

Primary form, a hexagonal prism. Cleavage lateral, in traces. Usual in clustered hexagonal prisms, forming crusts. Also in globules, or reniform, with a radiated structure.

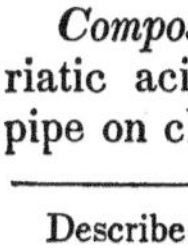

Color bright green or brown; sometimes fine orange-yellow, owing to an intermixture with chromate of lead. Streak white or nearly so. Luster more or less resinous. Nearly transparent to subtranslucent. Brittle. H=3·5—4. Gr=6·5—7·1.

Composition of a brown variety: oxyd of lead 78·58, muriatic acid 1·65, phosphoric acid 19·73. Before the blowpipe on charcoal fuses, and on cooling, the globule becomes

Describe pyromorphite. Of what does it consist?

* A subacetate is supposed to form first, and then to be immediately decomposed by the rising carbonic acid.

angular. In the inner flame, gives off fumes of lead. With boracic acid and iron, gives a phosphuret of iron and metallic lead.

Dif. Has some resemblance to beryl and apatite ; but is quite different in its action before the blowpipe, and much higher in specific gravity.

Obs. Leadhills, Wanlockhead, and other lead mines of Europe are foreign localities. In the United States, very handsome crystallized specimens occur at King's mine in Davidson county, N. C.: other localities are the Perkiomen and Phenixville mines, Pa.; the Lubec lead mines, Me.; Lenox, N. Y.; formerly, a mile south of Sing Sing, N. Y.; and the Southampton lead mine, Mass.

The name *pyromorphite* is from the Greek *pur*, fire, and *morphe*, form, alluding to its crystallizing on cooling from fusion before the blowpipe.

Mimetene. An arsenate of lead, resembling pyromorphite in crystallization, but giving a garlic odor on charcoal before the blowpipe. Color pale-yellow, passing into brown. H=2·75—3·5. Gr=6·41. From Cornwall and elsewhere ; Phenixville, Pa.

Hedyphane. An arseno-phosphate of lead and lime, containing 2 per cent. of chlorine. It occurs amorphous, of a whitish color, and adamantine luster. H=3·5—4. Gr=5·4—5·5. From Sweden.

CROCOISITE.—*Chromate of Lead.*

Occurs in oblique rhombic prisms, massive, of a bright red color and translucent. Streak orange-yellow. H=2·5—3. Gr=6.

Composition: chromic acid 31·85, protoxyd of lead 68·15. Produces a yellow solution in nitric acid. Blackens and fuses before the blowpipe, and forms a shining slag containing globules of lead.

Obs. Occurs in gneiss at Beresof in Siberia, and also in Brazil. This is the *chrome yellow* of the painters. It is made in the arts by adding to the chromate of potash in solution, a solution of acetate or nitrate of lead. The chromate of potash is usually procured by means of the ore chromic iron, which see, (p. 241.)

Melanochroite is another chromate of lead, containing 23·64 of chromic acid, and having a dark red color; streak brick red. Crystals usually tabular and reticulately arranged. Gr=5·75. From Siberia.

How is pyromorphite distinguished from beryl and apatite ? What is the color of chromate of lead ? its composition ? What is it called in the arts, and how used ?

Vauquelinite is a chromate of lead and copper, of a very dark green or pearly black color, occurring usually in minute irregularly aggregated crystals; also reniform and massive. H=2·5—3. Gr=5·5—5·8. From Siberia and Brazil. It has been found by Dr. Torrey at the lead mine near Sing Sing, in green and brownish-green mammillary concretions, and also nearly pulverulent.

Mendipite. Color white, yellowish or reddish, nearly opaque. Luster pearly. Gr=7—7·1. Contains chlorid of lead 38·4, oxyd of lead 61·6. From Mendip Hills, Somersetshire. ***Cotunnite*** is another chlorid of lead, occurring at Vesuvius in white acicular crystals. It contains 74·5 per cent. of lead.

Corneous lead. A chloro-carbonate of lead, occurring in whitish adamantine crystals. Gr=6—6·1. From Derbyshire and Germany. Also said to occur at the Southampton lead mine, Massachusetts.

Molybdate of lead. In dull-yellow octahedral crystals, and also massive. Luster resinous. Contains molybdic acid 34·25, protoxyd of lead 64·42. From Bleiberg and elsewhere in Carinthia; also Hungary. It has been found in small quantities at the Southampton lead mine, Mass., and in fine crystals, at Phenixville, Penn.

Selenate of lead. A sulphur-yellow mineral, occurring in small globules, and affording before the blowpipe on charcoal a garlic odor, and finally a globule of lead.

Vanadinite. A ***vanadate of lead***, occurring in hexagonal prisms like pyromorphite, and also in implanted globules. Color yellow to reddish brown. H=2·75. Gr=6·6—7·3. From Mexico; also from Wanlockhead in Dumfriesshire.

Tungstate of lead. In square octahedrons or prisms. Color green, gray, brown, or red. Luster resinous. H=2·5—3. Gr=7·9—8·1. Contains 51 of tungstic acid and 49 of lead.

Plumbo-resinite. In globular forms, having a luster somewhat like gum arabic, and a yellowish or reddish-brown color. H=4—4·5. Gr=6·3—6·4. Consists of protoxyd of lead 40·14, alumina 37·00, water 18·8. From Huelgoet in Brittany, and at a lead mine in Beaujeu; also from the Missouri mines, with black cobalt.

GENERAL REMARKS ON LEAD AND ITS ORES.

The lead of commerce is derived almost wholly from the sulphuret of lead or galena, the localities of which have already been mentioned.

This ore is reduced usually by heat alone in a reverberatory furnace. The process consists simply in burning out the sulphur after the ore is picked, pounded and washed. The galena is kept at a heat below that required for its fusion, and air is freely admitted to aid in the combustion. The sulphur is driven off, leaving the pure lead, or an oxyd formed in the process which passes to the state of a slag. The latter is heated again with charcoal, which separates the oxygen. A portion of quicklime is often added to stiffen the slag. In England, the whole operation of a smelting shift takes about 4½ hours, and four periods may be distinguished:—The *first fire* for roasting the ores, which requires

---

What is the source of the lead of commerce? How is the ore reduced?

very moderate firing, and lasts two hours; the *second* fire for smelting, requiring a higher heat with shut doors, and at the end the slags are dried up with lime, and the furnace is also allowed to cool a little; the *third* and *fourth fires*, also for smelting, requiring a still higher temperature.

A furnace for using the hot blast with lead has been contrived. The heated blast is made to diffuse itself equally through the whole "charge," carrying with it the flame of the burning fuel, and the reduction of the ore is effected with an economy and dispatch hitherto unknown in the processes of reducing this metal.*

According to another mode which has been practised in Germany and France, old iron (about 28 per cent.) is thrown into the melted ore, heated in a reverberatory furnace of small size; the iron acts by absorbing the sulphur, and the lead thus reduced flows into the bottom of the basin. There is here a gain of time and labor, but a total loss of the iron.

The mode of obtaining the silver from lead ore, is mentioned under Silver.

The principal mines of lead in the world are mentioned under Galena. The following is a statement of the approximate amount of lead produced by the mines of Europe:

Great Britain and Ireland,	1,200,000 cwt.	Sweden and Norway,	4,000 cwt.
Spain,	600,000 "	Prussia,	160,000 "
Austria,	140,000 "	Germany,	160,000 "
Russia and Poland,	6,000 "	Belgium,	20,000 "
France,	30,000 "	Piedmont and Switzerland,	10,000 "

According to the Statistical Tables of J. D. Whitney,† the mines of the Upper Mississippi and Missouri have afforded as follows:

	Upper Mississippi.	Missouri Mines.
1826,	428 tons.	1,343 tons.
1830,	5,331	1,832
1835,	8,469	3,227
1840,	11,987	2,793
1842,	13,992	3,348
1845,	24,328	
1847,	24,145	
1850,	17,768	1,500 ?
1853,	13,307	

The present yield of the Missouri mines is set down as not over 1500 tons. The proceeds of the western mines have been for many years on the decrease.

What other method is mentioned? What country affords the largest amount of lead at the present time, and how much? What is the yield of the mines of the Upper Mississippi. What of the Lower or Missouri mines?

* See Amer. Jour. Sci. xlii, p. 169.

† Whitney's Metallic Wealth of the United States, p. 421.

22. MERCURY

Mercury occurs native, alloyed with silver, and in combination with sulphur, chlorine, or iodine. Its ores are completely volatile, excepting the one containing silver.

NATIVE MERCURY.

Monometric; in octahedrons. Occurs in fluid globules scattered through the gangue. Color tin-white. Gr=13·6. Becomes solid and crystallizes at a temperature of—39° F.

Mercury, or quicksilver as it is often called, (a translation of the old name "argentum vivum,") is entirely volatile before the blowpipe, and dissolves readily in nitric acid.

Obs. Native mercury is a rare mineral, yet is met with at the different mines of this metal, at Almaden in Spain, Idria in Carniola, (Austria,) and also in Hungary and Peru. It is usually in disseminated globules, but is sometimes accumulated in cavities so as to be dipped up in pails.

Uses. Mercury is used for the extraction of gold and silver ores, and is exported in large quantities to South America. It is also employed for silvering mirrors, for thermometers and barometers, and for various purposes connected with medicine and the arts.

Native Amalgam. This mineral is a compound of mercury and silver, containing 64 to 72 per cent. of mercury, and occurring in silver-white dodecahedrons. H=2—2·5. Gr=10·5—14. Principally from the Palatinate; also from Hungary and Sweden. The *arquerite* of Berthier is an amalgam from Coquimbo, containing only 13·5 per cent. of silver.

CINNABAR.—*Sulphuret of Mercury.*

Rhombohedral. R : R=71° 47′. Cleavage transverse, highly perfect. Crystals often tabular, or six-sided prisms. Also massive, and in earthy coatings.

Luster unmetallic, adamantine in crystals; often dull. Color bright red to brownish-red, and brownish-black. Streak red. Subtransparent to nearly opaque. H=2—2·5. Gr=6·7—8·2. Sectile.

Composition: when pure, mercury 86·29, sulphur 13·71;

In what condition does mercury occur? What is a characteristic of its ores? Describe native mercury? Where is it found? For what is it used? What are the physical characters of cinnabar?

but often contains impurities. The *liver ore*, or *hepatic cinnabar*, contains some carbon and clay, and has a brownish streak and color. The pure variety volatilizes entirely before the blowpipe.

Dif. Distinguished from red oxyd of iron and chromate of lead by evaporating before the blowpipe ; from realgar by giving off on charcoal no allicaceous fumes.

Obs. Cinnabar is the ore from which the principal part of the mercury of commerce is obtained. It occurs mostly in connection with talcose and argillaceous shale, or other stratified deposits, both the most ancient and those of more recent date. The mineral is too volatile to be expected in any abundance in proper igneous or crystalline rocks, yet has been found sparingly in granite. The principal mines are at Idria in Austria, Almaden in Spain, in the Palatinate on the Rhine, and at Huanca Velica in Peru. Mercury occurs also at Arqueros in Chili, at various places in Mexico, in Hungary, Sweden, at several points in France, and at Ripa, in Tuscany ; also in China and Japan. A large mine has been discovered in Upper California. The ore there occurs in a ridge of the Sierra Azul, twelve miles south of San José, a few miles from the coast, and about half way from San Francisco to Monterey. The mouth of the principal mine (the mine of New Almaden) is a few yards down from the summit of the highest hill containing the ore, and is about 1200 feet above the neighboring plain. The prevailing rock is a greenish talcose rock. The ore is interspersed through the slate, in a yellow ochreous matrix, which forms a bed 42 feet in thickness. The richest ore is from the upper part of the bed. The supply is abundant, and of excellent quality, and one to two millions of pounds of mercury are now extracted annually. The rock is a metamorphic rock, much tilted and contorted ; but its exact age has not been ascertained.

Uses. This ore is the principal source of the mercury of commerce. It is also used as a pigment, and as a coloring ingredient for red sealing wax, and it is called in the shops *vermillion.* The vermillion of commerce is often adulterated with red lead, dragon's blood and realgar. Its entire volatility, without odorous fumes, will distinguish the pure material.

Horn Quicksilver, (chlorid of mercury.) A tough, sectile ore, of a

---

Of what does cinnabar consist ? Where are the principal mines ? For what is it used ?

light yellowish or grayish color, and adamantine luster, translucent or subtranslucent, crystallizing in secondaries to a square prism. H=1—2. Gr=6·48. It contains 85 per cent. of mercury.

Iodic Mercury is a still rarer ore from Mexico. Color reddish-brown.

Selenid of mercury, a dark steel-gray ore, which is wholly evaporated before the blowpipe. Occurs in Mexico near San Onofre.

GENERAL REMARKS ON THE ORES OF MERCURY.

The mines of Idria were discovered in 1497. The mining is carried on in galleries, as the rock is too fragile to allow of large chambers. The ore is obtained at a depth of about 750 feet, and is mostly a bituminous cinnabar, disseminated through the rock along with native mercury. The latter is in some parts so abundant that when the earthy rock is fresh broken, large globules fall out and roll to the bottom of the gallery. The pure mercury is first sifted out; the gangue is then washed, and prepared for reduction. For this purpose there is a large circular building, 40 feet in diameter by 60 in height, the interior of which communicates through small openings with a range of chambers around, each 10 or 12 feet square, and having a door communicating with the external air. The central chamber is filled with earthen pans, containing the prepared earth, the whole is closed up and heat is applied. The mercury sublimes and is condensed in the cold air of the smaller chambers, whence it is afterwards removed. After filtering, it is ready for packing. These mines afford annually 5000 cwt.

The above mode of reduction is styled by Ure "absolutely barbarous." He observes that the brick and mortar walls cannot be rendered either tight or cool; and that the ore ought to be pounded, and then heated in a series of cast-iron cylinder retorts, after being mixed with the requisite proportion of quicklime, (the lime aiding in the reduction of the cinnabar by taking its sulphur,) and the retorts should communicate with a trough through which a stream of water passes, for the purpose of condensing the mercury. An apparatus of this kind planned by Ure, is used at Landsberg, in Rhenish Bavaria.

The mines of the Palatinate, on the Rhine, and those of other parts of Germany, are stated by Burat to yield 7.600 quintals.

The mines of Almaden are situated near the frontier of Estremadura, in the province of La Mancha. They have been worked from a remote antiquity. According to Pliny, the Greeks obtained vermillion from them 700 years before our era, and afterwards imported annually 100,000 pounds. The mines are not over 300 yards in depth, although so long worked. The rock is argillaceous schist and grit, in horizontal beds, which are intersected by granitic and black porphyry eruptions. The mass of ore at the bottom of the principal vein, is 12 to 15 yards thick, and yields in the aggregate 10 per cent. of mercury. It is taken to the furnace without any kind of mechanical preparation. There are many veins in the vicinity, several of which have been explored. The furnaces of Almadenejos are fed almost exclusively by an ore obtained just east of the village, which is a black schist, strongly impregnated

---

What is said of the Idria mines? How is the ore reduced? What is a better process? What is said of the mines of Almaden?

with native mercury and cinnabar, with but little visible. These mines afford annually about 25.000 cwt. of mercury. The granitic and porphyritic eruptions of the region have been supposed to account for the presence of the mercury in the rocks: the heat produced exhalations of mercury and sulphur, which gave origin both to the cinnabar and the native mercury.

The mines of Huanca Velica, in Peru, have afforded a large amount of mercury for amalgamation at the Peruvian silver mines. Between the years 1570 and 1800, they are estimated to have produced 537.000 tons; and their present annual yield is 1800 quintals.

The Chinese have mines of cinnabar in Shensi, where the ore is reduced by the rude process of burning brushwood in the wells or pits dug out for the purpose, and then collecting the metal after condensation.

23. COPPER.

Copper occurs native in considerable quantities; also combined with oxygen, sulphur, selenium, and various acids.

The ores of copper vary in specific gravity from 3·5 to 8·5, and seldom exceed 4 in hardness. Many of the ores give to borax a green color in the outer flame, and an opaque dull-red in the inner. With carbonate of soda on charcoal, nearly all the ores are reduced, and a globule of copper obtained; borax and tin foil are required in some cases where a combination with other metals conceals the copper. When soluble in the acids, a clean plate of iron inserted in the solution becomes covered with copper, and ammonia produces a blue solution.

NATIVE COPPER.

Monometric. In octahedrons; no cleavage apparent. Often in plates or masses, or arborescent and filiform shapes.

Color copper-red. Ductile and malleable. H=2·5—3. Gr=8·58.

Native copper often contains a little silver, disseminated throughout it. Before the blowpipe it fuses readily, and on cooling it is covered with a black oxyd. Dissolves in nitric acid, and produces a blue solution with ammonia.

Obs. Native copper accompanies the ores of copper, and usually occurs in the vicinity of dikes of igneous rocks.

How does copper occur? How are copper ores distinguished? What are the characters of native copper?

Siberia, Cornwall, and Brazil, are noted for the copper they have produced. A mass supposed to be from Bahia, now at Lisbon, weighs 2616 pounds. The vicinity of lake Superior is one of the most extraordinary regions in the world for its native copper, where it occurs mostly in vertical seams in trap, and also in the enclosing sandstone. A mass weighing 3704 lbs. has been taken from thence to Washington city: it is the same that was figured by Schoolcraft, in the American Journal of Science, volume iii, p. 201. Masses from 1000 to 3700 pounds, from this region, have been exposed on the wharves of Boston, Mass. This is small compared with other pieces which have since been laid open. One large mass was quarried out in the "Cliff mine," whose weight has been estimated at 200 tons. It was 40 feet long, 6 feet deep, and averaged 6 inches in thickness. This copper contains intimately mixed with it about $\frac{3}{10}$ per cent. of silver. Besides this, perfectly pure silver, in strings, masses, and grains, is often disseminated through the copper, and some masses, when polished, appear sprinkled with large white spots of silver, resembling, as Dr. Jackson observes, a porphyry with its feldspar crystals. Crystals of native copper are also found penetrating masses of prehnite, and analcime, in the trap rock.

This mixture of copper and silver cannot be imitated by art, as the two metals form an alloy when melted together. It is probable that the separation, in the rocks, is due to the cooling from fusion being so extremely gradual as to allow the two metals to solidify separately, at their respective temperatures of solidification—the trap being an igneous rock, and ages often elapsing, as is well known, during the cooling of a bed of lava, covered from the air.

Small specimens of native copper have been found in the states of New Jersey, Connecticut, and Massachusetts, where the same formation occurs. One mass from near Somerville weighs 78 pounds, and is said originally to have weighed 128 pounds. Near New Haven, Conn., a mass of 90 pounds was formerly found. Near Brunswick, N. J., a vein or sheet of copper, from a sixteenth to an eighth of an inch thick, has been observed and traced along for several rods.

Where has native copper been found in the United States? What is said of its associations with silver? What explanation is given of this mixture of copper and silver?

COPPER GLANCE.—VITREOUS COPPER ORE.

Trimetric. Cleavage parallel to the faces of a right rhombic prism, but indistinct. M : M=119° 35′. Secondary forms, variously modified rhombic prisms. Also in compound crystals like arragonite ; often massive.

Color and streak blackish lead-gray, often tarnished blue or green. Streak sometimes shining. H=2·5—3. Gr=5·5—5·8.

Composition : sulphur 20·6, copper 77·2, iron 1·5. Before the blowpipe it gives off fumes of sulphur, fuses easily in the external flame, and boils. After the sulphur is driven off, a globule of copper remains. Dissolves in heated nitric acid, with a precipitation of the sulphur.

Dif. The vitreous copper ore resembles vitreous silver ore ; but the luster of a surface of fracture is less brilliant, and they afford different results before the blowpipe. The solution made by putting a piece of the ore in nitric acid, covers an iron plate (or knife blade) with copper, while a similar solution of the silver ore covers a copper plate with silver.

Obs. Occurs with other copper ores in beds and veins. At Cornwall, splendid crystallizations occur. Siberia, Hesse, Saxony, the Bannat, Chili, &c., afford this ore.

In the United States, a vein affording fine crystallizations occurs at Bristol, Conn. Other localities are at Wolcottville, Simsbury, and Cheshire, Conn. ; at Schuyler's Mines, and elsewhere, N. J. ; in the U. S. copper mine district, Blue Ridge, Orange county, Virginia ; between New Market and Taneytown, Maryland ; and sparingly at the copper mines of Michigan and the Western states ; also at some mines north of Lake Huron.

Blue Copper is a dull blue-black massive mineral. Gr=3·8—. It contains 65 per cent. of copper. It is named *Covelline.*

Harrisite is a copper glance, with cubic cleavage, from Canton mine, Ga., probably a pseudomorph after Galena.

COPPER PYRITES.—*Sulphuret of Copper and Iron.*

Dimetric. Crystals tetrahedral or octahedral ; sometimes

What are the physical characters of vitreous copper ? its constitution and chemical characters ? How does it differ from silver ores ?

compound. A : A=109° 53, and 108° 40′. Cleavage indistinct. Also massive, and of various imitative shapes.

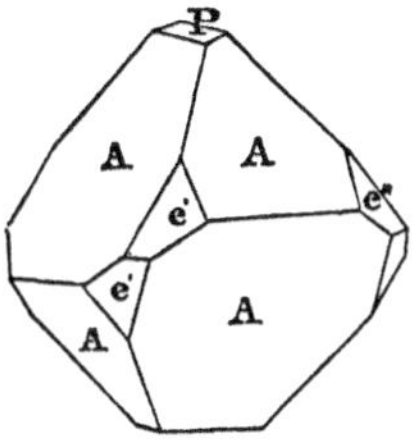

Color brass-yellow, often tarnished deep yellow, and also iridescent. Streak unmetallic, greenish-black, and but little shining. H=3·5—4. Gr=4·15—4·13.

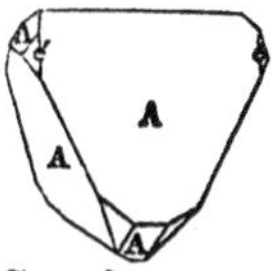

Composition: sulphur 34 9, copper 34·6, iron 30·5. Fuses before the blowpipe to a globule which is magnetic, owing to the iron present. Gives sulphur fumes on charcoal. With borax affords pure copper The usual effect with nitric acid.

Dif. This ore resembles native gold, and also iron pyrites. It is distinguished from gold by crumbling when it is attempted to cut it, instead of separating in slices; and from iron pyrites in its deeper yellow color and in yielding easily to the point of a knife, instead of striking fire with a steel.

Obs. Copper pyrites occurs in veins in granitic and allied rocks; also in graywacke, &c. It is usually associated with iron pyrites, and often with galena, blende, and carbonates of copper. The copper of Fahlun, Sweden, is obtained mostly from this ore, where it occurs with serpentine in gneiss. Other mines of this ore are in the Hartz, near Goslar; in the Bannat, Hungary, Thuringia, &c. The Cornwall ore is mostly of this kind, and 10 to 12,000 tons of pure copper are smelted annually. The ore for sale at Redruth is said to be by no means a rich ore. It rarely yields 12 per cent. and generally only 7 or 8, and occasionally as little as 3 to 4 per cent. of metal. In the latter case such poverty of ore is only made up by its facility of transport, the moderate expense of fuel, or the convenience of smelting. Its richness may generally be judged of from the color: if of a fine yellow hue, and yielding readily to the hammer, it is a good ore; but if hard and pale yellow it contains very largely of iron pyrites, and is of poor quality.

In the United States there are many localities of this ore.

What forms are presented by copper pyrites? What is its color and streak? its composition? How is it distinguished from iron pyrites and gold? What is said of the modes of occurrence of this ore and of its mines?

It occurs in Massachusetts, at the Southampton lead mines, at Turner's Falls on the Connecticut, at Hatfield and Sterling; in Vermont, at Strafford, where it was for a time worked, and at Shrewsbury, Corinth, Waterbury; in New Hampshire, at Franconia, Shelburn, Unity, Warren, Eaton, Lyme, Haverhill; in Maine, at the Lubec lead mines, and Dexter; in New York, at the Ancram lead mine, also near Rossie, and at Wurtzboro; in Pennsylvania, at Morgantown; in Virginia, at the Phenix copper mines, Fauquier county, and at the Walton gold mine, Luzerne county; in Maryland, in the vicinity of Liberty and New London, in Frederick Co., and at the Patapsco mines, near Sykesville; in North Carolina, in Davidson and Guilford counties. In Michigan, where native copper is so abundant, this is a rare ore; but it occurs at Presqu'isle, at Mineral Point, and in Wisconsin, where it is the predominating ore. In Tennessee, in Polk county, at the Hiwassee mines, where, however, only the overlying black copper is worked.

Uses. This ore, besides being mined for copper, is extensively employed in the manufacture of blue vitriol (sulphate of copper,) in the same manner that sulphate of iron (copperas) is obtained from iron pyrites.

Cuban is a sulphuret of copper and iron, containing sulphur 39·0, iron 38·0, copper 19·8, silica 2·3=99·12.

ERUBESCITE.—VARIEGATED COPPER PYRITES.

Monometric. Cleavage octahedral, in traces. Occurs in cubes and octahedrons. Also massive.

Color between copper-red and pinchbeck-brown. Tarnishes rapidly on exposure. Streak pale grayish-black and but slightly shining. Brittle. H=3. Gr=5.

Composition: specimen from Bristol, Conn., sulphur 25·7, copper 62·8, iron 11·6. Fuses before the blowpipe to a globule attractable by the magnet. On charcoal affords fumes of sulphur. Mostly dissolved in nitric acid.

Dif. This ore is distinguished from the preceding by its pale reddish-yellow color.

Obs. Occurs with other copper ores, in granitic and allied rocks, and also in secondary formations. The mines of Cornwall have afforded crystallized specimens, and it is there called from its color "horse-flesh ore." Other foreign

---

What is the appearance and composition of variegated copper pyrites? How is it distinguished from the preceding species?

localities of massive varieties are Ross Island, Killarney, Ireland; Norway, Hessia, Silesia, Siberia, and the Bannat.

Fine crystallizations occur at the Bristol copper mine, Conn., in granite; and also in red sandstone, at Cheshire, in the same state, with malachite and heavy spar. Massive varieties occur at the New Jersey mines, and in Pennsylvania.

TETRAHEDRITE.—GRAY COPPER.

Monometric. Occurs in modified tetrahedrons, and also in compound crystals. Cleavage octahedral in traces.

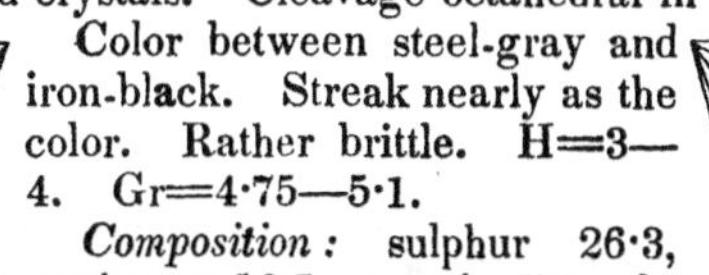

Color between steel-gray and iron-black. Streak nearly as the color. Rather brittle. H=3—4. Gr=4·75—5·1.

Composition: sulphur 26·3, copper 38·6, antimony 16·5, arsenic 7·2, along with some iron, zinc, and silver, amounting to 15 per cent. It sometimes contains 30 per cent. of silver in place of part of the copper, and is then called *argentiferous gray copper ore*, or *silver fahlerz*. The amount of arsenic varies from 0 to 10 per cent. One variety from Spain included 10 per cent. of platinum, and another from Hohenstein some gold; another from Tuscany 2·7 per cent. of mercury.

These varieties give off, before the blowpipe, fumes of arsenic and antimony, and after roasting yield a globule of copper. Dissolve, when pulverized, in nitric acid, affording a brownish-green solution.

Dif. Its copper reactions before the blowpipe and in solution in nitric acid, distinguish it from the gray silver ores.

Obs. The Cornish mines, Andreasberg in the Hartz, Kremnitz in Hungary, Freiberg in Saxony, Kapnik in Transylvania, and Dillenberg in Nassau, afford fine crystallizations of this ore. It is a common ore in the Chilian mines, and it is worked there and elsewhere for copper, and often also for silver.

Bournonite contains sulphur 20·3, antimony 26·3, lead 40·8, copper 12.7. Its crystals are modified rectangular prisms, of a steel-gray color and streak, and are often compounded into shapes like a cog-wheel, whence it is called *wheel-ore*. H=2·5—3. Gr=5·766. From the Hartz, Transylvania, Saxony, and Cornwall. Another allied ore, containing 47 per cent. of antimony, is called *antimonial copper;* it oc-

---

Describe gray copper ore. Mention its composition and blowpipe characters. How is it distinguished from silver ores?

curs in slender aggregated prisms, of a dark ead-gray color. Another containing also arsenic, is called *antimonial copper glance.*

Tennantite is a compound of copper, iron, sulphur, and arsenic. It occurs in dodecahedral crystals, brilliant, with a dark lead-gray color, and reddish-gray streak. From the Cornish mines near Redruth and St. Day. *Domeykite* is arsenical copper.

Selenid of Copper, is a silver-white ore, affording the horse-radish odor of selenium before the blowpipe. It contains 64 per cent. of copper. From Skrikerum, Sweden.

RED COPPER ORE.

Monometric. In regular octahedrons, and modified forms of the same. Cleavage octahedral. Also massive, and sometimes earthy.

1

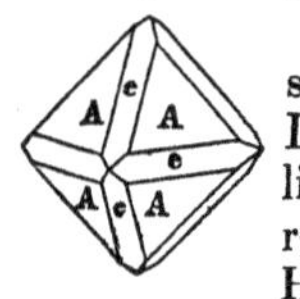

2

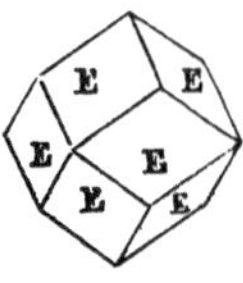

Color deep red, of various shades. Streak brownish-red. Luster adamantine or submetallic; also earthy. Subtransparent to nearly opaque. Brittle. H=3·5—4. Gr=6.

Composition: copper 88·8, oxygen 11·2. Before the blowpipe, on charcoal, it yields a globule of copper. Dissolves in nitric acid. The earthy varieties have been called *tile ore*, from the color.

Dif. From cinnabar it differs in not being volatile before the blowpipe; and from red iron ore, in yielding a bead of copper on charcoal, and copper reactions.

Obs. Occurs with other copper ores in the Bannat, Thuringia, Cornwall, at Chessy near Lyons, in Siberia, and Brazil. The octahedrons are often green, from a coating of malachite.

In the U. States, it has been observed crystallized and massive, at Schuyler's, Somerville, and the Flemington copper mines, N. J.; also near New Brunswick, N. J.; at Bristol, Ct.; also near Ladenton, Rockland county, N. Y.

Black Copper. Tenorite. An oxyd of copper, occurring as a black powder and in dull black masses, and botryoidal concretions, in veins or along with other copper ores. From Cornwall, and also the Vesuvian lavas. It is an abundant ore in some of the copper mines of the Mississippi valley, and yields 60 to 70 per cent. of copper. It results from the

---

What is the crystallization of red copper ore? Of what does it consist? How does it differ from cinnabar and red iron ore?

decomposition of the sulphurets and other ores. At the Hiwassee mine, Polk Co., Tennessee, it is abundant, and is worked as an ore. It was found of excellent quality in the Lake Superior copper region, but has there been exhausted.

The oxyds of copper are easily smelted by heating with the aid of charcoal alone. They may be converted directly into the sulphate or blue vitriol, by means of sulphuric acid, but are more valuable for the copper they afford.

BLUE VITRIOL.—*Sulphate of Copper.*

Triclinic. In oblique rhomboidal prisms. Also as an efflorescence or incrustation.

Color deep sky-blue. Streak uncolored. Subtransparent to translucent. Luster vitreous. Soluble, taste nauseous and metallic. H=2—2·5. Gr=2·21.

Composition: sulphuric acid 32·1, oxyd of copper, 31·8, water 36·1. A polished plate of iron in a solution becomes covered with copper.

Obs. Occurs with the sulphurets of copper as a result of their decomposition, and is often in solution in the waters flowing from copper mines. Occurs in the Hartz, at Fahlun in Sweden, and in many other copper regions.

Uses. Blue vitriol is much used in dyeing operations and in the printing of cotton and linen; also for various other purposes in the arts. It has been employed to prevent dry rot, by steeping wood in its solution: and it is a powerful preservative of animal substances; when imbued with it and dried, they remain unaltered. It is afforded by the decomposition of copper pyrites, in the same manner as green vitriol from iron pyrites. (p. 213.)

It is manufactured for the arts from old sheathing copper, copper turnings, and copper refinery scales. The scales are readily dissolved in dilute sulphuric acid at the temperature of ebullition; the solution obtained is evaporated to the point where crystallization will take place on cooling. Metallic copper is exposed in hot rooms to the atmosphere after it has been wet in weak sulphuric acid. By alternate wetting and exposure, it is rapidly corroded, and affords a solution which

---

What is blue vitriol? Describe it. What is said of its mode of occurrence? For what is it used? How is it manufactured in the arts? How is copper obtained from solutions in some mines? Describe green malachite.

is evaporated for crystals. 400 000 lbs. is the annual consumption of blue vitriol in the United States.

In Frederick county, Maryland, blue vitriol is made from a black earth which is an impure oxyd of copper with copper pyrites. The black oxyd of copper, which was found in the Lake Superior copper region, may be directly converted into blue vitriol.

In some mines, the solution of sulphate of copper is so abundant as to afford considerable copper, which is obtained by immersing clean iron in it, and is called *copper of cementation.* At the copper springs of Wicklow, Ireland, about 500 tons of iron were laid at one time in the pits; in about 12 months the bars were dissolved, and every ton of iron yielded a ton and a half, and sometimes nearly two tons, of a precipitated reddish mud, each ton of which produced 16 cwt. of pure copper. The Rio Tinto Mine in Spain, is another instance of working the sulphate in solution. These waters yield annually 1800 cwt. of copper, and consume 2400 cwt. of iron.

Brochantite. An insoluble sulphate of copper, containing 17·5 per cent. of sulphuric acid. Color emerald green. In tabular rhombic crystals, at Katherinenberg, in Siberia. Blackens before the blowpipe without fusing. *Krisuvigite* and *Konigite* are the same species.

MALACHITE.—*Green Carbonate of Copper.*

Monoclinic. Usual in incrustations, with a smooth tuberose, botryoidal or stalactitic surface; structure finely and firmly fibrous. Also earthy.

Color light green, streak paler. Usually nearly opaque; crystals translucent. Luster of crystals adamantine inclining to vitreous; but fibrous incrustations silky on a cross fracture. Earthy varieties dull. H=3·5—4. Gr=4.

Composition: carbonic acid 20, oxyd of copper 71·9, water 8·2. Dissolves with effervesence in nitric acid. Decrepitates and blackens before the blowpipe, and becomes partly a black scoria. With borax it fuses to a deep green globule, and ultimately affords a bead of copper.

Dif. Readily distinguished by its copper-green color and its association with copper ores. It resembles a siliceous ore of copper, chrysocolla, a common ore in the mines of the Mississippi valley; but it is distinguished by its complete

---

What is the composition of green malachite? How is it distinguished?

solution and effervescence in nitric acid. The color also is not the bluish-green of chrysocolla.

Obs. Green malachite usually accompanies other ores of copper, and forms incrustations, which when thick, have the colors banded and extremely delicate in their shades and blending. Perfect crystals are quite rare. The mines of Siberia, at Nischne Tagilsk, have afforded great quantities of this ore. A mass partly disclosed, measured at top 9 feet by 18; and the portion uncovered contained at least half a million pounds of pure malachite. Other noted foreign localities are Chessy in France, Sandlodge in Shetland, Schwartz in the Tyrol, Cornwall, and the island of Cuba.

The copper mine of Cheshire, Conn., has afforded handsome specimens; also Morgantown, Perkiomen and Phenixville, Penn.; Schuyler's mine, and the New Brunswick copper mine, N. J.: it occurs also in Maryland, between Newmarket and Taneytown, and in the Catoctin mountains; in the Blue Ridge, Penn., near Nicholson's Gap, and it is found more or less sparingly with all kinds of copper ores.

At Mineral Point, Wisconsin, a bluish silico-carbonate of copper occurs, which is for the most part chrysocolla, or a mixture of this mineral with the carbonate. An analysis of the rough ore afforded Mr. D. D. Owen, copper 35·7, carbonic acid 10·0, water 10·0, iron 15·7, oxygen 7, sulphur 8, silex 13·0. Specific gravity 3·69—3·87. The vein appears also to the northwest on Blue River, and southeast on the Peccatonica. This ore is abundant; it has been smelted on the spot and also exported to England.

Uses. This mineral receives a high polish and is used for inlaid work, and also ear-rings, snuff boxes, and various ornamental articles. It is not much prized in jewelry. Very large masses are occasionally obtained in Russia, which are worked into slabs for tables, mantel-pieces and vases, which are of exquisite beauty, owing to the delicate shadings of the radiations and zones of color. At Versailles, there is a room furnished with tables, vases, and other articles of this kind, and similar rooms are to be found in many European palaces. At Nischne Tagilsk, a block of malachite was obtained weighing 40 tons.

Malachite is sometimes passed off in jewelry as turquois, though easily distinguished by its shade of color and much

---

How does green malachite occur? What are its uses?

inferior hardness. It is a valuable ore when abundant; but it is seldom smelted alone, because the metal is liable to escape with the liberated volatile ingredient—carbonic acid.

AZURITE.—*Blue Carbonate of Copper.*

Monoclinic. In modified oblique rhombic prisms, the crystals rather short and stout; lateral cleavage perfect. Also massive. Often earthy.

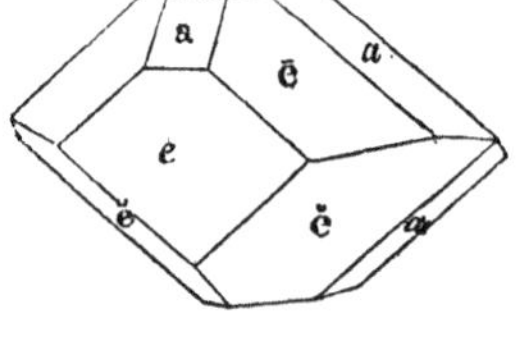

Color deep blue, azure or Berlin-blue. Transparent to nearly opaque. Streak bluish. Luster vitreous, almost adamantine.—Brittle. H=3·5—4·5. Gr=3·5—3·85.

Composition: carbonic acid 25·6, oxyd of copper 69·2, water 5·2. Before the blowpipe and in acids, it acts like the preceding.

Obs. Azurite accompanies other ores of copper. At Chessy, France, its crystallizations are very splendid. It is found also in Siberia, in the Bannat, and near Redruth in Cornwall; at Phenixville, Pa., in crystals.

As incrustations and rarely as crystals, it occurs near Singsing, N. Y.; near New Brunswick, N. J. Also near Nicholson's Gap, in the Blue Ridge, Penn.

Uses. When abundant it is a valuable ore of copper. It makes a poor pigment, as it is liable to turn green.

CHRYSOCOLLA.—*Silicate of Copper.*

Usually as incrustations; botryoidal and massive. Also in thin seams and stains; no fibrous structure apparent, nor any appearance of crystallization.

Color bright green, bluish-green. Luster of surface of incrustations smoothly shining; also earthy. Translucent to opaque. H=2—3. Gr=2—2·3. *Composition:*—

	SIBERIAN.		NEW JERSEY.		
	V. Kobell.		Berthier.	Bowen.	Beck.
Oxyd of copper	40·0		55·1	45·2	42·6
Silica	36·5		35·4	37·3	40·0
Water	20·2		28·5	17·0	16·0
Carbonic acid	2·1	loss	1·0	—	—
Oxyd of iron	1·0		—	—	1·4

Describe blue malachite. How does it differ from green malachite in composition? What is the appearance of chrysocolla? its composition?

The mineral varies much in the proportion of its constituents, as it is not crystallized. It blackens in the inner flame of the blowpipe without melting. With borax it is partly reduced. No effervescence nor complete solution in nitric acid, cold or heated.

Dif. Distinguished from green malachite as stated under that species.

Obs. Accompanies other copper ores in Cornwall, Hungary, the Tyrol, Siberia, Thuringia, &c. In Chili it is abundant at the various mines. In Wisconsin and Missouri it is so abundant as to be worked for copper. It was formerly taken for green malachite. It also occurs at the Somerville and Schuyler's mine, N. J., at Morgantown, Penn., and Wolcotville, Conn.

Uses. This ore in the pure state affords 30 per cent. of copper; but as it occurs in the rock will hardly yield one-third this amount. Still when abundant, as it appears to be in the Mississippi valley, it is a valuable ore. It is easy of reduction by means of limestone as a flux.

Dioptase is another silicate of copper, occurring in rhombohedral crystals and hexagonal prisms. R : R=126° 24′. Color emerald-green. Luster vitreous. Streak greenish. Transparent to nearly opaque. H=5. Gr=3·28. From the Kirghese Steppes of Siberia.

Besides the above salts of copper, there are the following species, which are of little use in the arts.

Arsenates of Copper.—*Euchroite* has a bright emerald-green color, and contains 33 per cent. of arsenic acid, and 48 of oxyd of copper. Occurs in modified rhombic prisms H=3·75. Gr=3·4. From Libethen, in Hungary. *Aphanesite* is of a dark verdigris-green inclining to blue, and also dark blue, H=2·5—3. Gr=4·19. It contains 30 per cent. of arsenic acid and 54 of oxyd of copper. From Cornwall. *Erinite* has an emerald-green color, and occurs in mammilated coatings. H=4·5—5. Gr=4·04. Contains 33·8 of arsenic acid and 59·4 of oxyd of copper. From Limerick, Ireland. *Liroconite* varies from sky-blue to virdigris-green, It occurs in rhombic prisms, sometimes an inch broad. H=2·5. Gr=2·8—2·9. Contains 14 per cent. of arsenic acid, 49 of oxyd of copper. *Olivenite* presents olive-green to brownish colors, and occurs in prismatic crystals or velvety coatings. H=3. Gr=4·2. Contains 36·7 per cent. of arsenic acid, to 56·4 of oxyd of copper. *Copper Mica* is remarkable for its thin foliated or mica-like structure. The color is emerald or grass-green. H=2. Gr=2·55. It contains 21 per cent. of arsenic acid, 58 of oxyd of copper, and 21 of water. From Cornwall and Hungary. *Copper Froth* is another arsenate of a pale apple-green and verdigris green color. It

---

How does chrysocolla differ from green malachite? Where is it abundant in the U. States? What is its use?

has a perfect cleavage. It contains 25 per cent of arsenic acid, 43·9 of oxyd of copper, 17·5 of water, with 13·6 of carbonate of lime. From Hungary, Siberia, the Tyrol, and Derbyshire. *Condurrite* has a brownish-black or blue color. From Cornwall. These different arsenates of copper give an allicaceous odor when heated on charcoal before the blowpipe.

Phosphates of Copper.—Pseudo-malachite occurs in very oblique crystals, or massive and incrusting, and has an emerald or blackish-green color. H=4·5—5. Gr=4·2. Contains 68 per cent. of oxyd of copper. From near Bonn, on the Rhine, and also from Hungary. *Libethenite* has a dark or olive-green color, and occurs in prismatic crystals and massive. H=4. Gr=3·6—3·8. Contains 64 per cent. of oxyd of copper. From Hungary and Cornwall. *Thrombolite* is a green phospate occurring massive in Hungary. Contains 39 per cent. of oxyd of copper. These phosphates give no fumes before the blowpipe; and have the reaction of phosphoric acid.

Chlorid of Copper.—Atacamite. Color green to blackish-green. Luster adamantine to vitreous. Streak apple-green. Translucent to subtranslucent. Occurs in right rhombic prisms and rectangular octahedrons, also massive. Consists of oxyd of copper 76·6, muriatic acid 10·6, water 12·8. Gives off fumes of muriatic acid before the blowpipe and leaves a globule of copper. From the Atacama desert, between Chili and Peru, and elsewhere in Chili; also from Vesuvius and Saxony. It is ground up in Chili, and sold as a powder for letters under the name of arenillo.

A *Sulphato-chlorid of Copper* has been observed in Cornwall, in blue acicular crystals, apparently hexagonal.

Vanadate of lead and copper. Reported as occurring in Chili. Color dark brown or brownish black; texture earthy, looking like a ferruginous earth. Occurs with other ores of lead and copper.

Vanadate of copper. Massive and foliated, or pulverulent; folia citron-yellow, pearly. From the Ural.

Buratite. A hydrous carbonate of copper, zinc, and lime, occurring in bluish radiating needles. Gr=3·2. From Chessy, France; the Altai mountains; and Tuscany. Probably same as Aurichalcite.

Velvet Copper Ore. In velvety druses or coatings, consisting of short fine fibrous crystallization. Color fine smalt blue.

GENERAL REMARKS ON COPPER AND ITS ORES.

The metal copper has been known since the earliest periods. It is obtained for the arts mostly from pyritous copper, the gray sulphurets, and the carbonate; also to some extent from the black oxyd, and from solutions of the sulphate, (page 296.)

Assay of Ores. For the assay of copper ores by the *dry way*, the following is a common method. A portion of the prepared ore, roasted in a closed tube, will show by the garlic or sulphurous smell of the fumes, and by the depositions on the tube, whether arsenic, sulphur, or both, be the mineralizers. If this last is the case, which often happens, 100 or 1000 grains of the ore are to be mixed with one half of its weight of sawdust, then imbued with oil, and heated moderately in a

---

What is the mode of assaying copper ores in the *dry* way?

crucible, till all the arsenical fumes are dissipated. The residuum, being cooled and triturated, is to be exposed in a shallow earthen dish, made of refractory material, to a slow roasting heat, and stirred till the sulphur and charcoal are burned away; what remains being ground and mixed with half its weight of calcined borax, or carbonate of soda, one-twelfth its weight of lamp black, (finely pulverized charcoal will answer,) and next, made into a dough with a few drops of oil, is then to be pressed down into a crucible, which is to be covered with a luted lid, and subjected in a powerful air-furnace, first to a dull red heat, then to vivid ignition for seven to twenty minutes. On cooling and breaking the crucible, a button of metallic copper will be obtained, which may be refined by melting again with borax in an open crucible. Its color and malleability indicate pretty well the quality, as does its weight the relative value, of the ore. It may be cupelled with lead to ascertain if it contain silver or gold; or it may be treated for the same purpose with nitric acid.

If the blowpipe trial show no arsenic, the first calcination may be omitted; and if neither sulphur nor arsenic are present, a portion of the pulverized ore should be dried and treated directly with borax, lamp-black, and oil.

The ores of copper, (the sulphuret as well as the oxyds, carbonates, &c.) may be reduced in the *wet way*, by solution in strong nitric acid. The solution, if made from the sulphuret, will contain sulphuric acid and free sulphur, as well as all the bases, (iron, nickel, cobalt, lead, silver, &c.) which may have been present in the original ore. If silver is present it will be found as a heavy white curdy precipitate, at the bottom, if the nitric acid employed contained any hydrochloric acid; and if the addition of this acid to the solution occasions no such precipitate, no silver is present. If the solution is free from lead, antimony, arsenic, and other metals precipitable by sulphureted hydrogen, the copper may be thrown down as sulphuret by means of a current of this gas, the black precipitate, collected on a filter washed with water, and redissolved in aqua regia, largely diluted, and finally precipitated by caustic potash, which throws down the black oxyd of copper. This dried and weighed will yield the true value of the ore in metallic copper. If only iron and copper are present, (which may be previously determined by the blowpipe,) they may be separated from their solutions in nitric acid by ammonia, which throws down the iron as hydrated peroxyd, but redissolves the copper precipitated by the first additions of ammonia. The determination of the weight of the iron may then give the amount of copper by the difference of weight, or the copper may again be thrown down by potash as before directed.

Reduction of Ores. Copper ores are reduced in England in a reverberatory furnace, and the process consists in alternate calcinations and fusions. The volatile ingredients are carried off by the calcinations, and any metals in combination with the copper are oxydized. The fusions serve to get rid of the various impurities, and finally bring out the pure metal.

The calcinations or roastings are performed either in a furnace, or by making piles in the open air. In this latter mode, which is in use

---

What is the mode of assaying copper ores in the *wet way?* How are copper ores reduced? Describe the process of calcination?

on the continent of Europe, the ore, after being pounded and assorted, is piled up in high pyramidal mounds, which mounds are covered with mortar, sod, &c., and have a chimney at the center. Hemispherical cavities are dug on the upper surface for the purpose of receiving the sulphur during the roasting, which arrives liquified at the surface. This process lasts about six months. In England, at Swansea, where the ores are carried for reduction, the calcinations are performed more rapidly in a reverberatory furnace; and this is especially necessary when the ores do not contain a sufficient proportion of iron pyrites to furnish enough sulphur to sustain the combustion. After calcination, the ore is black and powdery. In the Swansea establishments, the calcined ore is introduced into the furnace, (a reverberatory smaller than that used for calcination,) and is spread over the bottom, 1 cwt. at a time. The heat is raised, and the furnace closed. When fusion has taken place, the liquid mass is well rabbled or stirred, so as to allow of the complete separation of the slags from the metal; afterwards the slags are skimmed off. Then a second charge is added, and after a similar process, a third charge, if the furnace is deep enough to receive it without the metal's flowing from the door. After the last charge is reduced also, the tap-hole is opened, and the metal flows out into water, where it is granulated. The slags if not free from metal are again returned to the furnace, when other charges are put in. This granulated metal is usually about one-third copper; it contains sulphur, copper, and iron.

This coarse metal is next calcined, just as the ore was first calcined; by which the iron is oxydized. The charge remains in the furnace 24 hours, and is repeatedly stirred and turned.

It is then transferred to the furnace for melting, and there melted along with some slags from the previous fusion. The sulphur reduces any oxyd and the whole fuses down. The slags are skimmed off and the furnace tapped: the metal is again drawn off into water. In this state it contains about 60 per cent. of copper, and it is called fine metal. The fine metal is then calcined like the coarse metal; and next it is melted as before. It results in a coarse copper containing 80 to 90 per cent. of pure metal.

The coarse copper is then roasted in the melting furnace; the air drawing in large quantities over the copper in incipient fusion, oxydizes the iron and the volatile substances are driven off. The metal is fused toward the end of the operation, which is continued from 12 to 24 hours, and is then tapped into sand beds. The pigs formed are covered with black blisters and they are cellular within. The copper is then remelted in a melting furnace; it is heated slowly to allow of any farther oxydizing that may be necessary. The slag is removed and the metal is examined from time to time, by taking out some of it, and when it is in the right condition, it is next subjected to the process of toughening. It is now brittle, of a deep red color inclining to purple, with an open grain and a crystalline structure; the copper in this state is what is termed *dry*. The surface of the melted metal is first covered with charcoal; a pole, commonly of birch, is held in the liquid matter, causing considerable ebullition; and this poling is continued, with occasional additions of charcoal, till it is found in the assays taken

---

What are the several steps in reduction?

out that the crystalline grain has disappeared, and the copper when cut through has a silky polished appearance, and the color is light red. It is then ladeled out into moulds, usually 12 inches in width by 18 long. Lead is sometimes added in the purification, to aid by its own oxydation in the oxydation of the iron present.

The process of melting copper on the continent is done by blast furnaces instead of the reverberatory, and they are said to be more economical in fuel, and produce a less waste of copper in the slags. This mode is used at the works at Boston, while the Swansea mode has been adopted at the Baltimore furnaces, Maryland. At the Ha'ford works, South Wales, a furnace of three tiers of hearths has been introduced, which answers the double purpose of calcination and fusion at the same time.

Galvanism has been turned to account in the reduction of copper ores. The ore is converted into a sulphate by roasting with the free access of the atmosphere. From this sulphate the copper is deposited in a pure state by galvanic decomposition. See on this subject American Journal of Science, ii ser., volume iv, p. 276, or Franklin Journal, volume xi, p. 128.

Copper Mines. The principal mines of copper in the world are those of Cornwall and Devon, England ; of the island of Cuba ; of Copiapo, and other places in Chili; Chessy, near Lyons, in France; in the Erzgebirge, Saxony; at Eisleben and Sangerhausen, in Prussia ; at Goslar, in the Lower Hartz; at Schemmitz, Kremnitz, Kapnik, and the Bannat, in Hungary ; at Fahlun, in Sweden ; at Turinsk and Nischni-Tagilsk, and other places in the Urals ; also in China and Japan. Lately extensive mines have been opened in Southern Australia.

In the United States, considerable quantities have been raised from the mines of New Jersey, and those of Simsbury, Conn. At Bristol, Conn., is a fine vein of vitreous copper, now under profitable exploration. The Hiwassee mine, Tennessee, and the mine at Corinth, Vermont, are at present productive.

The most extensive deposits are those of Northern Michigan, near L. Superior. The Michigan mines are vertical veins mostly in the trap rock which intersects a red sandstone, probably identical in age with the red sandstone of Connecticut and New Jersey. The first discoveries of copper ore in this place were made at Copper harbor, where the chrysocolla and carbonate occur. Near Fort Wilkins the black oxyd was afterwards found in a large deposit, and 40,000 pounds of this ore were shipped to Boston. On farther exploration in the trap, the Cliff mine, 25 miles to the westward, was laid open, where the largest masses of native copper have been found, and which still proves to be highly productive. Other veins have since been opened in various parts of the region, at Eagle harbor, Eagle river, Grand Marais, Lac La Belle, Agate Harbor, Torch Lake, on the Ontonagon, in the Porcupine mountains, and elsewhere. At Mineral Point, Wisconsin, a blue siliceous carbonate is found. Other mines are opened in Missouri. The country north of Lakes Superior and Huron, also afford copper ores.

What is the process of reduction on the continent of Europe? Where are the principal foreign mines of copper? Where is copper found in the United States?

In the Lake Superior Region, the four larger mines have afforded, according to Whitney:

	Cliff Mine.	Minnesota.	N. American.	Northwest.
1845,	8·88 tons.			
1846,	16·79			
1847,	183·38			
1848,	444·85	4·00		
1849,	572·38	34.00	22.90	15.32
1850,	319·04	66·00	76·20	87·06
1851,	377·89	165·00	76·50	130·89
1852,	370·25	208·50	22·90	120·17
1853,	415·00	480·19	112·32	102·27

The total yield from all the mines in 1853, was 1,296·94 tons.

Mr. Whitney observes that various attempts have been made to work the mines associated with the trappean rocks north of Lake Superior and Lake Huron, but as yet with little success. On Spar Island, and also on Michipicoten Island, veins have been opened which are abandoned. The Bruce Mine, on Lake Huron, is the only one in those regions now worked with profit. It is situated about fifty miles below Saut Ste. Marie, and due north of the extremity of St. Joseph's Island. The ore is chiefly copper pyrites, with some variegated copper ore. During the year 1853, 1650 tons of ore were shipped. The Wallace Mine, 16 miles from La Clocke, a station of the Hudson's Bay Company, is said to furnish copper pyrites, like Bruce's Mine, and also nickel and cobalt ores.*

The amount of copper produced by different mining countries in Europe is as follows:

Great Britain,	300,000 cwt.†	Denmark,	8,500 cwt.
Russia,	130,000 "	Prussia,	30,000 "
Austria,	60,000 "	France,	2,000 "
Germany,	12,000 "	Spain,	8,000 "
Sweden and Norway,	40,000 "		

Other countries afford in 1853 (Whitney's Met. Wealth):

Africa, ...	1,200 cwt.
Asia, ...	60,000 "
Australia and New Zealand, .	60,000 "
Chili, ...	280,000 "
South America, exclusive of Chili,	26,000 "
Cuba, ...	50,000 "
United States and Canada, . .	40,000 "

Making the total for the world of about 1,150,000 cwt., or 55,750 tons.

What three countries are most productive in copper? Where are the principal mines in the United States?

---

* Whitney's Metallic Wealth of the United States.

† 5-6ths of the whole from Cornwall.

What will be ultimately the proceeds of the copper region of Lake Superior, cannot now be fully determined. But there is every prospect that the country will prove boundless in its resources.

Uses. The metal copper was known in the earliest periods and was used mostly alloyed with tin, forming *bronze.* The mines of Nubia and Ethiopia are believed to have produced a great part of the copper of the early Egyptians. Eubæa and Cyprus are also mentioned as affording this metal to the Greeks. It was employed for cutting instruments and weapons, as well as for utensils; and bronze chisels are at this day found at the Egyptian stone-quarries, that were once employed in quarrying. This bronze, (*chalkos* of the Greeks, and *æs* of the Romans,) consisted of about 5 parts of copper to 1 of tin, a proportion which produces an alloy of maximum hardness. Nearly the same material was used in early times over Europe; and weapons and tools have been found consisting of copper, edged with iron, indicating the scarcity of the latter metal. Similar weapons have also been found in Britain; yet it is certain that iron and steel were well known to the Romans and later Greeks, and to some extent used for warlike weapons and cutlery.

Copper at the present day is very various in its applications in the arts. It is largely employed for utensils, for the sheathing of ships, and for coinage. Alloyed with zinc it constitutes brass, and with tin it forms bell-metal as well as bronze.

The best *brass* contains 2 parts of copper to 1 of zinc; the proportion of 4 of copper to 1 of zinc, makes a good brass. *Pinchbeck* contains 5 of copper to 1 of zinc; and *tombac* and *Dutch gold*, are other allied compounds. *Bath metal* consists of 9 of zinc to 32 of brass. A whitish metal used by the button-makers of Birmingham, and called *platina*, is made of 5 pounds of zinc to 8 of brass.

Bronze is an alloy of copper with 7 to 10 per cent. of tin. This is the material used for cannon. With 8 per cent. of tin, it is the bronze for medals. With 20 of tin, the material for cymbals. With 30 to 33 parts of tin, it forms *speculum metal*, of which the mirrors for optical instruments are made. Lord Rosse used for the speculum of his great telescope, 126 parts of copper to 57½ parts of tin.

The brothers Keller, celebrated for their *statue castings*, used a metal consisting of 91·4 per cent. of copper, 5·53 of zinc, 1·7 of tin, and 1·37 of lead. An equestrian statue of Louis XIV, 21 feet high, and weighing 53,263 French pounds, was cast by them in 1699, at a single jet.

Bell-metal is made of copper, with a third to a fifth as much tin by weight, the proportion of tin varying according to the size of the bell and sound required. The *Chinese gong* contains 80 parts of copper to 20 or 25 of tin; to give it its full sonorousness, it must be heated and suddenly cooled in cold water.

Sheet copper is made by heating the copper in a furnace and rolling it between iron rollers. Copper is also worked by forging and casting. In casting, it will not bear over a red heat without burning.

How did the ancients use copper? What is the proportion of alloy in the ancient bronze?

2. NOBLE METALS.

1. PLATINUM.—IRIDIUM.—PALLADIUM.

NATIVE PLATINUM.

In flattened or angular grains or irregular masses, the masses occasionally large. Crystalline form cubic, but seldom observed. Cleavage none

Color and streak pale or dark steel-gray. Luster metallic, shining. Ductile and malleable. H=4—4·5. Gr=16—19.

Composition. Platinum is usually combined with more or less of the rare metals Iridium, Rhodium, Palladium, and Osmium, besides copper and iron, which give it a darker color than belongs to the pure metal, and increase its hardness. A Russian specimen afforded, platinum 78·9, iridium 5·0, osmium and iridium 1·9, rhodium 0·9, palladium 0·3, copper 0·7, iron 11·0=98·75.

Platinum is soluble in heated aqua regia. It is one of the most infusible substances known, being wholly unaltered before the blowpipe. It is very slightly magnetic, and this quality is increased by the iron it may contain.

Dif. Platinum is at once distinguished by its malleability and extreme infusibility.

Obs. Platinum was first detected in grains in the alluvial deposits of Choco and Barbaçoa in South America, where it received the name *platina*, derived from the word *plata*, meaning *silver*. Although before known, an account by Ulloa, a Spanish traveller in America in 1735, directed attention in Europe, in 1748, to the metal. It has since been found in the Urals, on Borneo, in the sands of the Rhine, and in those of the river Jocky, St. Domingo; and recently traces have been observed in the United States, in North Carolina.

The Ural localities of Nischne Tagilsk, and Goroblagodat, have afforded much the larger part of the platinum of commerce. It occurs, as elsewhere, in alluvial beds; but the courses of platiniferous alluvium have been traced to a great extent up Mount La Martiane, which consists of crystalline

---

What is the condition and appearance of native platinum? What is said of its crystallization? What is its specific gravity? With what is it usually combined? Where and when was it first found? Where else does it occur?

rocks, and is the origin of the detritus. *One* to *three* pounds are procured from 3700 pounds of sand.

Though commonly in small grains, masses of considerable size have occasionally been found. A mass weighing 1088 grains was brought by Humboldt from South America and deposited in the Berlin Museum. Its specific gravity was 18·94. In the year 1822, a mass from Condoto was deposited in the Madrid museum, measuring 2 inches and 4 lines in diameter, and weighing 11,641 grains. A more remarkable specimen was found in the year 1827 in the Urals, not far from the Demidoff mines, which weighed 11½ (more accurately, 11·57) pounds troy; and similar masses are now not uncommon. The largest yet discovered weighed 21 pounds troy; it is in the Demidoff cabinet.

Russia affords annually about 80 cwt. of platinum, which is nearly ten times the amount from Brazil, Columbia, St. Domingo, and Borneo. Borneo affords six or eight hundred pounds per year.

The North Carolina platinum was found with gold in Rutherford county. It was a single reniform granule, weighing 2·54 grains. Other instances are reported from the southern gold region, and from Point Orford and elsewhere in California.

Uses. The infusibility of platinum and its resistance to the action of the air, and moisture and most chemical agents, renders it of great value for the construction of chemical and philosophical apparatus. The large vessels employed in the concentration of sulphuric acid are now made of platinum, as it is unaffected by this corrosive acid. It is also used for crucibles and capsules in chemical analysis; for galvanic batteries; as foil or worked into cups or forceps for supporting objects before the blowpipe. It alloys readily when heated with iron, lead, and several of the metals, and is also attacked by caustic potash, and phosphoric acid, in contact with carbon; and consequently there should be caution when heating it not to expose it to these agents.

It is employed for coating copper and brass; also for painting porcelain and giving it a steel luster, formerly highly prized. It admits of being drawn into wire of extreme tenuity: Dr. Wollaston obtained a wire not exceeding a two-thousandth of an inch in diameter.

Platinum is coined in Russia, but is not a legal tender.

What are the uses of platinum?

The coins have the value of 11 and 22 rubles each. The amount coined from 1826 to 1844 equals 2½ millions of dollars.

For many years after its discovery, platinum was almost a useless metal on account of the difficulty of obtaining it in masses. The grains weld when heated, but because of their small size, this was interminable labor, and moreover the metal was not pure. Dr. Wollaston introduced the process now in use, which consists in dissolving the metal in nitro-muriatic acid, and throwing down from the solution an orange precipitate by means of muriate of ammonia. This precipitate (a double chlorid of platinum and ammonium) is then heated and thus reduced to the metallic state ; the platinum is now in an extremely minute state of division. This black powder ("spongy platinum") is next compressed in steel moulds by the aid of heat and strong pressure ; and when sufficiently compact, is forged under the hammer and then reduced at last to solid masses.

This metal fuses readily before the "compound blowpipe ;" and Dr. Hare succeeded in 1837 in melting twenty-eight ounces into one mass.* The metal was almost as malleable and as good for working as that obtained by the other process ; it had a specific gravity of 19·8. He afterwards succeeded in obtaining from the ore masses which were 90 per cent. platinum, and as malleable as the metal in ordinary use, though somewhat more liable to tarnish, owing to some of its impurities.

Platin-iridium. Grains of iridium have been obtained at Nischne Tagilsk, consisting of 76·8 iridium, and 19·64 platinum, with some palladium and copper. A similar platin-iridium has been obtained at Ava in the East Indies. Another from Brazil contained 27·8 iridium, 55·5 platinum, and 6·9 of rhodium.

Iridosmine. A compound of iridium and osmium from the platinum mines of Russia, South America, the East Indies and California. The crystals are pale steel-gray hexagonal prisms: occurs usually in flat grains. H=6·7. Gr=19·5—21·1. Malleable with difficulty.

The *composition* varies. One variety contains iridium 46·8, osmium 49·3, rhodium 3·2, iron 0·7. Another, iridium 25·1, osmium 74·9 ; another, iridium 20, osmium 80. They are distinguished by their superior hardness from the grains of platinum, and also by the peculiar odor of osmium when heated with niter. Iridosmine is common with the gold of California, and injures its quality for jewelry. It is proposed to separate it by keeping the gold melted for a short time, to allow the grains of iridosmine to settle.

What is the value of Russian platinum coins? How is platinum worked into masses?

---

* American Jour. Sci., xxxiii, 195 ; xxxvlii, 155, 163, and ii ser. iv, 39.

The metal iridium is extremely hard, and is used as well as rhodium for nibs to gold pens. Its specific gravity is 21·8. Rhodium (1 to 2 per cent.) gives great hardness to steel, and would be a useful metal were it more abundant.

NATIVE PALLADIUM.

In regular octahedrons. Also in hexagonal tables. Occurs mostly in grains, apparently composed of divergent fibers. Color steel-gray, inclining to silver-white. Ductile and malleable. H. above 4·5. Gr=11·8—12·2.

Consists of palladium, with some platinum and iridium. Fuses with sulphur, but not alone.

Obs. Occurs in Brazil with gold, and is distinguished from platinum with which it is associated by the divergent structure of its grains. *Selenpalladite* is nothing but the native palladium.

Uses. This metal is malleable, and when polished has a splendid steel-like luster which does not tarnish. A cup weighing 3¼ pounds was made by M. Breant in the mint at Paris, and is now in the *garde-meuble* of the French crown. In hardness it is equal to fine steel. 1 part fused with 6 of gold forms a white alloy; and this compound was employed, at the suggestion of Dr. Wollaston, for the graduated part of the mural circle, constructed by Troughton for the Royal Observatory at Greenwich. Palladium has been employed also for certain surgical intruments.

Quite large masses of the metal palladium are brought from Brazil. It is extracted from the auriferous sands by first fusing it with silver, and consequently forming a quaternary alloy of gold, palladium, silver and copper, which is granulated by projecting it into water. By means of nitric acid all but the gold is dissolved; and from the solution, the silver is first precipitated by common salt as an insoluble chlorid, and then, after separating the chlorid, the palladium and copper are precipitated by plates of zinc. This precipitate is redissolved in nitric acid, an excess of ammonia added, and then hydrochloric acid sufficient to saturate; a double chlorid of palladium and ammonia is deposited as a crystalline yellow powder, which on calcination produces spongy palladium.

Describe native palladium? Where and how does it occur? How is it used?

2. GOLD.

Gold occurs mostly native, being either pure or alloyed with silver and other metals. It is occasially found mineralized by tellurium.

NATIVE GOLD.

Monometric. In cubes, without cleavage. Also in grains, thin laminæ and masses ; sometimes filiform or reticulated.

Color various shades of gold-yellow ; occasionally nearly silver-white, from the silver present. Very ductile and malleable. H=2·5—3. Gr=12—20, varying according to the metals alloyed with the gold.

Composition. Native gold usually contains silver, and in very various proportions. The finest native gold from Russia yielded gold 98·96, silver 0·16, copper 0·35, iron 0·05 ; Gr=19·099. A gold from Marmato afforded only 73·45 per cent. of gold, with 26·48 per cent. of silver ; Gr=12·666. This last is in the proportion of 3 of gold to 1 of silver. The following proportions also have been observed : 3½ to 1 ; 5 to 1 ; 6 to 1 ; 8 to 1, and this is the most common ; 12 to 1, also of frequent occurrence.

Copper is often found in alloy with gold, and also palladium and rhodium. A *rhodium-gold* from Mexico gave the specific gravity 15·5—16·8, and contained 34 to 43 per cent. of rhodium.

Dif. Iron and copper pyrites are often mistaken for gold by those inexperienced in ores. Gold is at once distinguished by being easily cut in slices and flattening under a hammer. The pyrites when pounded are reduced to powder ; iron pyrites is too hard to yield at all to a knife, and copper pyrites affords a dull greenish powder. Moreover, the pyrites give off sulphur when strongly heated, while gold melts without any such odor.

Obs. Native gold is mostly confined to those subcrystalline slaty or schistose rocks that abound in quartz veins, and more especially to talcose and chloritic slates. It occurs

---

In what condition does gold occur in nature ? What is the crystallization of native gold ? What are its common forms in the rocks ? Mention its characters. With what is it alloyed ? How is gold distinguished from iron and copper pyrites ? From what rocks is gold obtained ?

sparingly in granite, gneiss or mica slate, for the veins of these more highly crystalline rocks are commonly feldspathic or granitic rather than quartzose, and granitic veins seldom afford gold. The quartz veins often intersect the slaty rocks in great numbers, are generally very irregular in size, and often lie as beds conformable to the lamination. The quartz is frequently rather cellular, containing cavities in which it is crystallized. It generally contains more or less pyrites, and sometimes galena and other minerals. The decomposition of the pyrites leaves the quartz very cavernous, and somewhat rusty in appearance; and occasionally a little sulphur lines the cavities, derived from the removed pyrites. The rock in this cavernous state, (as it is very liable to be near the surface,) is rather easily quarried out; but deep below, where the minerals are not removed by decomposition, mining is far more difficult.

The pyrites itself is nearly as hard as quartz, when unaltered, and readily strikes fire with a steel. This pyrites is often very abundant, and contains throughout it considerable gold; but the gold is so finely distributed, that little of it can be removed by the ordinary process of crushing and amalgamation, and nature's way of decomposing the pyrites and thereby making it drop its load, is the only effectual one. This is accomplished by exposing the pyrites in heaps, with moisture and perhaps a little heat, by which it changes to copperas, which may be dissolved and the gold obtained. The galena of a gold region is also usually auriferous.

Gold sometimes occurs in the slate rocks adjoining the veins, though mostly confined to the latter. The quartz may contain gold when none is visible to the naked eye.

The minerals most common in gold regions are platinum, iridosmine, magnetic or titanic iron, iron pyrites, galena, copper pyrites, blende, tetradymite, zircon, rutile, heavy spar; also in some cases, brookite, monazite and diamond. Platinum and iridosmine accompany the gold of the Urals, Brazil and California; and diamonds are found in the gold region of Brazil, and occasionally in the Urals and Eastern United States.

Gold is widely distributed over the globe. It occurs in Brazil (where formerly a greater part of that used was obtained) along the chain of mountains which runs nearly parallel with the coast, especially near Villa Rica, and in the province of Minas Geraes; in New Grenada at Antioquia, Choco, and Giron; in Chili; sparingly in Peru and Mexico; in the southern of the United States. In Europe, it is most abundant in Hungary at Konigsberg, Schemnitz and Felsobanya, and in Transylvania at Kapnik, Vöröspatak and Offenbanya; it occurs also in the sands of the Rhine, the Reuss and the Aar; on the southern slope of the Pennine Alps from the Simplon and Monte Rosa to the valley of Aosta; in Piedmont; in Spain, formerly worked in Asturias; in the county of Wicklow, Ireland; in Sweden at Edelfors.

In the Urals are valuable mines at Beresof, and other places on the eastern or Asiatic flank of this range, and the comparatively level portions of Siberia; also in the Altai mountains. Also in the Cailas mountains in Little Thibet.

There are mines in Africa at Kordofan, between Darfour and Abyssinia; also south of Sahara in the western part of Africa, from the Senegal to Cape Palmas; also along the coast opposite Madagascar, between the 22d and 35th degrees south latitude, supposed to have been the *Ophir* of the time of Solomon. Other regions are China, Japan, Formosa, Ceylon, Java, Sumatra, western coast of Borneo, the Philippines, Australia, Van Diemen's land and New Zealand.

The present total yield of the gold mines of the world is not less than 195 tons. Much the larger part of this (about 175 tons) comes from Asiatic Russia, South America, Australia and California.

The Russian mines till recently ranked first in productiveness. They are principally alluvial washings, and these washings seldom yield more than 65 grains of gold for 4000 pounds of soil; never more than 120 grains. The alluvium is generally most productive where the loose material is most ferruginous. The mines of Ekaterinenberg are in the parent rock—a quartz constituting veins in a half decomposed granite called "beresite," which is connected with talcose and chloritic schists. The shafts are sunk vertically in the beresite, seldom below 25 feet, and from them lateral galleries are run to the veins. These mines afforded between the years 1725 and 1841, 679 poods of gold, or about 30,000 pounds troy. The whole of the Russian mines yielded in 1842, 970 poods of gold, or 42,000 pounds troy, half

---

What is said of the distribution of gold over the globe? What countries afford the greatest part of the gold of commerce?

of which was from Siberia, east of the Urals. In 1843, the yield was nearly 60,000 pounds troy, or about $13,000,000; in 1845 it amounted to $13,250,000; and in 1846, to 1722·7·16 poods, equal to 75,353 troy pounds, and $16,500,000.*

At the Transylvania mines of Vöröspatak, the gold is obtained by mining, and these mines have been worked since the time of the Romans.

The annual yield of Europe, exclusive of Russia, is not above $1,000,000. Austria afforded in 1844, 6785 marks. The sands of the Rhone, Rhine and Danube contain gold in small quantities. The Rhine has been most productive between Bale and Manheim; but at present only $9000 are extracted annually. The sands of the richest quality contain only about 56 parts of gold in a hundred millions; sands containing less than half this proportion are worked. The whole amount of gold in the auriferous sand of the Rhine is estimated at $30,000,000, but it is mostly covered by soil under cultivation.

Africa yields annually at least 4500 pounds troy, ($850,000,) and Southern Asia and the East Indies 25,000 pounds.

The mines of South America and Mexico were estimated by Humboldt to yield annually about $11,500,000; but the amount is now not over $10,000,000. Brazil of late has furnished about 6000 pounds troy; New Grenada, etc., 15,000; Peru 1900; Bolivia 1200; Chili 3000; in all for South America 27,100 pounds. Mexico yields about 10,000 pounds annually. It is estimated that between 1790 and 1830, Mexico produced $31,250,000 in gold, Chili $13,450,000, and Buenos Ayres $19,500,000, making an average annual yield of $16,050,000.

The whole product of Europe, Asia, Africa and South America, is not far from 125,000 pounds troy, annually; and this is far less than is derived at the present time from either Australia or the United States.

The gold mines of Australia afford at this time about

What amount was furnished by Russia in 1846? What is the annual yield of the other mines of Europe?

---

* The value of gold, silver and platinum, coined in Russia from 1644 to 1844, at present rates equals 545,360,317 silver rubles, or 409,020,000 dollars, in addition to which, during the same period, the value of 37,500,000 dollars in copper was coined.

250,000 pounds. These mines occur in eastern and south-eastern Australia, about the mountains called the Australian Alps, and their continuation north beyond the Blue Mountains. They were first made known to the world in 1851. The localities discovered were on Summer Hill Creek and the Lewis Pond River, (near lat. 33° N., long. 149°—150° E.,) streams which run from the northern flank of the Coriobolas down to the river Macquarie, a river flowing westward and northward. It was afterwards found on the Turon river, which rises in the Blue Mountains; and finally a region of country 1000 miles in length, north and south, was proved to be auriferous. The country is a region of metamorphic rocks, granite and slates, and in some parts abounds in quartz veins. The gold has been obtained mainly from alluvial washings.

Van Diemen's Land or Tasmania, and New Zealand, also afford the precious metal.

The mines of California yield per year about 200,000 pounds troy, or $50,000,000. The first discovery was made early in the spring of 1848, on the American Fork, a tributary to the Sacramento, near the mouth of which Sutter's establishment was situated. Soon Feather river, another affluent, 18 or 20 miles north, was also proved to abound in gold about its upper portions; and it was not long after before each stream in succession, north and south, along the western slope of the Sierra Nevada was found to flow over auriferous sands. The gold as now developed extends along that chain, through the whole length of the great north and south valley which holds the rivers and plains of the Sacramento and San Joaquin. It continues south nearly to the Tejon pass, in latitude 35°, and north beyond the Shasty mountains to the Umpqua, and less productively into Oregon and Washington territories. Gold also occurs in some places in the coast range of mountains. Even the very site of San Francisco has been found to contain traces. Beyond the Shasty mountains there are important mines on the Klamath and the Umpqua, and some of the best on the sea-shore between Gold Bluff, in 41° 30′ south of the Klamath (30 miles south of Crescent City) to the Umpqua. What once was Rogue river is now called Gold river.

The gold of the Sierra Nevada occurs mainly about the upper parts of the tributaries to the Sacramento and other rivers, rather high up among the mountains; and not only along the streams where the torrents perform annually the washing process, but also in the gravelly material or drift that covers the country, and over the slopes of the valley. At places along the valley where the descending waters meet an obstacle or a projecting rock, both in the river bed and on the declivities, "pockets" of gold are found. Certain layers of the drift are especially rich in the metal. This drift material is explored by turning the streams across it by artificial channels, where nature has not prepared the way, and thus the gold is separated and gathered.

The gold is mostly in thin scales or grains, usually of quite small size, and sometimes in plates or lumps; occasionally in masses of fifteen or twenty pounds, mixed more or less with quartz. Each region is generally distinguished by some peculiarity in the form or size of the scales, or their color, the lighter colored containing the most silver. Some of the plates are beautifully crystallized in dentritic or plumose forms made of united crystals. A few simple crystals of large size have been found. The annexed figure represents one of natural size, figured and described by Mr. Alger, of Boston.

The gold of the alluvial washings, as in other cases, has been derived from gold-bearing rocks. By some long action of denudation, those rocks have been extensively worn down to gravel and sand, and the gold is distributed along the water courses or on the slopes. As the metal is so very heavy—seven times heavier than the gravel—it has mostly been dropped by the waters high up the streams. The smaller scales have been carried farthest away, and no doubt minute traces exist throughout the Sacramento valley. The forms of the scales have arisen partly from the original lamellar form in the rocks, and partly from the process of wear.

Quartz veins, rich in gold, have been found in many parts of the country, and great efforts have been made to work them, especially in Nevada, Tuolumne, and Placer counties. For a knowledge of particular localities in California, see an article by W. P. Blake, in the American Journal of Science, volume xx, page 72, 1855.

Other gold mines exist in Lower California, the Great Basin and New Mexico.

The gold mines of the Eastern United States have produced of late less than a million of dollars. They are mostly confined to the states of Virginia, North and South Carolina, and Georgia, or along a line from the Rappahannock to the Coosa in Alabama. But the region may be said to extend north to Canada; for gold has been found at Canaan, N. H., Bridgewater, Vt., Dedham, Mass., Albion and Madrid in Maine, and on the Chaudiére river and elsewhere in Canada. Gold also occurs in Arkansas and Texas.

In *Virginia*, the principal deposits are in Spotsylvania county, on the Rappahannock, at the United States mines and at other places to the southwest; in Stafford county, at the Rappahannock gold mines, ten miles from Falmouth; in Culpepper county, at the Culpepper mines, on Rapidan river: in Orange county, at the Orange grove gold mine, and at the Greenwood gold mines; in Goochland county, at Moss and Busby's mines; in Louisa county, at Walton's gold mine; in Buckingham county, at Eldridge's mine. In *North Carolina*, the gold region is mostly confined to the counties of Montgomery, Cabarrus, Mecklenberg and Lincoln, which are situated about in a line running N. E. and S. W., parallel nearly with the coast. The mines at Mecklenburg are principally vein deposits; those of Burke, Lincoln, McDowell and Rutherford, are mostly in alluvial soil. In *Georgia*, gold mines occur in Habersham county, and at many places in Rabun and Hall counties, and the Cherokee country. In *South Carolina*, the gold regions are the Fairforest in Union district, and the Lynch's creek and Catawba regions, chiefly in Lancaster and Chesterfield districts; also in Pickens county, adjoining Georgia. The only mine not deserted is the Dorn mine in the Abbeville district. There is gold also in eastern Tennessee.

Viewing the gold region of the eastern United States as a whole, it is perceived that it ranges along the Appalachians, particularly the Eastern slope, from Maine to Alabama, having nearly a northeast and southwest course.

Masses of gold of considerable size have been found in North Carolina. The largest was discovered in Cabarrus county; it weighed twenty-eight pounds avoirdupois, ("steelyard weight," equals 37 lbs. troy,) and was 8 or 9 inches long by 4 or 5 broad, and about an inch thick. In Paraguay, pieces from 1 to 50 pounds weight were taken from a mass of rock which fell from one of the highest mountains. Several specimens weighing 16 pounds have been found in the Ural, and one of 27 pounds: and in the valley of Taschku-Targanka, in 1842, a mass was detached weighing very nearly 100 pounds troy. This mass is now in the museum of the Institute of Mining Engineers at St. Petersburg.

The largest mass yet discovered in any part of the world, is one from California, weighing 134 pounds 7 ounces, and affording 109 pounds 11 ounces of pure gold: it sold for £5,532. Another of 27½ pounds, here figured,

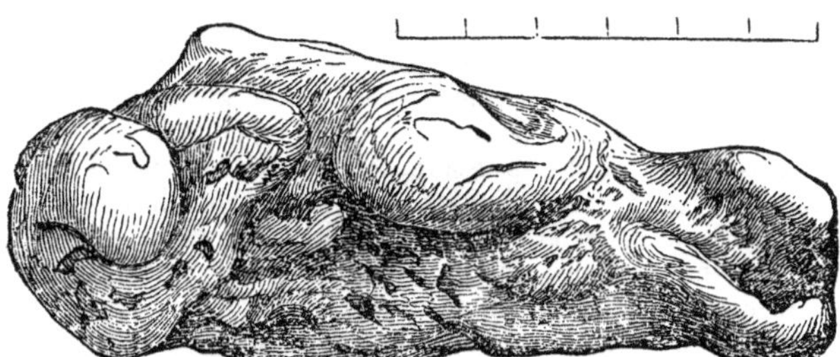

was found at Forest Creek, Mount Alexander, in the colony of Victoria. It was 11 inches long and 5 in breadth at its broadest part.

The origin of gold veins, or rather of the gold in the veins, is little understood. The rocks, as has been stated, are metamorphic slates that have been crystallized by heat; and they are the talcose and argillaceous, that have been but imperfectly crystallized, rather than the mica schist and gneiss which are well crystallized: and the veins of quartz which contain the gold, occupy fissures through the slates, and openings among the layers, which must have been made when the metamorphic change or crystallization took place. It was a period, for each gold region, of long continued heat, (occupying, probably, a prolonged age,) and also of vast up-

---

What is said of the gold rock of the United States?

liftings and disturbances of the beds; for the beds are tilted at various angles, and the veins show where were the fractures of the layers, or the separations and gapings of the tortured strata. The heat appears not to have been of the intensity required for the better crystallization of the more perfectly crystalline schists. The quartz veins could not have been filled from below, by injection,—a view not now accepted for the generality of mineral veins. They must have been filled either laterally or from above. In all such conditions of continued heat beneath an ocean, the hot water would dissolve silica freely within the rocks or from them, (as happens at the Geysirs of Iceland and elsewhere,) so that the region would become one of hot siliceous solutions permeating and overlying the upheaving strata. Thus silica would be free to consolidate or metamorphose the strata, and to fill up all rents or openings, whether they were no thicker than a sheet of paper, or rods in width. The waters would work laterally into these fissures, as this would be the tendency of the internal flow or movement, and they would carry mineral material of various kinds with them; besides, the superficial waters might deposit what mineral matter they contained along with the silica; and at the same time vapors might rise from below along the lines of rents, and be still a third source of metallic or mineral material. Between these methods appears to lie the process by which the gold was introduced into the quartz veins, and it remains for further research to ascertain the particular facts in the case. The pyrites formed in the veins is usually auriferous, showing that they were crystallized under the same circumstances as the depositing of the gold in strings, crystals and grains. Murchison has stated, that in the Urals the gold diminishes on descending in a vein; but this is not yet regarded as an established truth. The time when the gold veins were formed may differ in different regions. Along our eastern coast it appears to have been after the coal period.

An *examination of a gold rock* for gold is a simple process. The rock is first pounded up fine and sifted; a certain quantity of the sand thus obtained is washed in a shallow iron pan, and as the gold sinks, the material above is allowed to pass off into some receptacle. The largest part of the gold is thus left in the angle of the pan; by a repetition of the process a further portion is obtained; and when the bulk the bulk of sand is thus reduced to a manageable quantity, the gold is amalgamated with clean mercury; the amalgam is next strained to separate any excess of mercury, and finally is heated and the mercury expelled, leaving the gold. In this way by successive trials with the rock, the proportion of gold is quite accurately ascertained. It is the same process used with the larger washings, though on a small scale. Mercury unites readily with gold, and thus separates it from any associated rock or sand; and it is employed in all extensive gold minings, though much gold may be often obtained by simple washing without amalgamation.

The operation of hand washing is called in Virginia *panning*. With a small iron pan, they wash the earth in a tub or in some brook, and thus extract much gold from the gravel or soil, which is said to *pan well* or *pan poorly* according to the result. Masses of quartz, with no external indications of gold, examined in the above way at a Virginia mine, afforded an average of more than eight dollars to the bushel of gold rock.

When gold is alloyed with copper or silver, the mode of assay for separating the copper depends on the process of cupellation; and that for separating the silver, on the power of nitric acid to dissolve silver without acting on the gold.

The *process of cupellation* consists in heating the assay in a small cup (called a cupel,) made of bone ashes, (or in a cavity containing bone ashes,) *while the atmosphere has free access*. The heated metal is oxydated by the air passing over it, and the oxyd formed sinks into the porous cup, leaving the precious metal behind. The shape of the cupel is shown in fig. 1. In order to fuse the alloy and still have the atmosphere circulating over it, the cupel is placed in a small oven-shaped vessel, called a muffle (fig. 2:) it is of infusible stone ware, and has a number of oblong holes, through which to admit the flame from the fire, and give exit to the atmosphere which passes into it. The muffle is inserted in a hole fitting it in the side of a vertical furnace, with the open mouth out-

1

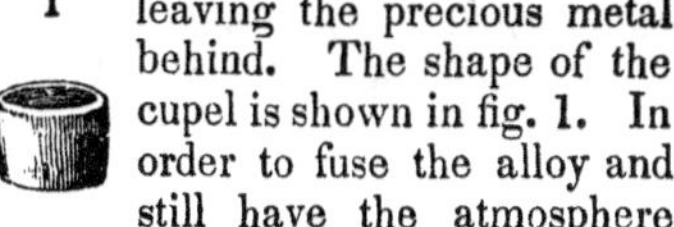

2

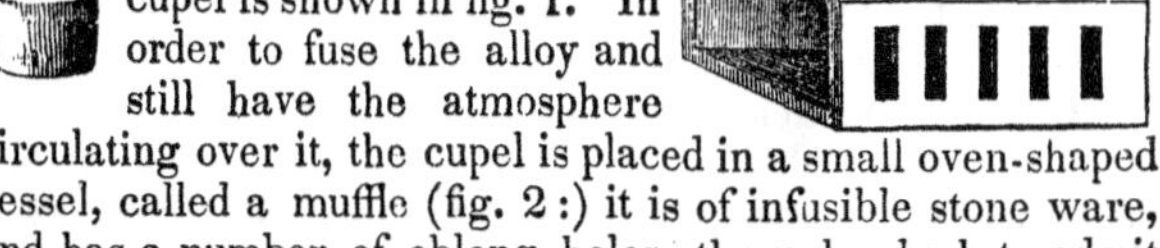

How is a rock examined for gold? What are the processes for separating gold from silver or copper? Describe the process of cupellation.

ward and even nearly with the exterior surface of the furnace. The fire is made within the furnace, below, around, and above; and after heating up, the cupel is put in the muffle with the assay in its shallow cup-shaped cavity. It thus has the heat of the furnace to fuse the assay, and the air at the same time is drawn in over it through the large opening of the muffle. The oxygen of the atmosphere unites with the lead of the assay, and produces an oxyd, which oxyd sinks into the cupel, leaving the silver or gold behind. The completion of the process is at once known by the change of the assay suddenly to a bright shining globule.

In the cupellation of gold containing copper, lead is melted with the assay. The lead on being fused in a draft of air oxydizes, and also promotes the oxydation of the copper, and both oxyds disappear in the pores of the cupel leaving the gold behind, and the silver alloyed with it. In this process the gold is melted with three times its weight of silver, (a *quartation* as it is termed, the gold being one part out of four of the alloy,) in order by its diffusion to effect a more complete removal of the silver as well as the contained copper. The cupel is placed in the heated furnace, and the gold, silver, and lead, on the cupel; the heat is continued until the surface of the metal is quiet and bright, when the cupellation is finished; the metal then is slowly cooled and removed. The button obtained, after annealing it by bringing it to a red heat, is rolled out into a thin plate and boiled in strong nitric acid. This process is repeated two or three times with a change of the acid each time, and the silver is thus finally removed. At the United States mint, half a gramme of the gold is submitted to assay. The assay-gold and quartation-silver are wrapped in a sheet of lead weighing about ten times as much as the gold under assay. After cupellation, the plate of gold and silver, loosely rolled into a coil, is boiled for 20 minutes in $4\frac{1}{4}$ oz. of nitric acid, of 20 to 22° Beaumé; the acid is then poured off and another portion of stronger acid is added, about half the former quantity, and boiled 10 minutes; then the same again. The gold thus purified is washed and exposed to a red heat, for the purpose of drying and annealing it, and then weighed.

Uses. The uses of gold are well known; and also that it owes a great part of its value to its extreme malleability, and the fact of its not tarnishing on exposure. Although a costly metal, it is one of the cheapest means of ornament, on account of the thinness of the leaves into which it is beaten. A grain of the metal may be made to cover 56¾ square inches of surface, and the thinnest leaf is but 1-280,000th of an inch thick.

Perfectly pure gold is denominated gold of 24 *carats*, or *fine* gold. If it contains 22 parts of pure gold to 2 of silver, or to 1 of copper and 1 of silver, it is said to be 22 carats fine; so also for 20 carats fine, it contains 20 parts of pure gold. The carat is divided into $\frac{1}{4}$, $\frac{1}{8}$, $\frac{1}{16}$, $\frac{1}{32}$ parts, for a more minute specification of the quality of gold.

The standard gold of the United States consists of 900 parts of gold to 100 of an alloy of copper and silver. The eagle (10 dollars) contains 232 grains of fine gold.

Aurotellurite, called also *Sylvanite*, is a grayish or silver-white mineral, containing gold combined with Tellurium.

3. SILVER.

Silver occurs native and alloyed; also mineralized with sulphur, selenium, arsenic, chlorine, bromine, or iodine, and in combination with different acids.

The ores of silver fuse easily and decompose before the blowpipe, affording a globule of silver either alone or with soda; the globule is known to be silver by its flattening out readily under a hammer, and also by its sectility. The species vary in specific gravity from 5·5 to 10·5.

NATIVE SILVER.

Monometric. In octahedrons. No cleavage apparent. Occurs often in filiform and arborescent shapes, the threads having a crystalline character; also in laminæ.

Color and streak silver-white and shining. Sectile. Malleable. H=2·5—3. G.=10·3—10·5.

Composition: native silver is usually an alloy of silver and copper, the latter ingredient often amounting to 10 per cent. It is also alloyed with gold, as mentioned under that metal. A *bismuth silver* from Copiapo, S. A., contained 16 per cent. of bismuth.

What surface may a grain of gold be made to cover? How much pure gold is there in the American eagle? What is the use of the term carat? What is the condition of silver in nature? Describe native silver.

Before the blowpipe it fuses easily and affords a globule which becomes angular on cooling. Dissolves in nitric acid, from which it is precipitated by putting in a clean piece of copper.

Dif. Distinguished by being malleable ; from bismuth and other white native metals by affording no fumes before the blowpipe ; by affording a solution with muriatic acid, which becomes black on exposure.

Obs. Native silver occurs in masses and string-like arborescences, penetrating rocks, and is found in igneous rocks and in sedimentary strata, in the vicinity of dikes of trap and porphyry.

The mines of Norway, at Kongsberg, formerly afforded magnificent specimens of native silver, but they are now mostly under water. One specimen from this locality, at Copenhagen, weighs five hundred pounds. Other European localities are in Saxony, Bohemia, the Hartz, Hungary, Dauphiny. Peru and Mexico also afford native silver. A Mexican specimen from Batopilas, weighed when obtained, 400 pounds; and one from Southern Peru, (mines of Huantajaya,) weighed over 8 cwt. In the United States, elegant specimens are associated with the native copper of Lake Superior. The silver generally penetrates the copper in masses and strings, and is very nearly pure, notwithstanding the copper about it.

Much of the galena of the west contains a very small per centage of silver, and that of Monroe, Conn., yields nearly 3 per cent.

Native silver has also been observed near the Sing Sing state prison ; at the Bridgewater copper mines, N. J. ; and in handsome specimens at King's mine, Davidson county, North Carolina.

Uses. The uses of silver are, for the manufacture of various articles of luxury, for plating other metals, for philosophical instruments, for coinage, and also various purposes in the arts. For coins, it is alloyed in this country with copper, and is thus rendered harder and more durable ; 1000 parts of the coin contains 100 parts of copper. When this alloy is boiled with a solution of cream of tartar and sea-salt, or scrubbed with water of ammonia, the superficial

---

How is native silver distinguished ? How does it occur and in what rocks ? Where does silver occur in the U. States, and how ? What are the uses of silver ?

particles of copper are removed, and a surface of fine silver is left. Silver is much less malleable than gold, and cannot be beaten into unbroken leaves less than 160,000th part of an inch thick.

In expressing in the arts the purity of silver, if absolutely pure, it is said to be silver of 12 pennyweights; if it contain $\frac{1}{12}$ of its weight of alloy it is called silver of 11 pennyweights; if 2-12ths be alloy, it is called silver of 10 pennyweights, and so on.

SILVER GLANCE.—*Sulphuret of Silver.*

Monometric. In dodecahedrons more or less modified. Fig. 22*a*, page 30, and also other modifications. Cleavage sometimes apparent parallel to the faces of the dodecahedron. Also reticulated and massive.

Luster metallic. Color and streak blackish lead-gray; streak shining. Brittle. H=2—2·5. Gr=7·19—7·4.

Composition: when pure, silver 87·04, sulphur 12·96. Before the blowpipe it intumesces, gives off an odor of sulphur, and finally affords a globule of silver. Soluble in dilute nitric acid.

Dif. Resembles some ores of copper and lead, and other ores of silver, but is distinguished as a sulphuret by giving the odor of sulphur before the blowpipe, and as an ore of silver by affording a globule of this metal, by heat alone. Its specific gravity is much higher than any copper ores, and it is sectile.

Obs. This important ore of silver occurs in Europe, principally at Annaberg, Joachimstahl, and other mines of the Erzgebirge; at Schemnitz, and Kremnitz, in Hungary, and at Freiberg in Saxony. It is a common ore at the Mexican silver mines, and also in the mines of South America.

A mass of sulphuret of silver, is stated by Troost, to have been found in Sparta, Tennessee. It also occurs with native silver and copper in Northern Michigan.

Uses. This is a common and highly valuable ore of silver.

Besides this sulphuret of silver there are two others, which contain also sulphuret of iron or copper.

What is the appearance of vitreous silver? What is its composition? What is its value? How is it distinguished?

Stromeyerite. This is a steel-gray sulphuret of silver and copper, containing 52 per cent. of silver. Gr=6·26. Before the blowpipe it fuses and gives an odor of sulphur; but a silver globule is not obtained except by cupellation with lead. A solution in nitric acid covers a plate of iron with copper, and a plate of copper with silver indicating the copper and silver present. From Peru, Siberia, and Europe.

Sternbergite. A sulphuret of silver and iron containing 33 per cent. of silver. It is a highly foliated ore resembling graphite, and like it leaving a tracing on paper; the thin laminæ are flexible and may be smoothed out by the nail. Luster metallic, color pinchbeck brown. Streak black. It affords the odor of sulphur and a globule covered with silver on charcoal, before the blowipe. With borax a globule of silver is obtained. From Joachimstahl, in Bohemia.

BRITTLE SILVER ORE.—*Sulphuret of Silver and Antimony.*

Trimetric. In modified right rhombic prisms. M : M= 115° 39′. No perfect cleavage. Often in compound crystals. Also massive.

Luster metallic; streak and color iron-black. H=2—2·5. Gr=6·27.

Composition: Sulphur 16·4, antimony 14·7, silver 68·5, copper 0·6. Before the blowpipe it gives an odor of sulphur and also fumes of antimony, and yields a dark metallic globule from which silver may be obtained by the addition of soda. Soluble in dilute nitric acid, and the solution indicates the presence of silver by silvering a plate of copper.

Dif. The black color of this ore distinguishes it from the preceding; and more decidedly the fumes of antimony given off before the blowpipe. By the trial with nitric acid as well as by soda and the blowpipe, it is ascertained to be an ore of silver.

Obs. It occurs with other silver ores at Freiberg, Schneeberg, and Johanngeorgenstadt, in Saxony; also in Bohemia, and Hungary. It is an abundant ore in Chili, Peru, and Mexico. It is sometimes called *black silver.*

An antimonial sulphuret of silver is said to occur with native silver and native copper, at the copper mines in Michigan.

Uses. This is a very important ore for obtaining silver, especially at the South American mines.

Besides this there are other antimonial, and also arsenical and seleniferous ores of silver.

What is the composition of brittle silver ore? its color and appearance? For what is it valued?

Antimonial Silver, consists simply of silver and antimony, (77 parts to 23,) and has nearly a tin-white color. Gr=9·4—9·8. Before the blowpipe gray fumes of antimony pass off, leaving finally a globule of silver. Called also *Discrasite.*

Polybasite is near brittle silver ore in color, specific gravity, and composition, but contains some arsenic and copper, with 75·2 per cent. of silver. The crystals are usually in tabular hexagonal prisms, without cleavage. From Mexico and Peru.

Miargyrite is an antimonial sulphuret of silver, containing but 36·5 per cent. of silver, and having a *dark cherry-red streak,* though iron-black in color. Before the blowpipe gives off fumes of antimony and an odor of sulphur ; and with soda, a globule is left which finally yields a button of pure silver.

Dark Red Silver Ore, and *Light Red Silver Ore,* are two allied ores rhombohedral in their crystals. The former contains silver (59 per cent.,) antimony, and sulphur, and has a color varying from black to cochineal red, a metallic adamantine luster, *and a red streak.* H=2·5. Gr=5·7—5·9.

The latter consists of silver, (65·4 per cent.) arsenic, and sulphur. Its *color and streak* are *cochineal red.* H=2—2·5. Gr=5·4—5·6. Before the blowpipe these species fuse easily, give off fumes, one of antimony, the other of arsenic ; and finally a globule of silver is obtained. They are abundant ores in Mexico, and occur also in Saxony, Hungary, and Bohemia. These ores have been called *ruby silver.*

Eucairite is a *seleniferous* ore of silver and copper occurring in black metallic films. It gives before the blowpipe fumes of selenium, having an odor like that of decaying horse-radish. From Sweden. Another seleniferous ore, from the Hartz, called *selensilver,* contains silver and selenium, with a little lead, and crystallizes in cubes.

Telluric Silver is a Russian ore, of a steel-gray color, containing silver 62·8, and tellurium 37·2. Another variety contains 18 per cent. of gold. Gr=8·3—8·8. With soda, silver is obtained.

Xanthocone is another silver ore, containing silver (66·2 per cent.) combined with sulphur and arsenic. Color dull red to clove-brown ; powder yellow.

HORN SILVER.—*Chlorid of Silver.*

Monometric. In cubes, with no distinct cleavage. Also massive, and rarely columnar ; often incrusting.

Color gray, passing into green and blue, and looking somewhat like horn or wax. Luster resinous, passing into adamantine. Streak shining. Translucent to nearly opaque. Cuts like wax or horn.

Composition : when pure, silver 75·3, chlorine 24·7. Fuses in the flame of a candle, and emits acrid fumes. Affords silver easily on charcoal. The surface of a plate of iron rubbed with it is silvered.

Describe horn silver. Of what does it consist ?

Obs. A very common ore and extensively worked in the mines of South America and Mexico, where it occurs with native silver. It also occurs at the mines of Saxony, Siberia, Norway, the Hartz, and in Cornwall.

Iodic Silver. Bromic Silver. Silver also occurs in nature united with iodine and bromine. These rare ores occur with the preceding in Mexico, and the latter in Chile, and at Huelgoet, in Brittany.

Embolite. A chlorobromid of silver, resembling the chlorid or horn silver. Color asparagus to olive green. Contains 51 of chlorid of silver, to 49 of bromid. This ore is not less common in Chili than the chlorid. It has also been found in Chihuahua, Mexico.

REMARKS ON SILVER AND ITS ORES.

The ores from which the silver of commerce is mostly obtained are the *vitreous silver, brittle or black silver ore, red silver ore* and *horn silver*, in addition to native silver. Besides these, silver is obtained in large quantities from galena, (lead ore,) and from different ores of copper : and some galenas are so rich in silver that the lead is neglected for the more precious metal. This metal occurs in rocks of various ages, in gneiss, and allied rocks, in porphyry, trap, sandstone, limestone, and shales ; and the sandstone and shales may be as recent as the middle secondary, as is the case in Prussia, and probably also in our own Michigan mining region. The silver ores are associated often with ores of lead, zinc, copper, cobalt, and antimony, and the usual gangue is calc spar or quartz, with frequently fluor spar, pearl spar, or heavy spar.

The silver of South America is derived principally from the horn silver, brittle silver ores, including arseniuretted silver ore, vitreous silver ore, and native silver. Those of Mexico are of nearly the same character. Besides, there are earthy ores called *colorados*, and in Peru *pacos*, which are mostly earthy oxyd of iron, with a little disseminated silver ; they are found near the surface where the rock has undergone partial decomposition. The sulphurets of lead, iron, and copper, of the mining regions, generally contain silver, and are also worked.

The mines of Mexico are most abundant between 18° and 24° north latitude, on the back or sides of the Cordilleras and especially the west side ; and the principal are those of the districts of Guanaxuato, Zacatecas, Fresnillo, Sombrerete, Catorce, Oaxaca, Pachuca, Real del Monte, Moran, and Pasco. The veins traverse very different rocks in these regions. The vein of Guanaxuato, the most productive in Mexico, intersects argillaceous and chloritic shale, and porphyry ; it affords one-fourth of all the Mexican silver. The Valencian mine is the richest in Guanaxuato, and has yielded for many years, from one to two millions of dollars annually. In the district of Zacatecas the veins are in gray-

---

Where is horn silver a common ore ? From what ores is the silver of commerce mostly obtained ? How do they occur ? What are the common ores of South America ?

wacke. In Sombrerete they occur in limestone; and there are extensive veins of the antimonial sulphuret, one of which gave in six months 700,000 marcs, (418,000 lbs. troy) of silver. The Pachuca, Real del Monte, and Moran districts, are near one another. Four great parallel veins transverse these districts, through a decomposed porphyry. From the vein Biscaina, in Real del Monte, $5,000,000 were realized by the Count de Regla, in twelve years.

In South America the Chilian mines are on the western slope of the Cordilleras, and are connected mostly with stratified deposits, of a shaly, sandstone, or conglomerate, character, or with their intersections with porphyries. The chlorids and native amalgams are found in regions more towards the coast, while the sulphurets and antimonial ores abound nearer the Cordilleras. The mountains north of the valley of Huasco contain the richest silver mines of Chili. The mines of Mt. Chanarcillo produces at the present time more than 80,000 marcs of silver per year. The veins abound in horn silver, and begin to yield arsenio-sulphurets at a depth of about 500 feet. The mines of Punta Brava, in Copiapo, which are nearer the Cordilleras, afford the arseniuretted ores.

In Peru, the principal mines are in the districts of Pasco, Chota, and Huantaya. Those of Pasco are 15,700 feet above the sea, while those of Huantaya are in a low desert plain, near the port of Yquique, in the southern part of Peru. The ores afforded are the same as in Chili. The mines of Huantaya are noted for the large masses of native silver they have afforded.

The Potosi mines in Buenos Ayres, occur in a mountain of argillaceous shale, whose summit is covered by a bed of argillaceous porphyry. The ore is the red silver, the vitreous ore along with native silver. It has been estimated that they have afforded since their discovery $1,300,000,000. These mines have diminished in value, though they still rank next to those of Guanaxuato.

In Europe the principal mines are those of Spain, of Kongsberg in Norway, of Saxony, the Hartz, Austria, and Russia. The mines of Kongsberg occur in gneiss and hornblende slate, in a gangue of calc spar. They were especially rich in native silver, but are now nearly exhausted. The silver of Spain is obtained mostly from galena, and principally in the Sierra Almagrera in Grenada.

The mines of Saxony occur mostly in gneiss, in the vicinity of Freyberg, Ehrenfriedensdorf, Johangeorgenstadt, Annaberg and Schneeberg.

The ores of the Hartz are mostly argentiferous copper pyrites and galena, yet the red silver, vitreous silver ore, brittle silver ore, and arsenical silver, occur, especially at Andreaskreutz, and the mines of that vicinity. The rock intersected by the deposits is mostly an argillaceous shale. Carbonate of lime is the usual gangue, though it is sometimes quartz

In the Tyrol, Austria, sulphuret of silver, argentiferous gray copper, and mispickel occur in a gangue of quartz, in argillaceous schist. The Hungarian mines at Schemmitz and Kremnitz, occur in syenite and hornblende porphyry, in a gangue of quartz, often with calc spar or heavy spar, and sometimes fluor. The ores are sulphuret of silver

---

Where are the principal mines in Europe?

gray copper, galena, blende, pyritous copper and iron; and the galena, and copper ores are argentiferous.

The Russian mines of Kolyvan in the Altai, and of Nertchinsk in the Daouria mountains, Siberia, (east of Lake Baikal,) are increasing in value, and yield annually at this time, 58,000 troy pounds of silver. The Daouria mines afford an argentiferous galena which is worked for its silver. It occurs in a crystalline limestone. The silver ores of the Altai occur in silurian schists in the vicinity of porphyry, which contain besides silver ores, gold, copper, and lead ores.

In England argentiferous galena is worked for its silver. 40,000 tons of the ore were reduced in 1837, one half of which contained 8 to 8½ oz. of silver to the ton of lead, and the other half only 4 to 5 oz. of silver.

In the United States, the Washington silver mine, in Davidson county, N. Carolina, had afforded up to 1845, 30,000 dollars of silver. The native silver of Michigan is associated with copper in trap and sandstone. These mines promise to be highly productive.

The silver mines of the world have been estimated to yield at the present time $50,000,000 annually.

The annual product of the several countries of Europe is nearly as follows:—

	Pounds troy.		Pounds troy.
British Isles,	70,000	Saxony, the Hartz, and other parts of Germany,	120,000
France,	5,000		
Austria,	90,500	Belgium,	410
Sweden and Norway,	20,000	Piedmont, Switzerland and Russia,	58,000
Spain,	130,000		

making in all nearly 500,000 troy pounds, or about 7,750,000 dollars annually. This is small compared with the amount from America, which at the beginning of the present century equaled 2,100,000 pounds, or 31½ millions of dollars, nearly six times the above sum; and it is probable that these mines will again yield this amount when properly worked. The annual amount from Mexico is set down at 1,750,000 pounds, and from Chili, Peru and Bolivia, near 700,000 pounds. The whole sum from Russia, Europe and America, makes nearly 3,000,000 pounds troy.

The common modes of *reducing silver ores* in the large way are two; by *amalgamation*, and by *smelting*. Both mercury and lead have a strong affinity for silver, and these reducing processes are based on this fact. In amalgamation, the silver ore is brought to the state of a chlorid by a mixture of the powdered ore (or "schlich,") with about ten per cent. of common salt; the chlorid is reduced by means of salts or sulphurets of iron, or metallic iron in filings, and at the same time mercury which has been added, combines with the liberated silver, and thus separates it in the condition of an amalgam, (a compound of mercury and silver.) The mixture of salt and "schlich" requires several days to become complete. Heat is employed at the Saxon mines, but not at those of Mexico, where the climate is tropical. After the mercury is put in, (6 or 8 parts to 1 of silver,) the mixture is kept in constant agi-

---

Where are the Russian mines? What is the yield of the silver mines of the world? What was afforded by South America at the beginning of this century? Describe the process of amalgamation.

tation until the process is finished. In the best arrangements, as in Saxony, this agitation is performed in revolving barrels, and the result is accomplished in a few hours; but in Mexico it is effected by the treading of mules or oxen, and requires two or three weeks or more. The amalgam, separated from the muddy mass, by a current of water or washing, is then filtered of the excess of mercury; as a last step it is subjected to heat in a distilling furnace, by which the silver is left behind, the mercury passing off in a state of vapor to be condensed in a condensing chamber or receptacle. The loss of mercury by the process is often large.

In case of the ordinary sulphurets and arseniurets of silver, or the chlorid, in Mexico and South America, the poorer ores are first fused with a flux, and the result, (called the "matt") is then roasted to expel the sulphur; afterwards it is mixed with better ores, again fused, and on cooling, again roasted. This fusion and roasting is again repeated with the best ores. The result from this fusion is next mixed thoroughly with melted lead; the lead separates the silver; and the impurities which float on the surface, are removed in plates as a crust cools, to be again melted with new ores, as the slag is apt to contain some of the silver.

When the argentiferous galena is the ore, it is reduced by roasting in a reverberatory furnace in the ordinary way for lead ore; the resulting lead contains also the silver.

The accompanying sketch represents the essential characters of a reverberatory furnace. It is a transverse section. *a* is the grate on which the fire is made, and from which the flame proceeds through the horizontal chamber or general cavity of the furnace, (usually very low,) to the flue at *e*. *b* is the sole of the hearth, for rereceiving the ore or assay, having an elliptical or circular form according to the shape of the furnace; *c* is the fire bridge, separating the fire from the sole; *d* is the arched roof. The flame plays horizontally over the charge of ore, and as the air may be made to pass freely with it, we may have in such a furnace a combined effect derived from the heat and the presence of the atmosphere; the ore, or its metal, if capable of uniting with the oxygen of the atmosphere, may be oxydated by the process, precisely as in the *outer* or *oxydating* flame of the blowpipe. In an ordinary blast furnace, (page 233,) the ore and its flux are confined from the atmosphere, (except the air that enters with the blast,) and the result is the reduction of an ore or its *deoxydation*, as in the *inner* or *reducing* flame of the blowpipe. This latter effect may in many cases be obtained also with a reverberatory furnace, when the atmosphere is excluded except what is essential to feeding the fire.

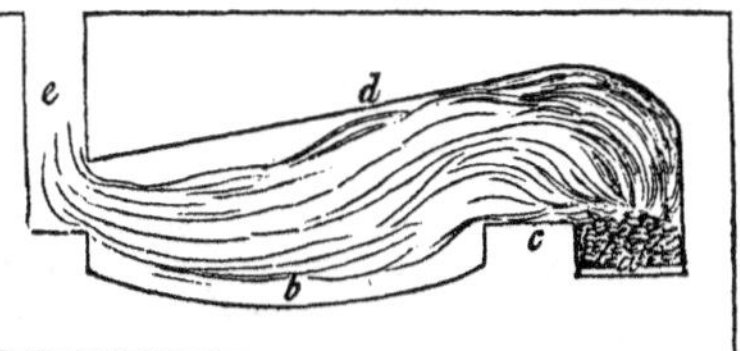

In the reverberatory furnace, there is a small door near the fire-grate, *a*, for putting in fuel. There is also an opening either at top, or on the

Describe a reverberatory furnace.

side, for introducing the charge ; also there may be one or more doors on each side for working the charge while exposed to the heat. There may also be a tap hole for drawing off the reduced metal into one or more pots attached for the purpose ; another in some cases for the escape of slag as in cupellation, and where there is a vaporizable ingredient to be condensed, one or two flues leading to a condensing chamber. In large establishments several of these reverberatory furnaces connect with a single chimney. They are actually like large elliptical or circular ovens, of brick or stone, communicating with a common flue.

In reverberatory furnaces adapted for melting metals, the hearth is a gently inclined plane, sloping to a spot towards one end, in order that the fused metal may flow down together and be convenient for drawing off. For many other purposes, the *sole* is flat, and the depth is greater than in the above figure.

To separate the silver from the lead, the lead is heated in a reverberatory furnace, the hearth of which is covered with wood ashes and clay, so as to give it the nature of a cupel. The air received through an aperture on one side, passes over the metal in fusion, in a constant current, oxydizing it and changing it to litharge, which is from time to time drawn out ; finally the lead is thus removed, and the silver remains nearly pure. The completion of the process is known by the metal becoming brilliant. It is again subjected to another similar operation, and thus rendered quite pure. The litharge from the latter part of the process is also subjected to another operation for the silver it usually contains.

According to Pattinson's new process, adopted in England, the silver is separated by melting the lead, and, as it begins to cool, straining out the crystals with an iron strainer. The portion left behind contains nearly all the silver. This is several times repeated, each time the remaining lead becoming richer in silver. This is then cupelled. An ore containing only 3 ounces of silver to the ton of lead, (or but 1-10,000th part,) may thus be profitably worked, and with little loss of lead.

When the ore containing silver is a copper ore, as is often the case with gray copper ore, the calcined ore is mixed with lead or lead ore, and fused and calcined, and the resulting products are either liquated to *sweat* out the silver or cupelled. In liquation, the copper is run into pigs, (called *liquation* cakes,) and kept above a red heat for two or three days ; the lead first melts and flows in drops into cast iron troughs, carrying with it the silver, which is afterwards obtained by cupelling. The copper still contains some of the lead.

In *trials* by cupellation, a piece of lead of known weight is placed in a cup of bone-ashes, and this is subjected to heat in a small air chamber or oven, and placed in a furnace so that the air shall have free access. The lead is oxydized, and the oxyd sinks into the cupel, leaving a globule of silver behind. The globule being then weighed, and compared with the weight of lead, the proportion of silver is ascertained. Silver may thus be found in almost any specimen of the lead of com-

---

What is the process of amalgamation with an argentiferous lead ore ? What is the mode of trial by cupellation ?

merce, however small the proportion. The weight of the globule, especially when quite minute, may be also ascertained by measurement, according to a scale given by Prof. W. W. Mather, in the American Journal of Science, volume iii, second series, page 414. Much that has been mentioned in the preceding pages on the American mines of silver, has been derived from an article by Prof. Mather, in volume xxiv of the same Journal.

Other modes of reducing silver ores without quicksilver, have been proposed. According to one, the ore is calcined with common salt, as in Mexico, and converted thus to a chlorid. It is then removed to some proper vessel, and a hot solution of salt poured over it; this takes up the chlorid of silver and holds it in solution. The liquid is transferred to another vessel, and by means of metallic copper the silver is deposited.

Another process consists in roasting the sulphurets and converting them in a reverberatory furnace to sulphates; then by boiling water, dissolving the sulphates in a proper vessel, and finally precipitating as above by copper. This process requires the presence of a good deal of sulphur, and is the best when there is much iron and copper pyrites present.

In the assay to separate copper from silver, the alloy is dissolved in nitric acid, and the silver precipitated in the state of a chlorid by common salt. The amount of silver may then be ascertained by weighing the precipitated chlorid, and observing that 75·33 per cent. of the chlorid is pure silver.

SUPPLEMENT.

Forms of Gems.—Gems are cut either by cleaving, by sawing with a wire armed with diamond dust, or by grinding. Some remarks on the cutting of the diamond are given on page 83. The harder stones, as the sapphire and topaz, are cut on a copper wheel with diamond powder soaked with olive oil, and are afterwards polished with tripoli. For other gems, less hard, a lead wheel with emery and water is first used, and then a tin or zinc wheel with putty of tin or rotten stone and water.

The following are some of the common forms. It will be remembered that the upper truncated pyramid is called the *table*, the lower part or pyramid, the *collet*, and the line of junction between the two parts, the *girdle*. Figures 1 and 2 represent the *brilliant*, the best form of the *diamond*, used also for other stones, as well as pastes. Figs. 3 and 4 are views of a variety of the *rose* diamond. Figs. 5 and 6 the same of an *emerald*. The cut in steps is called the *pavilion* cut. Fig. 7 is an upper view of a mode of cutting the *sapphire*. A side view would be nearly like figure 6, except that the collet is more like that of figure 8. Fig. 8 represents a side view of an *oriental topaz*. The table has the *brilliant* cut, like figs. 1 and 2. Figure 9 represents a *Bohemian garnet*, which is made thin because its color is deep. The *common topaz* is cut like figure 8; often also like figure 9, but much thicker, and frequently having the table bordered by two or more rows of triangular facets. Figure 10 is a very simple table. Figures 11 and 12 represent the form "*en cabochon*" given the *opal;* and figures 12 and 13, "*en cabochon*" *with facets*, a mode of cutting the *chrysoberyl.*

The following are some of the common forms. It will be remembered that the upper truncated pyramid is called the *table*, the lower part or pyramid, the *collet*, and the line of junction between the two parts, the *girdle*. Figures 1 and 2 represent the *brilliant*, the best form of the *diamond*, used also for other stones, as well as pastes. Figs. 3 and 4 are views of a variety of the *rose* diamond. Figs. 5 and 6 the same of an *emerald*. The cut in steps is called the *pavilion* cut. Fig. 7 is an

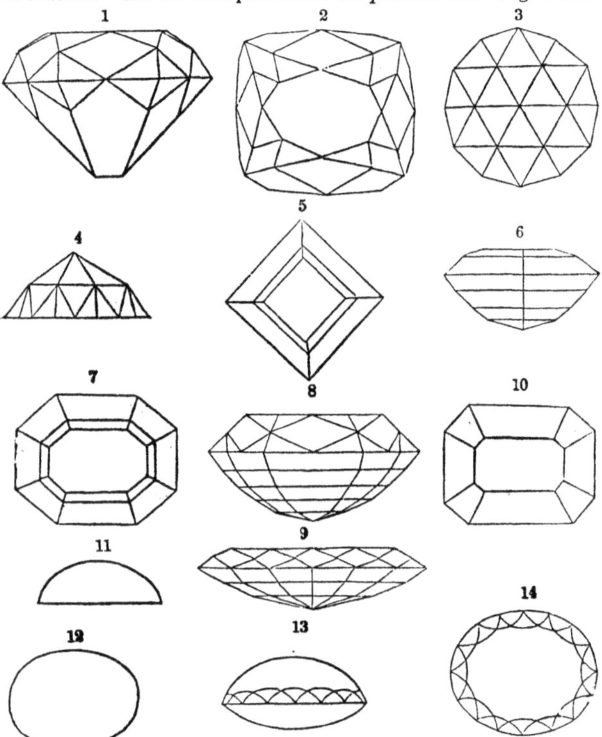

upper view of a mode of cutting the *sapphire*. A side view would be nearly like figure 6, except that the collet is more like that of figure 8. Fig. 8 represents a side view of an *oriental topaz*. The table has the *brilliant* cut, like figs. 1 and 2. Figure 9 represents a *Bohemian garnet*, which is made thin because its color is deep. The *common topaz* is cut like figure 8; often also like figure 9 but much thicker, and frequently having the table bordered by two or more rows of triangular facets. Figure 10 is a very simple table. Figures 11 and 12 represent the form "*en cabochon*" given the *opal;* and figures 12 and 13, "*en cabochon*" *with facets*, a mode of cutting the *chrysoberyl*.

The form of the diamond after being cut anew, is represented in the following figures, which are of natural size :

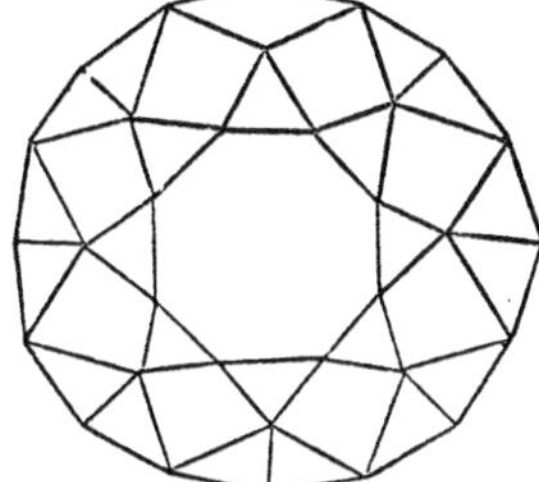

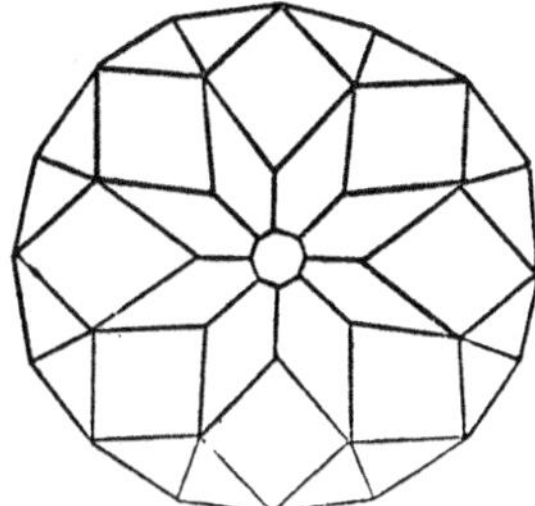

This re-cutting has diminished the weight of the diamond over one third, a sacrifice of magnificence to mere ornamental brilliancy.

CHAPTER VII.

CHEMICAL COMPOSITION AND FORMULAS OF MINERALS.

On a former page a brief explanation is given of the constitution of minerals. The following table contains the names of all the elements thus far discovered by Chemistry, together with the abbreviations or symbols by which they are indicated in chemical formulas, and the combining or atomic weights. Thus Al stands for the element *Aluminium*, Sb for *Antimony* (derived from *Stibium*, the Latin name for Antimony). As all the elements combine with oxygen, and oxyds (as such compounds are called), are the most common of all compounds, Berzelius proposed to use a dot over a letter for oxygen, that is, *one* dot for one proportion of oxygen, *two* for two proportions, *three* for three proportions, &c. In this way, $\dot{\text{B}}\text{a}$ means that one part of oxygen is combined with one of Baryum; the compound is protoxyd of baryum. So $\overset{\cdot\cdot\cdot\cdot\cdot}{\text{P}}$ signifies that *five* parts of oxygen are combined with *one* of phosphorus; the compound is phosphoric acid. Again to express two of Aluminium, a bar crosses the letter A, so that $\dddot{\bar{\text{A}}}\text{l}$ means a compound of *three* of oxygen and *two* of aluminium; $\dddot{\bar{\text{F}}}\text{e}$ means *three* of oxygen and *two* of iron, or a sesquioxyd of iron. Besides the atomic weights of the elements, the principal occurring oxyds are given, with their atomic weights, and also the percentage of oxygen in each.

Table of Atomic Weights.

ALUMINIUM, Al,	171·25
Alumina, $\dddot{\bar{\text{A}}}\text{l}$,	642·5 (O, 46·7)
ANTIMONY (Stibium), Sb,	1612·5
ARSENIC, As,	937·5
BARYUM, Ba,	856·25
Baryta, $\dot{\text{B}}\text{a}$,	956·25 (O, 10·45)
BISMUTH, Bi,	2600
BORON, B,	136·2
Boracic acid, $\dddot{\text{B}}$,	436·2 (O, 68·8)
BROMINE, Br,	1000

CADMIUM, Cd,	696·8
CALCIUM, Ca,	250
Lime, $\dot{C}a$,	350 (O, 28·57)
CARBON, C,	75
Carbonic acid, $\ddot{C}$,	275
CERIUM, Ce,	587·5
Protoxyd of C., $\dot{C}e$,	687·5 (O, 14·55)
CHLORINE, Cl,	443·3
Hydrochloric acid, HCl,	455·8
CHROMIUM, Cr,	333·75
Oxyd of Ch., $\dddot{\bar{C}}r$,	967·5 (O, 31·0)
Chromic acid, $\dddot{C}r$,	633·75 (O, 47·3)
COBALT, Co,	368·65
COLUMBIUM, Cb,	2300 ?
Columbic acid, $\bar{\ddot{C}}b$,	2600 (O, 11·5)
COPPER (Cuprum), Cu,	396·25
Oxyd of Copper, $\dot{\bar{C}}u$,	892·5 (O, 11·2)
Oxyd of Copper, $\dot{C}u$,	496·25 (O, 20·15)
DIDYMIUM, D.	
ERBIUM, Eb.	
FLUORINE, F,	237·5
Hydrofl. acid, HF,	250·0
GLUCINUM (Beryllium), Be,	58·75
Glucina, $\dddot{\bar{B}}e$,	476·25 (O, 63)
GOLD (Aurum), Au,	1231·25
HYDROGEN, H,	12·5
Water, $\dot{\bar{H}}$,	112·5 (O, 88·89)
IODINE, I,	1587·5
IRIDIUM, Ir,	1237·5
IRON (Ferrum), Fe,	350
Protoxyd of I., $\dot{F}e$,	450 (O, 22·22)
Sesquioxyd of I., $\dddot{\bar{F}}e$,	1000 (O, 30)
LANTHANUM, La,	587·5
Protoxyd of L., $\dot{L}a$,	687·5 (O, 14·5)
LEAD (Plumbum), Pb,	1294·6
Protoxyd of Lead, $\dot{P}b$,	1394·6 (O, 7·17)
LITHIUM, Li,	81·6
Lithia, $\dot{L}i$,	181·6 (O, 55)
MAGNESIUM, Mg,	150
Magnesia, $\dot{M}g$,	250 (O, 40)

MANGANESE, Mn,	344·7
Protoxyd of M., $\dot{\mathrm{Mn}}$,	444·7 (O, 22·5)
Sesquioxyd of M., $\dddot{\overline{\mathrm{Mn}}}$,	989·4 (O, 30·3)
MOLYBDENUM, Mo,	575
Molybdic acid, $\dddot{\mathrm{Mo}}$,	875
NICKEL, Ni,	369·3
Protoxyd of Ni., $\dot{\mathrm{Ni}}$,	469·3 (O, 21·3)
NITROGEN, N,	175
Nitric acid, $\overline{\overset{\cdot\cdot\cdot\cdot\cdot}{\mathrm{N}}}$,	675 (O, 74)
OSMIUM, Os,	1243·6
OXYGEN, O,	100
PALLADIUM, Pd,	665·5
PHOSPHORUS, P,	387·5
Phosphoric acid, $\overline{\overset{\cdot\cdot\cdot\cdot\cdot}{\mathrm{P}}}$,	887·5 (O, 56·34)
PLATINUM, Pt,	1237·5
POTASSIUM (Kalium), K,	488·9
Potassa, $\dot{\mathrm{K}}$,	588·9 (O, 16·98)
QUICKSILVER (Hydrargyrum), Hg,	1250
RHODIUM, Rd,	652
RUTHENIUM, Ru,	652
SELENIUM, Se,	493·75
SILICIUM, Si,	266·25
Silica, $\dddot{\mathrm{Si}}$,	566·25 (O, 52·98)
SILVER (Argentum), Ag,	1350
SODIUM (Natrium), Na,	287·5
Soda, $\dot{\mathrm{Na}}$,	387·5 (O, 25·8)
STRONTIUM, Sr,	547·5
Strontia, $\dot{\mathrm{Sr}}$,	647·5 (O, 15·44)
SULPHUR, S,	200
Sulphurous acid, $\ddot{\mathrm{S}}$,	400
Sulphuric acid, $\dddot{\mathrm{S}}$,	500 (O, 60)
TANTALUM, Ta,	2300
Tantalic acid, $\dddot{\overline{\mathrm{Ta}}}$,	2600
TELLURIUM, Te,	801·8
TERBIUM, Tb.	
THORIUM, Th,	743·9
TIN (Stannum), Sn,	725
Oxyd of Tin, $\ddot{\mathrm{Sn}}$,	925 (O, 21·6)
TITANIUM, Ti,	312·5
Oxyd of Titanium, $\dddot{\overline{\mathrm{Ti}}}$	925 (O, 32·4)
Titanic acid, $\ddot{\mathrm{Ti}}$,	512·5 (O, 39)

TUNGSTEN (Wolfram), W,	1150
Tungstic acid, $\dddot{W}$,	1450
URANIUM, U,	750
Protoxyd of U., $\dot{U}$,	850
Peroxyd of U., $\ddot{\overline{U}}$,	1800
VANADIUM, V,	856·9
YTTRIUM, Y,	402·5
Yttria, $\dot{Y}$,	502·5 (O, 19·9)
ZINC, Zn,	406·6
Oxyd of Zinc, $\dot{Zn}$,	506·6 (O, 19·74)
ZIRCONIUM, Zr,	419·7
Zirconia, $\ddot{\overline{Zr}}$,	1139·5 (O, 26·3)

By *atomic weights* is understood the combining proportions of the elements. For example, when *iron* and *oxygen* combine, they unite in the proportions of 350 parts by weight of iron to 100 of oxygen, or in some simple multiple of these numbers. The protoxyd contains one part or atom of each, and has therefore the atomic weight 450; the peroxyd (more precisely *sesquioxyd*) contains 2 of iron (2×350=700) to 3 of oxygen (3×100=300), and therefore has the atomic weighs 1000 (700+300=1000). To ascertain the per-centage of oxygen in this oxyd, we have 300 of oxygen in 1000 parts; hence the ratio,—1000 are to 300 as 100 to the number of parts in 100; therefore dividing 300×100 by 1000 give the oxygen per-centage. Hence too if we multiply the percentage of oxygen by the atomic weight of the oxyd, we obtain as a result, after dividing by 100, the oxygen amount in the compound. For alumina, 46·7×642·5÷100=300, the amount of oxygen; and in this way the correctness of the oxygen per-centage may be verified.

The mode of deducing chemical formulas may be illustrated by two or three examples.

1. We have an analysis of Red Silver Ore as follows:

Silver 59·02, antimony 23·49, and sulphur 17·49 per cent

It is desired to ascertain the relative number of atoms of each element in the compound. This number must depend on the weights of the atoms, as compared with the quantity of each, for the less the weight, the greater the number of atoms. The rule consequently is,—*Divide the per-centage of each element by the atomic weight of the same;* as, 5·02 by

1350, the atomic weight of silver, and so on. (See preceding table.) This process gives the relation,

0·0437 : 0·0146 : 0·0875,

and *dividing each by the smallest*, to simplify it, it becomes

3 : 1 : 6,

which is therefore the number of atoms of each, silver, antimony, and sulphur. The formula 3Ag+1Sb+6S, or Ag^3 Sb S^6, expresses this relation.

As chemistry makes known a sulphuret of silver consisting of 1 of sulphur to 1 of silver, and also a sulphuret of antimony containing 3 of sulphur to 1 of antimony, the ingredients are regarded as thus combined in the compound. The 3Ag take 3S, making 3AgS, and leave 3S for the 1Sb to form $1(SbS^3)$; so that Red Silver Ore is supposed to be represented by the formula

$$3AgS+SbS^3; \text{ or } \overset{\prime}{A}g^3\overset{\prime\prime\prime}{S}b,$$

if the mark (') be used, as is common, for sulphur.

2. An analysis of Feldspar gives in 100 parts,

Silica 64·78, Alumina 18·38, Potash 16·84.

Now if we ascertain the proportion of *oxygen* in these constituents we learn the ratio of the constituents, since we know that silica contains 3 of oxygen, alumina 3, and potash 1. From the above table we find that 100 of silica contain 52·98 of oxygen; consequently if 100 give 52·98, the amount in 64·78 parts will be found by multiplying 52·98 by 64·78, and dividing by 100; or what is equivalent, multiplying 0·5298 by 64·78. So in 100 parts of alumina the oxygen is 46·7; hence the oxygen in 18·38 parts of alumina will equal 18·38×46·7÷100. In this way we ascertain that

64·78 of silica contain	34·32 oxygen,	or, dividing each by the smallest,	12
18·38 of alumina "	8·58 "		3
16·84 of potash "	2·86 "		1

Hence the amount of oxygen in the potash, alumina, and silica, is as 1:3:12. Now as each atom of silica contains 3 of oxygen, 12 atoms of oxygen correspond to 4 of silica: so also 3 of oxygen for the alumina correspond for a like reason to 1 of alumina; and 1 of oxygen for the potash to 1 atom of potash. The compound therefore contains 4 parts of silica to 1 of alumina and 1 of potash.

The next step in the usual method, is to determine how these constituents are combined; how much of the silica with the potash, and how much with the alumina. Reference is made to the possibility or probability of certain compounds, which Chemistry alone can teach; but aid is found in the principle, that the number of atoms of oxygen in each acid and base is usually some simple multiple, the one of the other. If in the above compound, 1 of silica be united with 1 of potash, the ratio alluded to is 1 to 3; and if the alumina be combined with the remaining 3 atoms of silica, the same ratio holds. This is the mode of combination commonly adopted; it is expressed in the following formula, the dots as explained, indicating the oxygen:

$$\dot{K}\,\dddot{Si}+\dddot{\overline{Al}}\,\dddot{Si}^{3}.$$

The index ^{3} expresses the number of atoms of silica: had the 3 been written as a prefix, thus, $3\dddot{\overline{Al}}\,\dddot{Si}$, it would have meant 3 atoms of a compound of silica and alumina.

The formula might also be written with equal precision, and without dividing the silica between the bases, as follows:

$$(\dot{K}+\dddot{\overline{Al}})\,\dddot{Si}^{4}.$$

In a similar manner, an analysis of Garnet affords the ratio of ingredients as follows:

3 of lime ($3\dot{Ca}$), 1 of alumina ($1\dddot{\overline{Al}}$), 2 of silica ($2\dddot{Si}$),

corresponding to the oxygen ratio 3 : 3 : 6 or 1 : 1 : 2. Apportioning the silica to the bases, we have the formula

$$\dot{Ca}^{3}\dddot{Si}+\dddot{\overline{Al}}\dddot{Si},$$

in which the oxygen ratio for each member is 1 : 1. Idocrase and Meionite afford other simple examples.

3. In Feldspar, above cited, the protoxyd portion is often not potash alone, but part soda or lime. Again, in Garnet, the protoxyds, instead of being all lime, may be part magnesia, protoxyd of iron, &c. In each case, however, all the protoxyds added together, make up the same specific number of atoms as if there were but one alone. So the peroxyd portion may not be all of it alumina, but part peroxyd of iron, the amount of this peroxyd of iron being just equivalent to the deficiency in the alumina; that is, an equivalent not in actual weight, but in atomic weight. In Garnet, as stated above, the oxygen ratio is 1 : 1 : 2; and whatever the

peroxyds or protoxyds, the ratio still holds. Suppose an analysis of Garnet affords the per-centage, Silica 39·6, alumina 22·5, lime 32·6, protoxyd of iron 5·3: we ascertain the oxygen in each constituent in the manner explained, (in the *Silica*, by multiplying 0·5298 by 39·6,—in the *alumina*, by multiplying 0·467 by 22·5,—in the *lime* by multiplying 0·2857 by 32·6,—in the *protoxyd of iron* by multiplying 0·2222 by 5·3); then on *adding* the oxygen of the protoxyd of iron to that of the lime, the amount just equals that of the alumina, as the oxygen ratio requires. Moreover the oxygen of all the protoxyds and peroxyds together equals the oxygen of the silica.

As different protoxyds may thus replace one another, and as different peroxyds likewise admit of mutual replacement, it is common to write $\dot{R}$ as a general symbol for the protoxyds of a compound, and $\dddot{\bar{R}}$ for the peroxyds. It is also common to write the special symbols of the protoxyds which replace one another, in parentheses, with a comma between them. Thus in the Garnet referred to, in which lime and protoxyd of iron replace one another, the general formula may be either

$$\dot{R}^3\dddot{Si}+\dddot{\bar{A}l}\dddot{Si};\ \text{or}\ (\dot{C}a,\dot{F}e)^3\dddot{Si}+\dddot{\bar{A}l}\dddot{Si}.$$

The proportions of the lime and protoxyd of iron are not here stated, but may be; if 1 : 2, the formula becomes

$$(\tfrac{1}{3}\dot{C}a+\tfrac{2}{3}\dot{F}e)^3\dddot{Si}+\dddot{\bar{A}l}\dddot{Si}.$$

Again, the formula $(\dot{C}a,\dot{M}g)\ddot{C}$, signifies that the compound is a carbonate of lime and magnesia, in definite or indefinite proportions; $(\tfrac{1}{2}\dot{C}a+\tfrac{1}{2}\dot{M}g)\ddot{C}$ that the proportion is 1 : 1; $(\tfrac{2}{5}\dot{C}a+\tfrac{3}{5}\dot{M}g)\ddot{C}$ that the proportion is 2 : 3, $\tfrac{2}{5}$ and $\tfrac{3}{5}$ having this ratio, and together equaling a unit. $\dot{R}\ddot{C}$ is a general expression for a carbonate of any protoxyd.

4. The formula for Garnet, $\dot{R}^3\dddot{Si}+\dddot{\bar{A}l}\dddot{Si}$, may also be written with equal precision as follows:

$$(\tfrac{1}{2}\dot{R}^3+\tfrac{1}{2}\dddot{\bar{A}l})\dddot{Si},$$

the ratio 1 : 1 : 2 being still retained, and the fact being also presented to the eye that the oxygen of all the oxyds is to that of the silica as 1 : 1. In Gehlenite, another silicate of alumina and lime, the oxygen ratio is 3 : 3 : 4, which gives 6 : 4 or 3 : 2 for the ratio of the oxyds and silica (that is, the oxygen of the silica is *two-thirds* that of all the oxyds), while that of the protoxyds and peroxyds is 1 : 1, The formula may hence be

$$\dot{R}^3\dddot{Si}^{\frac{2}{3}}+\dddot{\bar{R}}\dddot{Si}^{\frac{2}{3}}\text{; or }(\tfrac{1}{2}\dot{R}^3+\tfrac{1}{2}\dddot{\bar{R}})\dddot{Si}^{\frac{2}{3}}.$$

In the first of these formulas each of the two members has the same oxygen ratio 3 : 2; in the second this ratio is also retained, and is more briefly expressed, without the hypothetical idea that the silica in the compound is divided off between the protoxyd and peroxyd bases.

5. To deduce the per-centage atomic relations from a formula, the process above described is reversed. For example: for Feldspar we have 4 of silica, 1 of alumina, 1 of potash. In the preceding table the atomic weight of silica is 566·25, and four times this is 2265. Setting this down and the atomic weights of alumina and potash below it, and adding, we have

4 of silica, - - - - - -	2265
1 of alumina, - - - - -	642·5
1 of potash, - - - - -	588·9
Total atomic weight of the feldspar,	3496·4

Now if this amount (3496·4) of feldspar contains 2265 of silica, what will 100 parts contain? Hence, to obtain the per-centage, we divide the atomic weight of each constituent in succession by the sum of the whole, and this gives the percentage relation for each; viz. silica 64·78, alumina 18·38, potash 16·84.

The following are the formulas of the more common mineral species following the order of the book.

Table of Chemical Formulas of Minerals.

Sal Ammoniac (100),	NH^4Cl
Niter (101),	$\dot{K}\overset{\cdot\cdot\cdot\cdot\cdot}{N}$
Glauber Salt (102),	$\dot{N}a\dddot{S}+10\dot{H}$
Nitratine (Nit. Soda, 103),	$\dot{N}a\overset{\cdot\cdot\cdot\cdot\cdot}{N}$
Natron (103),	$\dot{N}a\ddot{C}+10\dot{H}$
Trona (103),	$\dot{N}a^2\ddot{C}^3+4\dot{H}$
Common Salt (104),	$NaCl$
Borax (107)	$\dot{N}a\dddot{B}^2+10\dot{H}$
Barytes (Heavy Spar, 108),	$\dot{B}a\dddot{S}$
Witherite (109),	$\dot{B}a\ddot{C}$

Celestine (110),	$Sr\dddot{S}$
Strontianite (111),	$Sr\ddot{C}$
Gypsum (112),	$\dot{Ca}\dddot{S}+2\dot{H}$
Anhydrite (114),	$\dot{Ca}\dddot{S}$
Calcite (115),	$\dot{Ca}\ddot{C}$
Aragonite (118),	$\dot{Ca}\ddot{C}$
Dolomite (119),	$(\dot{Ca}, \dot{Mg})\ddot{C}$
Apatite (120),	$\dot{Ca}^3\overset{\cdot\cdot\cdot\cdot\cdot}{\bar{P}}+\frac{1}{3}Ca(Cl, F)$
Fluor spar (121),	CaF
Epsomite (Epsom salt, 124),	$\dot{Mg}\dddot{S}+7\dot{H}$
Magnesite (124),	$\dot{Mg}\ddot{C}$
Brucite (125),	$\dot{Mg}\dot{H}$
Boracite (126),	$\dot{Mg}^3\dddot{B}^4$
Potash alum (127),	$\dot{K}\dddot{S}+\dddot{\bar{Al}}\dddot{S}^3+24\dot{H}$
Soda alum (128),	$\dot{Na}\dddot{S}+\dddot{\bar{Al}}\dddot{S}^3+24\dot{H}$
Alunite (129),	$\dot{K}\dddot{S}+3\dddot{\bar{Al}}\dddot{S}+6\dot{H}$
Wavellite (130),	$(\dddot{\bar{Al}}^4\overset{\cdot\cdot\cdot\cdot\cdot}{\bar{P}}^3+18\dot{H})+\frac{1}{3}AlF^3$
Gibbsite (131),	$\dddot{\bar{Al}}\dot{H}^3$
Quartz (132),	$\dddot{Si}$
Opal (139),	$\dddot{Si}(+Aq.)$
Wollastonite (141),	$\dot{Ca}^3\dddot{Si}^2$
Datholite (142).	$(\dot{Ca}^3, \dot{H}^3, \dddot{B})\dddot{Si}^{\frac{2}{3}}$
Talc (143),	$\dot{Mg}^6\dddot{Si}^5+2\dot{H}$
Chlorite (145),	$(\dot{R}^3, \dddot{\bar{R}})\dddot{Si}^{\frac{3}{4}}+1\frac{1}{2}\dot{H}$
Ripidolite (145),	$(\dot{R}^3, \dddot{\bar{R}})\dddot{Si}^{\frac{2}{3}}+1\frac{1}{2}\dot{H}$
Serpentine (145),	$\dot{Mg}^3\dddot{Si}^2+1\frac{1}{2}\dot{Mg}\dot{H}^2$
Meerschaum (148),	$\dot{Mg}\dddot{Si}+\dot{H}$?
Clintonite (148),	$(\dot{R}^3, \dddot{\bar{R}})(\dddot{Si}, \dddot{\bar{Al}})^{\frac{2}{3}}+\frac{1}{2}\dot{H}$
Pyroxene (150),	$(\dot{Mg}, \dot{Ca}, \dot{Fe}, \dot{Mn})^3\dddot{Si}^2$
Hornblende (152),	$(\dot{Mg}, \dot{Ca}, \dot{Fe}, \dot{Mn})^4\dddot{Si}^3$
Spodumene (156),	$(\dot{Li}, \dot{Na})^3\dddot{Si}^2+4\dddot{\bar{Al}}\dddot{Si}^2$
Chrysolite (157),	$(\dot{Mg}, \dot{Fe}, \dot{Ca})^3\dddot{Si}$
Chondrodite (157),	$(\dot{Mg}, \dot{Fe})^4\dddot{Si}$
Corundum (158),	$\dddot{\bar{Al}}$
Spinel (160),	$(\dot{Mg}, \dddot{\bar{Al}})$
Automolite (var. of Spinel, 161),	$\dot{Zn}\dddot{\bar{Al}}$
Hercinite, "	$\dot{Fe}\dddot{\bar{Al}}$
Ceylanite, "	$(\dot{Mg}, \dot{Fe})\dddot{\bar{Al}}$
Dysluite, "	$(\dot{Zn}, \dot{Mn})(\dddot{\bar{Al}}, \dddot{\bar{Fe}})$

Kreittonite, (var. of Spinel), $(\dot{Z}n, \dot{F}e)(\ddot{\bar{A}}l, \ddot{\bar{F}}e)$

Halloysite (161), $\ddot{\bar{A}}l\dddot{S}i+3\dot{H}$ (in part)

Heulandite (164), $\dot{C}a\dddot{S}i+\ddot{\bar{A}}l\dddot{S}i^3+5\dot{H}$

Stilbite (165), $\dot{C}a\dddot{S}i+\ddot{\bar{A}}l\dddot{S}i^3+6\dot{H}$

Apophyllite (165), $\dot{C}a$, $\dot{K}$, $\dddot{S}i$, $\dot{H}$, F.

Laumontite (166), $\dot{C}a^3\dddot{S}i^2+3\ddot{\bar{A}}l\dddot{S}i^2+12\dot{H}$

Natrolite (166), $\dot{N}a\dddot{S}i+\ddot{\bar{A}}l\dddot{S}i+2\dot{H}$

Scolecite (167), $\dot{C}a\dddot{S}i+\ddot{\bar{A}}l\dddot{S}i+3\dot{H}$

Thomsonite (167), $\dot{R}^3\dddot{S}i+3\ddot{\bar{A}}l\dddot{S}i+7\dot{H}$

Harmotome (168), $\dot{B}a\dddot{S}i+\ddot{\bar{A}}l\dddot{S}i^2+5\dot{H}$

Analcime (168), $\dot{N}a^3\dddot{S}i^2+3\ddot{\bar{A}}l\dddot{S}i^2+6\dot{H}$

Chabazite (169), $\dot{R}^3\dddot{S}i^2+3\ddot{\bar{A}}l\dddot{S}i^2+18\dot{H}$

Prehnite (170), $2\dot{C}a^3\dddot{S}i+3\ddot{\bar{A}}l\dddot{S}i+\dot{H}^3\dddot{S}i$

Sillimanite (172), $\ddot{\bar{A}}l\dddot{S}i^{\frac{2}{3}}$; (partly $\ddot{\bar{A}}l\dddot{S}i$?)

Kyanite (173), $\ddot{\bar{A}}l\dddot{S}i^{\frac{2}{3}}$

Andalusite (174), $\ddot{\bar{A}}l\dddot{S}i^{\frac{2}{3}}$

Staurotide (174), $\ddot{\bar{A}}l\dddot{S}i^{\frac{1}{2}}$

Leucite (175), $\dot{K}^3\dddot{S}i^2+3\ddot{\bar{A}}l\dddot{S}i^2$

Orthoclase (176), $(\dot{K}+\ddot{\bar{A}}l)\dddot{S}i^4=\dot{K}\dddot{S}i+\ddot{\bar{A}}l\dddot{S}i^3$

Albite (177), $(\dot{N}a+\ddot{\bar{A}}l)\dddot{S}i^4=\dot{N}a\dddot{S}i+\ddot{\bar{A}}l\dddot{S}i^3$

Labradorite (178), $(\dot{C}a+\ddot{\bar{A}}l)\dddot{S}i^2=\dot{C}a\dddot{S}i+\ddot{\bar{A}}l\dddot{S}i$

Nepheline (179), $(\dot{N}a+\ddot{\bar{A}}l)\dddot{S}i^{\frac{3}{2}}=\dot{N}a^2\dddot{S}i+2\ddot{\bar{A}}l\dddot{S}i$

Scapolite (180), $(\frac{1}{3}\dot{R}^3+\frac{2}{3}\ddot{\bar{A}}l)\dddot{S}i^{\frac{4}{3}}=\dot{R}^3\dddot{S}i+\ddot{\bar{A}}l^2\dddot{S}^3$

Meionite (181), $(\frac{1}{3}\dot{C}a^3+\frac{2}{3}\ddot{\bar{A}}l)\dddot{S}i=\dot{C}a^3\dddot{S}i+2\ddot{\bar{A}}l\dddot{S}i$

Petalite (182), $(\frac{1}{5}\dot{R}^3+\frac{4}{5}\ddot{\bar{A}}l)\dddot{S}i^4=\dot{R}^3\dddot{S}i^4+4\ddot{\bar{A}}l\dddot{S}i^4$

Epidote (182), $(\frac{1}{3}\dot{R}^3+\frac{2}{3}\ddot{\bar{A}}l)\dddot{S}i=\dot{R}^3\dddot{S}i+2\ddot{\bar{A}}l\dddot{S}i$

Allanite (183, 207), $(\frac{1}{2}\dot{R}^3+\frac{1}{2}\ddot{\bar{A}}l)\dddot{S}i=\dot{R}^3\dddot{S}i+\ddot{\bar{A}}l\dddot{S}i$

Idocrase (184), $(\frac{3}{5}\dot{R}^3+\frac{2}{5}\ddot{\bar{A}}l)\dddot{S}i=3\dot{R}^3\dddot{S}i+2\ddot{\bar{A}}l\dddot{S}i$

Garnet (184), $(\frac{1}{2}\dot{R}^3+\frac{1}{2}\ddot{\bar{A}}l)\dddot{S}i=\dot{R}^3\dddot{S}i+\ddot{\bar{A}}l\dddot{S}i$

Tourmaline (187), $(\dot{R}^3, \ddot{\bar{A}}l, \dddot{B})\dddot{S}i^{\frac{3}{4}}$

Axinite (190), $(\dot{R}^3, \ddot{\bar{R}}, \dddot{B})\dddot{S}i$

Iolite (190), $(\frac{1}{4}\dot{R}^3+\frac{3}{4}\ddot{\bar{A}}l)\dddot{S}i^{\frac{5}{4}}=\dot{R}^3\dddot{S}i^{\frac{5}{4}}+3\ddot{\bar{A}}l\dddot{S}i^{\frac{5}{4}}$

Muscovite (191), $(\frac{1}{13}\dot{R}^3+\frac{12}{13}\ddot{\bar{R}})\dddot{S}i^{\frac{5}{4}}=\dot{R}^3\dddot{S}i^{\frac{5}{4}}+12\ddot{\bar{R}}\dddot{S}i^{\frac{5}{4}}$

Biotite (193), $(\frac{1}{2}\dot{R}^3+\frac{1}{2}\ddot{\bar{A}}l)\dddot{S}i=\dot{R}^3\dddot{S}i+\ddot{\bar{A}}l\dddot{S}i$

Phlogopite (193), $(\frac{3}{5}\dot{R}^3+\frac{2}{5}\ddot{\bar{A}}l)\dddot{S}i=3R^3\dddot{S}i+2\ddot{\bar{A}}l\dddot{S}i$

Topaz (194), $\ddot{\bar{A}}l\dddot{S}i^{\frac{2}{3}}$ with F replacing part of O

Beryl (197), $(\frac{1}{2}\ddot{\bar{B}}e+\frac{1}{2}\ddot{\bar{A}}l)\dddot{S}i^2=\ddot{B}e\dddot{S}i^2+\ddot{\bar{A}}l\dddot{S}i^2$

Euclase (199),	$(\frac{1}{2}\dddot{\bar{B}}e+\frac{1}{2}\dddot{\bar{A}}l)\dddot{S}i^{\frac{3}{4}}=\dddot{\bar{B}}e\dddot{S}i^{\frac{3}{4}}+\dddot{\bar{A}}l\dddot{S}i^{\frac{3}{4}}$
Chrysoberyl (199),	$\dddot{\bar{B}}e+\dddot{\bar{A}}l^{3}$
Zircon (200),	$\dddot{\bar{Z}}r\dddot{S}i$
Yttrocerite (206),	$(Ca, Ce, Y)F$
Monazite (206),	$(\dot{C}e, \dot{L}a, \dot{T}h)^{3}\overset{\cdot\cdot\cdot\cdot\cdot}{\bar{P}}$
Allanite (see above).	
Rutile (210),	$\ddot{T}i\cdot$
Sphene (211),	$(\dot{C}a+\ddot{T}i)\dddot{S}i^{\frac{2}{3}}=(\dddot{\bar{R}})\dddot{S}i^{\frac{2}{3}}$
Tin Pyrites (213),	$\bar{C}uS(Sn^{2}S^{3}, Fe^{2}S^{3})$
Tin Ore (214),	$\ddot{S}n$
Molybdenite (217),	MoS^{2}
Molybdic ocher (218),	$\dddot{M}o$
Tungstic ocher (218),	$\dddot{W}$
Scheelite (Tungstate of lime, 219),	$\dot{C}a\dddot{W}$
Gray Antimony (222),	$Sb^{2}S^{3}$
White Antimony (224),	$Sb^{2}O^{3}$
White Arsenic (226),	$As^{2}O^{3}$
Orpiment (226),	$As^{2}S^{3}$
Realgar (226),	AsS
Uranite (228),	$\dot{C}a^{2}\overset{\cdot\cdot\cdot\cdot\cdot}{\bar{P}}+\dddot{\bar{U}}_{4}\overset{\cdot\cdot\cdot\cdot\cdot}{\bar{P}}+16\dot{H}$
Chalcolite (229),	$\dot{C}u^{2}\overset{\cdot\cdot\cdot\cdot\cdot}{\bar{P}}+\dddot{\bar{U}}_{4}\overset{\cdot\cdot\cdot\cdot\cdot}{\bar{P}}+16\dot{H}$
Pyrites (231),	FeS^{2}
Pyrrhotine (233),	$Fe^{7}S^{3}$; FeS.
Mispickel (234),	$Fe(As, S)^{2}$
Magnetite (235),	$\dot{F}e\dddot{\bar{F}}e$
Specular Iron (237),	$\dddot{\bar{F}}e$
Limonite (239),	$\dddot{\bar{F}}e^{2}\dot{H}^{3}$
Göthite (240),	$\dddot{\bar{F}}e\dot{H}$
Franklinite (240),	$\dot{Z}n(\dddot{\bar{F}}e, \dddot{\bar{M}}n)$
Ilmenite (241),	$(\dddot{\bar{F}}e, \dddot{\bar{T}}i)$
Chromic Iron (241),	$\dot{F}e(\dddot{\bar{C}}r)$
Columbite (243),	$\dot{F}e^{3}\dddot{C}b^{2}$
Tantalite (ferrotantalite, 244),	$\dot{F}e\dddot{T}a$
Wolfram (244),	$(\dot{F}e, \dot{M}n)\dddot{W}$
Copperas (246),	$\dot{F}e\dddot{S}+7\dot{H}$
Spathic Iron (247),	$\dot{F}e\ddot{C}$
Vivianite (248),	$\dot{F}e^{3}\overset{\cdot\cdot\cdot\cdot\cdot}{\bar{P}}+8\dot{H}$
Rhodonite (258),	$\dot{M}n^{3}\dddot{S}i^{2}$
Pyrolusite (259),	$\ddot{M}n$

Psilomelane (259), $\dot{R}\ddot{M}n^{2}+\dot{H}$?
Wad (260), $\dot{R}\ddot{M}n+\dot{H}$?
Triplite (260), $(\dot{M}n, \dot{F}e)^{4}\overset{\cdot\cdot\cdot\cdot\cdot}{P}$
Copper Nickel (263), NiAs
Chloanthite (263), $(Ni, Co)As^{2}$
Safflorite (263), $(Ni, Co, Fe)As_{2}$
Nickel Glance (263), $Ni(S, As)^{2}$
Capillary Pyrites or Millerite (264), NiS
Emerald Nickel (264), $Ni\ddot{C}+2\dot{N}i\dot{H}^{3}$
Smaltine (266), $(Co, Ni)As_{2}$
Erythrine (267), $\dot{C}o^{3}\overset{\cdot\cdot\cdot\cdot\cdot}{As}+8\dot{H}$
Blende (269), ZnS
Zincite (270), $\dot{Z}n$
Sulphate of Zinc (271), $\dot{Z}n\dddot{S}+7\dot{H}$
Smithsonite (272), $\dot{Z}n\ddot{C}$
Calamine (272), $\dot{Z}n^{3}\dddot{S}i+1\frac{1}{2}\dot{H}$
Galena (277), PbS
Minium (280), $Pb^{3}O^{4}$
Anglesite (281), $\dot{P}b\dddot{S}$
Cerusite (281), $\dot{P}b\ddot{C}$
Pyromorphite (283), $\dot{P}b^{3}\overset{\cdot\cdot\cdot\cdot\cdot}{P}+\frac{1}{3}PbCl$
Crocoisite (284), $\dot{P}b\dddot{C}r$
Cinnabar (287), HgS
Copper Glance (292), ȻuS
Copper Pyrites (292), $\text{Ȼ}uS+Fe^{2}S^{3}$
Erubescite (294), (Fe, Ȼu)S
Tetrahedrite (295), $\text{Ȼ}uS+\frac{1}{4}(Sb, As)^{2}S^{3}$
Red Copper (296), $\dot{C}u$
Blue Vitriol (297), $\dot{C}u\dddot{S}+5\dot{H}$
Malachite (298), $\dot{C}u^{2}\ddot{C}+\dot{H}$
Azurite (300), $2\dot{C}u\ddot{C}+\dot{C}u\dot{H}$
Chrysocolla (300), $\dot{C}u^{3}\dddot{S}i_{2}+6\dot{H}$
Dioptase (301), $\dot{C}u^{3}\dddot{S}i^{2}+3\dot{H}$
Silver Glance (325), AgS
Brittle Silver Ore (326), $AgS+\frac{1}{6}Sb_{2}S^{3}$
Dark Red Silver Ore (327), $AgS+\frac{1}{3}Sb^{2}S^{3}$
Light Red Silver Ore (327), $AgS+\frac{1}{3}As^{2}S^{3}$
Horn Silver (327), AgCl
Bromic Silver (327), AgBr
Embolite (327, Ag(Cl, Br)

CHAP. VIII.—ROCKS OR MINERAL AGGREGATES.

General Nature of Rocks. In the early part of this volume it is stated that the rocks of the globe are mineral in their nature, and consist either of a single mineral in a massive state, or of intimate combinations of different minerals. Limestone, when pure, is a single mineral,—it is the species calcite or carbonate of lime; common *granite* is a compound or aggregate of three minerals, quartz, feldspar, and mica. *Sandstones* may consist of grains of quartz alone, like the sands of many sea-coasts, being such a rock as these sands would make if agglutinated; it is common to find along with the quartz, grains of feldspar, and sometimes mica. *Clay slates* consist of quartz and feldspar or clay, with sometimes mica, all so finely comminuted, that often the grains cannot be observed. *Conglomerates* or *puddingstones*, may be aggregates of pebbles of any kind: of granite pebbles, of quartz pebbles, of limestone pebbles, or of mixtures of different kinds, cemented together by some cementing material, such as silica, oxyd of iron, or carbonate of lime.

Texture or structure of Rocks.—Rocks differ also in texture. In some, as granite, or syenite, the texture is crystalline: that is, the grains are more or less angular, and show faces of cleavage; the aggregation was the result of a cotemporaneous crystallization of the several ingredients. Common statuary or white building marble, consists of angular grains, and is crystalline in the same manner. But a pudding-stone is evidently not a result of crystallization; it consists only of adhering pebbles of other rocks with a cementing material which is often not apparent. Sandstones also are an agglutination of grains of sand,—just such rocks as would be made from ordinary sand by compacting it together; and clay slates are often just what would result from solidifying a bed of clay. There are therefore *crystalline* and *uncrystalline* rocks. It should be remembered, however, that in each kind of rock the grains themselves are crystalline, as all solid matter becomes solid by crystallization. But the *former* kind is a crystalline aggregation of grains, the *latter* a mechanical aggregation.

In crystalline rocks it is not always possible to distinguish the grains, as they may be so minute, or the rock so compact, that they are not visible. Much of the crystalline rock called basalt is thus compact.

Positions, or modes of occurrence of Rocks. A great part of the rocks of the earth's surface constitute extensive beds or layers, lying one above the other, and varying in thickness from a fraction of an inch to many scores of yards. There are compact limestones, beds of sandstone, and shales or clay slates, in many and very various alternations. In some regions, certain of these rocks, or certain parts of the series, may extend over large areas or underlie a whole country, while others are wholly wanting or present only in thin beds. The irregularities in their geographical arrangement and in the order of superposition are very numerous, and it is one object of geology to discover order amid the apparent want of system. Thus in Pennsylvania, over a considerable part of the state, there are sandstones, shales, and limestones, connected with beds of coal. In New York there are other sandstones, shales, and limestones, without coal; and the geologist ascertains at once by his investigations, (as was observed in the remarks on coal,) that no coal can be expected to be found in New York. These rocks contain each its own peculiar organic remains, and these are one source of the confident decision of the geologist. The stratified rocks bear evidence in every part—in their regular layers, their worn sand or pebbles, and their fossils,—that they are the result of gradual accumulations beneath water, marine or fresh, or on the shores of seas, lakes or rivers.

Besides the *stratified rocks* alluded to, there are others which, like the ejections from a volcano, or an igneous vent, form beds, or break through other strata and fill fissures often many miles in length. The rock filling such fissures, is called a dike. Such are the *trap dikes* of New England and elsewhere; they are fissures filled by trap. Porphyry dikes, and many of the veins in rocks, are of the same kind. Similar rocks may also occur as extensive layers; for the lavas of a single volcanic eruption are sometimes continuous for 40 miles. They may appear underlying a wide region of country, like granite.

The stratified rocks, or such as consist of material in regular layers, are of two kinds. The worn grains of which they are made are sometimes distinct, and the remains of shells farther indicate that they are the result of gradual accumulation. But others, or even certain parts of beds that elsewhere contain these indications, have a crystalline texture. A limestone bed may be compact in one part, and granular or crystalline, like statuary marble, in another. Here is an effect of heat on a portion of the bed; heat, which has acted since the rock was deposited. Other rocks, such as mica slate, gneiss, and probably some granites, have thus been crystallized. They are called *metamorphic rocks.*

In these few general remarks on the structure of the globe, we have distinguished the following general facts:

1. The great variety of alternations of sandstone, conglomerates, clay shale, and limestones.
2. The existence of igneous rocks in beds and intersecting dikes or veins.
3. The mechanical structure of sandstone, conglomerate, and shales.
4. The crystalline character of igneous rocks.
5. The crystalline character of many stratified or sedimentary rocks, arising from the action of heat upon the beds of rock themselves, after they were first formed.

We follow this comprehensive survey of the arrangement and general nature of rocks, with descriptions of the more prominent varieties and a mention of their applications in the arts.*

---

* One of the most important uses of stone is for architectural purposes. The character of the material depends not only upon its durability, but also its contraction or expansion from changes of temperature. This latter cause occasions fractures or the opening of seams, and produces in cold climates serious injuries to structures. The following table, by Mr. A. J. Adie, gives the rate of expansion in length for different materials, for a change of temperature of 180° F.—*Proc. Roy. Soc. Edinb.*, i, 95, 1835.

Granite,	·0008968—·0007894
Sicilian white marble,	·00110411
Carrara marble,	·0006539
Black marble, from Galway, Ireland,	·00044519
Sandstone, (Craigleith quarry, Scotland,)	·0011743
Slate, Penryhn quarry, Wales,	·0010376
Greenstone,	·0008089
Best brick,	·0005502
Fire brick,	·0004928
Cast iron,	·00114676—·001102166
Rod of wedgewood ware,	·00045294

GRANITE. SYENITE.

Granite consists of the three minerals, quartz, feldspar, and mica. It has a crystalline granular structure, and usually a grayish-white, gray, or flesh-red color, the shade varying with the color of the constituent minerals. When it contains hornblende in place of mica, it is called *syenite;* hornblende resembles mica in these rocks but the laminæ separate much less easily and are brittle.

Granite is said to be *micaceous, feldspathic*, or *quartzose*, according as the *mica, feldspar*, or *quartz*, predominates.

It is called *porphyritic* granite, when the feldspar is in large crystals, and appears over a worn surface like thickly scattered white blotches, often rectangular in shape.

Graphic granite has an appearance of small oriental characters over the surface, owing to the angular arrangement of the quartz in the feldspar, or of the feldspar in the quartz.

Graphic Granite.

When the mica of the granite is wanting, it is then a granular mixture of feldspar and quartz, called *granulite* or *leptynite.*

When the feldspar is replaced by albite, it is called *albite* granite. The albite is usually white, but otherwise resembles feldspar. When replaced by talc, it is called *protogine.*

Granite is the usual rock for veins of tin ore. It contains also workable veins of pyritous, vitreous, and gray copper ore, of galena or lead ore, of zinc blende, of specular and magnetic iron ore, besides ores of antimony, cobalt, nickel, uranium, arsenic, titanium, bismuth, tungsten, and silver, with rarely a trace of mercury. The rare cerium and yttria minerals are found in granite, and mostly frequently in

---

The experiments of W. H. C. Bartlett, Lieut. U. S. Engineers, led to the following results.—*Amer. J. Sci.*, xxii, 136, 1832.

	For 1° F.	For 180° F.
Granite,	·000004825	·00086904
Marble,	·000005668	·00102024
Sandstone,	·000009532	·00171596
Hammered copper,	·000009440	.00169920

albitic granite. It also contains emerald, topaz, corundum, zircon, fluor spar, garnet, tourmaline, pyroxene, hornblende, epidote, and many other species.

Diorite is a rock of the granitic series, consisting of hornblende and feldspar. Color dark green or greenish-black Crystalline texture distinct.

Granite is one of the most valuable materials for building. The rock selected for this purpose, should be fine and even in texture, as the coarser varieties are less durable; it should especially be pure from pyrites or any ore of iron, which on exposure to the weather will rust and destroy, as well as deface, the stone. The only certain evidence of durability, must be learned from examining the rock in its native beds; for some handsome granites which have every appearance of durability, decompose rapidly from some cause not fully understood. The more feldspathic are less enduring than the quartzose, and the syenitic (or hornblendic) variety more durable than proper granite itself. The rock, after removal from the quarry, hardens somewhat, and is less easily worked than when first quarried out.

Massachusetts is properly the granite state of the union. New Hampshire and Maine also afford a good material. The Quincy quarries in Massachusetts, south of Boston, have for many years been celebrated. Besides this locality, there are others in the eastern part of this state, between cape Ann and Salem, in Gloucester, at Fall River, in Troy, in Danvers; also south between Quincy and Rhode Island, where it is wrought in many places, as well as in Rhode Island, even to Providence. The so-called Chelmsford granite comes from Westford and Tyngsborough, beyond Lowell, and an excellent variety is obtained at Pelham, a short distance north in New Hampshire. Masses 60 feet in length are obtained at several of the quarries. They are worked into columns for buildings, many fine examples of which are common in Boston, New York, and other cities.

Good granite is also quarried in Waterford, Greenwich, and elsewhere, in Connecticut.

The granite is detached in blocks by drilling a series of holes, one every few inches, to a depth of three inches, and then driving in wedges of iron between steel cheeks. In this manner masses of any size are split out. There is a choice of direction, as the granite has certain directions of easiest fracture. Masses are often got out in long narrow strips, a foot wide, for fence posts.

Granite is also used for paving, in small rectangular blocks neatly fitted together, as in London and in some parts of New York and other cities. The feldspathic granite is of great value in the manufacture of porcelain, as remarked upon under *Feldspar*.

Granite was much used by the ancients, especially the Egyptians, where are obelisks that have stood the weather for 3000 years.

GNEISS.

Gneiss has the same constitution as granite, but the mica is more in layers, and the rock has therefore a stratified appearance. It generally breaks out in slabs a few inches to a foot thick. It is hence much used both as a building material and for flagging walks. The quarries in the vicinity of Haddam, Conn., on the Connecticut river, are very extensively opened, and a large amount of stone is annually taken out and exported to the Atlantic cities, even as far as New Orleans. There are also quarries at Lebanon and other places, in Connecticut; at Wilbraham, Millbury, Monson, and many other places in Massachusetts.

MICA SLATE.

Mica slate has the constituents of gneiss, but is thin slaty, and breaks with a glistening or shining surface, owing to the large proportion of mica, upon which its foliated structure depends. It contains less feldspar and much more mica than gneiss.

The thin even slabs of the more compact varieties of mica slate are much used for flagging, and for door and hearth stones; also for lining furnaces. The finer arenaceous varieties make good *scythe stones*.

It is quarried extensively of fine quality, in large even slabs, at Bolton in Connecticut; also in the range passing through Goshen and Chesterfield, Mass. It is worked into whetstones in Enfield, Norwich, and Bellingham, Mass., and extensively at Woonsocket Hill, Smithfield, R. I. The south part of Chester, Vt., affords a slate like that of Bolton. Mica slate is used at Salisbury, Conn., for the inner wall of the iron furnace.

Hornblende slate resembles mica slate, but has not as glistening a luster, and seldom breaks into as thin slabs. It is more tough than mica slate, and is an excellent material for flagging.

TALCOSE SLATE.—TALCOSE ROCK.

Talcose slate resembles mica slate, but has a more greasy feel, owing to its containing talc instead of mica. It is usually light gray or dark grayish-brown. It breaks into thin slabs, but is generally rather brittle, yet it often makes good fire-stones.

A talcose slate in Stockbridge, Vt., is worked for scythe stones and hones, and is of excellent quality for this purpose.

Talcose rock is a hard and tough compact rock, containing more or less talc, and often quite compact. It is usually very much intersected by veins of white quartz. Much of it contains chlorite (an olive-green mineral) in place of talc, here and there disseminated.

Chlorite slate has a dark green color, and is similar in general characters to talcose slate. These slaty rocks are to a great extent the gold rocks of the world, especially the quartzose veins, as mentioned under Gold. Platinum, iridosmine, pyrites and many other minerals, occur in them, or in associated beds.

STEATITE, OR SOAPSTONE.

Steatite is a soft stone, easily cut by the knife and greasy in its feel. Its color is usually grayish-green; but when smoothed and varnished it becomes dark olive-green. It occurs in beds, associated generally with talcose slate.

Owing to the facility with which soapstone is worked, and its refractory nature, it is cut into slabs for fire stones and other purposes, as stated on page 144. The powder is employed for diminishing friction, and for mixing with blacklead in the manufacture of crucibles. It is also used, as observed by Dr. C. T. Jackson, for the sizing rollers in cotton factories, one of which is 4½ feet long and 5 to 6 inches in diameter. The most valuable quarries in Massachusetts are at Middlefield, Windsor, Blanford, Andover, and Chester; in Vermont, at Windham and Grafton; in New Hampshire, at Francestown and Oxford; in Orange county, North Carolina. The Francestown soapstone sells at Boston at from 36 to 42 dollars the ton, or from 3 to 3½ dollars the cubic foot.*

Steatite often contains disseminated crystals of magnesian carbonate of lime, (dolomite,) and brown spar; also crystals of pyrites and actinolite.

* Geol. N. H., by C. T. Jackson, 1844; p. 168.

Potstone is a compact steatite. Rensselaerite is another compact variety, (page 144,) found in Jefferson and St. Lawrence counties, N. Y., and used for inkstands.

SERPENTINE.

This dark green rock is usually associated with talcose rocks, and often also with granular limestones. It has been described on page 145, where its uses are alluded to. It often contains disseminated a foliated green variety of hornblende called diallage. A compound rock consisting of diallage and feldspar, has been called *diallage rock* or *euphotide.*

TRAP.—BASALT.

Trap is a dark greenish or brownish black rock, heavy and tough. Specific gravity 2·8—3·2. It has sometimes a granular crystalline structure, and at other times it is very compact without apparent grains. It is an intimate mixture of feldspar and augite. It is often called *dolerite.* The feldspar in this rock is usually the kind called labraborite. (p. 178.)

Amygdaloid, (from the Latin *amygdalum,* an almond,) is a trap containing small almond-shaped cavities, which are filled with some mineral: usually a zeolite, quartz or chlorite.

Porphyritic trap is a trap containing, like porphyritic granite, disseminated crystals of feldspar.

Basalt resembles trap, but consists of augite, olivine and feldspar. It varies in color from grayish to black. In the lighter colored, which are sometimes denominated *graystone,* feldspar predominates: and in the darker, iron, or a ferruginous augite. The chrysolite (or olivine) it contains is in small grains of a bottle-glass appearance. Magnetic or titanic iron are also frequently present in the rock. When feldspar crystals are coarsely disseminated, it is called *porphyritic basalt;* and when containing minerals in small nodules, it is *amygdaloidal basalt.* Basalt is a common product of volcanic action.

Wacke or *toadstone* is an earthy basalt, or a sedimentary rock of trap or basaltic material.

Both trap and basalt occur in columnar forms, as at the Giant's Causeway and other similar places.

Trap and basalt are excellent materials for macadamizing roads, on account of their toughness. Trap is also used for buildings. It breaks into irregular angular blocks, and is employed in this condition. For a Gothic building it is well fitted, on account of an appearance of age which it has.

PORPHYRY.—CLINKSTONE.—TRACHYTE.

Porphyry consists mainly of compact feldspar, with disseminated crystals of feldspar. Red or brownish-red and green, are common colors; but gray and black are met with. The feldspar crystals are from a very small size to half or three quarters of an inch in length, and have a much lighter shade of color than the base, or are quite white. It breaks with a smooth surface and conchoidal fracture. The specific gravity and other characters of the rock are the same nearly as for the mineral feldspar; the hardness is usually a little higher than in that mineral.

Porphyry receives a fine polish, and has been used for columns, vases, mortars, and other purposes. Green porphyry is the *oriental verd antique* of the ancients, and was held in high esteem. The red porphyry of Egypt is also a beautiful rock. It has a clear brownish red color, and is sprinkled with small spots of white feldspar.

Clinkstone or *Phonolite* is a grayish-blue rock, consisting, like porphyry, mainly of feldspar. It passes into gray basalt, and is distinguished by its less specific gravity. It rings like iron when struck with a hammer, and hence its name.

Trachyte is another feldspathic rock, distinguished by breaking with a rough surface, and showing less compactness than clinkstone. It sometimes contains crystals of hornblende, mica, or some glassy feldspar mineral. It occurs in volcanic regions.

LAVA.—OBSIDIAN.—PUMICE.

The term *lava* is applied to any rock material which has flowed in igneous fusion from a volcano. Basalt is one kind of lava; and when containg cellules, it is called *basaltic lava.* Trachyte is also a lava. There are thus both feldspathic and basaltic lavas. The feldspathic are light colored, and of low specific gravity, (not exceeding 2·8); the basaltic vary from grayish-blue to black, and are above 2·8 in specific gravity. The general term *basaltic* sometimes includes *doleritic* lava, which is closely allied. Chrysolite is present in basaltic lavas; and such lavas are not unfrequently porphyritic, or contain disseminated crystals of feldspar.

The light cellular ejections of a volcano are called *scoria* or *pumice*.

Pumice is feldspathic in constitution ; it is very porous, and the fine pores lying in one direction make the rock appear to be fibrous. It is so light as to float on water. It is much used for polishing wood, ivory, marble, metal, glass, etc., and also parchment and skins. The principal localities are the islands of Lipari, Ponza, Ischia, and Vulcano, in the Mediterranean between Sicily and Naples. Both scoria and pumice are properly the scum of a volcano.

Volcanic ashes are the light cinders, or minute particles of rock, ejected from a volcano in the course of an eruption.

Obsidian is a volcanic glass. It resembles ordinary glass. Black and smoky tints are the common colors. In Mexico, it was formerly used both for mirrors, knives and razors. *Pitchstone* is less perfectly glassy in its character, and has a pitch-like luster. Otherwise it resembles obsidian. *Pearlstone* has a grayish color and pearly luster. *Spherulite* is a kind of pearlstone, occurring in small globules in massive pearlstone. *Marekanite* is a pearl-gray translucent obsidian from Marekan in Kamschatka.

ARGILLACEOUS SHALE, OR CLAY SLATE.—ARGILLITE.

Slate is an argillaceous rock, breaking into thin laminæ ; *shale* a similar rock, with the same structure usually less perfect and often more brittle ; *schist* includes the same varieties of rock, but is extended also to those of a much coarser laminated structure. The ordinary clay slate has the same constitution as mica slate ; but the material is so fine that the ingredients cannot be distinguished. The two pass into one another insensibly. The colors are very various, and always dull or but slightly glistening.

Roofing slate is a fine grained argillaceous variety, commonly of a dark dull blue or bluish-black color, or somewhat purplish. To be a good material for roofing, it should split easily into even slates, and admit of being pierced for nails without fracturing. Moreover, it should not be absorbent of water, either by the surface or edges, which may be tested by weighing, after immersion for a while in water. It should also be pure from pyrites and every thing that can undergo decomposition on exposure.

Roofing slates occur in England, in Cornwall and Devon, Cumberland, Westmoreland.

In the United States, a good material is obtained in Maine at Barnard, Piscataquies, Kennebec, Bingham and elsewhere; also in Massachusetts, in Worcester county, in Boylston, Lancaster, Harvard, Shirley, and Peperell; in Vermont, at Guilford, Brattleborough, Fairhaven, and Dummerston; in Hoosic, New York; on Bush creek and near Unionville, Maryland; at the Cove of Wachitta, Arkansas. At Rutland Vt., is a manufactory of slate pencils, from a greenish slate.

These slate rocks are also used for gravestones; and we cannot go through New England cemetries without frequent regret that a material which is sure to fall to pieces in a few years, should have been selected for such records.

Drawing slate is a finer and more compact variety, of bluish and purplish shades of color. The best slates come from Spain, Italy, and France. A good quality is quarried in Maine and Vermont.

Novaculite, *hone-slate*, or *whet-stone*, is a fine grained slate, containing considerable quartz, though the grains of this mineral are not perceptible. It occurs of light and dark shades of color, and compact texture. It is found in North Carolina, 7 miles west of Chapel Hill, and elsewhere; in Lincoln and Oglethorpe counties, Georgia; on Bush creek, and near Unionville, Maryland; at the Cove of Wachitta, Arkansas.

Argillite is a general term given to argillaceous or clay slate rocks. Many shales or argillites crumble easily, and are unfit for any purpose in the arts, except to furnish a clayey soil.

Alum shale is any slaty rock which contains decomposing pyrites, and thus will afford alum or sulphate of alumina on lixiviation. (See under Alum, page 128.)

Bituminous shale is a dark colored slaty rock containing some bitumen, and giving off a bituminous odor.

Plumbaginous schist is a clay slate containing plumbago or graphite, and leaving traces like black lead.

The *Pipestone* of the North American Indians was in part a red claystone or compacted clay from the Coteau de Prairies. It has been named *catlinite*. A similar material, now accumulating, occurs on the north shore of Lake Superior, at Nepigon bay. Another variety of pipestone is a dark grayish compact argillite; it is used by the Indians of the northwest coast of America.

Agalmatolite is a soft mineral, impressible by the nail,

and waxy in luster when polished, presenting grayish and greenish colors and other shades. Gr=2·8—2·9. It has a greasy feel. It consists of silica 55·0, alumina 30·0, potash 7·0, water 3 to 5 per cent., with a trace of oxyd of iron. It is carved into images, and is hence called *figure-stone.*

QUARTZ ROCK.

Quartz rock is a compact rock consisting of quartz, and often appearing granular. Its colors are light gray, reddish or dull bluish; also sometimes brown.

When the granular quartz contains a little mica, it often breaks in slabs like gneiss or mica slate. The *itacolumite* of Brazil, with which gold and topaz are associated, is a micaceous granular quartz rock of this kind.

Flexible sandstone is an allied rock of finer texture. It owes its flexibility to the mica present, and its looseness of aggregation. Granular quartz graduates into the proper sandstones, which are treated of for convenience on a following page.

Granular quartz is one of the most refractory of rocks. It is consequently used extensively for hearthstones, for the lining of furnaces, and for lime kilns. At Stafford, Conn., a loose grained micaceous quartz rock is highly valued for furnaces; it sells at the quarry for 16 dollars a ton.*

Granular quartz is also used for flagging, and a fine quarry is opened in Washington, near Pittsfield, Mass.; it also occurs of good quality at Tyringham and Lee, Mass. In the shape of cobble stones, it is a common paving material.

A highly important use of this rock is in the manufacture of glass and sandpaper, and for sawing marble. In many places it occurs crumbled to a fine sand, and is highly convenient for these purposes. In Cheshire, Berkshire county, Mass., and in Lanesboro', Mass., it occurs of superior quality, and in great abundance. It is also in demand for the manufacture of glass and pottery. In Unity, N. H., a granular quartz is ground for sandpaper and for polishing powder; the latter is a good material for many purposes.

A fine variety of granular quartz is a material much valued for whet-stones.

BUHRSTONE.

Buhrstone is a quartz rock containing cellules. It is as hard and firm as quartz crystal, and owes its peculiar value

* Rep. on Connecticut, by C. U. Shepard—p. 78.

to this quality and the cellules, which give it a very rough surface. In the best stones for wheat or corn the cavities about equal in space the solid part. The finest quality comes from France, in the basin of Paris and some adjoining districts.

The stones are cut into wedge-shaped parallelopipeds called panes, which are bound together by iron hoops into large millstones. The Paris buhrstone is from the tertiary formation, and is therefore of much more recent origin than the quartz rock above described.

Buhrstone of good quality is abundant in Ohio, and others of the western states. It is associated there with proper sandstones, as more particularly mentioned on page 346.

The quartz rock of Washington, near Pittsfield, Mass., is in some parts cellular, and makes good millstones.

A buhrstone occurs in Georgia, about 40 miles from the sea, near the Carolina line; also in Arkansas, near the Cove of Wachitta.

SANDSTONES.—GRIT ROCKS.—CONGLOMERATES.

Sandstones consist of small grains, aggregated into a compact rock. They have a harsh feel, and every dull shade of color from white through yellow, red and brown to black. Many sandstones are very compact and hard, while others break or rub to pieces in the fingers. They usually consist of siliceous sand; but grains of feldspar are often present. In many compact sandstones there is much clay, and the rock is then an argillaceous sandstone.

Sandstones are of all geological ages, from the lower Silurian to the most recent period. The older rocks are in general the most firm and compact. The "old red" sandstone is a sandstone below the coal in age; while the so called "new red" is more recent than the coal. But these terms are of indefinite application out of Great Britain, and are not now used in this country. Red sandstone, when used as a building material, is often called freestone.

Grit rock. When the sandstone is very hard and harsh, and contains occasional siliceous pebbles, it is called a grit rock, or millstone grit.

Conglomerates. Conglomerates consist mostly of pebbles compacted together. They are called *pudding stone* when the pebbles are rounded, and *breccia* when they are angular. They may consist of pebbles of any kinds, as of granite, quartz, limestone, etc., and they are named accordingly *granitic, quartzose, calcareous, conglomerates.*

The use of sandstone as a building material is well known. For this purpose it should be free, like granite, from pyrites or iron sand, as these rust and disfigure the structure. It should be firm in texture, and not liable to peel off on exposure. Some sandstones, especially certain argillaceous varieties, which appear well in the quarry, when exposed for a season where they will be left to dry, gradually fall to pieces. The same rock answers well for structures beneath water, that is worth nothing for buildings. Other sandstones which are so soft as to be easily cut from their bed without blasting, harden on exposure, (owing to the hardening of silica in the contained moisture,) and are quite durable. These are qualities which must be tested before a stone is used. Moreover it should be considered that in frosty climates, a weak absorbent stone is liable to be destroyed in a comparatively short time, while in a climate like that of Peru, even sunburnt bricks will last for centuries.

Mr. Ure observes, that "such was the care of the ancients to provide strong and durable materials for their public edifices, that but for the desolating hands of modern barbarians, in peace and in war, most of the temples and other public monuments of Greece and Rome would have remained perfect at the present day, uninjured by the elements during 2000 years. The contrast in this respect of the works of modern architects, especially in Great Britain, [much more true of the United States,] is very humiliating to those who boast so loudly of social advancement; for there is scarcely a public building of recent date which will be in existence a thousand years hence." Many splendid structures are monuments (not endless) of folly in this respect. He observes also that the stone intended for a durable edifice ought to be tested as to its durability by immersion in a saturated solution of sulphate of soda, and exposure to the air for some days: the crystallization within the stone will cause the same disintegration that would result in time from frost.

The dark red sandstone (*freestone*) of New Jersey and Connecticut, when of fine gritty texture and compact, is generally an excellent building material. Trinity Church in New York is built of the stone from Belville, New Jersey. At Chatham, on the Connecticut, is a large quarry, which supplies great quantities of stone to the cities of the coast;

and there are numerous others in the Connecticut valley, both in Connecticut and Massachusetts. A variety in North Haven, at the east end of Mount Carmel, has been spoken of as excellent for ornamental architecture. That of Longmeadow and Wilbraham, in Massachusetts, is a very fine and beautiful variety and is much used. A *freestone* occurs also at the mouth of Seneca creek, Maryland, convenient for transportation by the Chesapeake and Ohio canal; white and colored sandstones occur also at Sugarloaf mountain, Maryland.

The sandstone of the Capitol at Washington, is from the Potomac; it is a poor material.

Sandstones when splitting into thin layers, form excellent *flagging stones*, and are in common use.

Hard, gritty sandstones and the grit rocks are used for the *hearths of furnaces*, on account of their resistance to heat. They are also much used for *millstones*, and when of firm texture, make a good substitute for the buhrstone.

The true *buhrstone* has been described as a cellular siliceous rock, without an apparent granular texture. The buhrstone of Ohio approaches this character; it is in part a true sandstone containing fossils in some places, and overlying the coal. Much of it contains lime; and it is possible that the removal of the lime by solution, since its deposition, may have occasioned its cellular character. It has an open cellular structure where quarried for millstones. It occurs in Ohio, in the county of Muskingum, and the counties south and west of south, on the Raccoon river and elsewhere. The manufacture commenced in this region in 1807, and in Richland, Elk, and Clinton, and in Hopewell, the manufacture is now carried on extensively. Stones 4 feet in diameter bring $150.*

The "green sand" of the cretaceous formation contains grains of silicate of iron and potash, to which it owes its greenish tint. It occurs abundantly in New Jersey as a soft rock, and is much used for improving lands: a value it owes mostly to the alkali it contains.

Pudding stones and breccias are fitted, in general, only for the coarser uses of stone, as for foundations, butments of bridges. Occasionally when of limestone, they make handsome marbles, as the "Potomac breccia marble" on the

* S. P. Hildreth, Geol. Report, Ohio.

Monocacey, of which the columns in the Hall of Representatives at Washington.

Porphyry couglomerates, *basaltic conglomerates*, *pumiceous conglomerates*, consist respectively of pebbles or fragments of porphyry, basalt, pumice.

Tufa is a sandrock consisting of volcanic material, either cinders or the comminuted lavas. Pozzuolana is a kind of tufa found in the vicinity of Rome, Italy. It consists of silica 34·5, alumina 15, lime 8·8, magnesia 4·7, potash 1·4, soda 4·1, oxyds of iron and titanium 12, water 9·2. *Peperino* is a coarse sandrock, made up of volcanic cinders or fine fragments of scoria, partially agglutinated.

LIMESTONES.

Limestones consist essentially of carbonate of lime, and belong to the species calcite, (p. 115,) or of the carbonates of lime and magnesia. They are distinguished by being easily scratched with a knife, and by effervescing with an acid. They are either *compact* or *granular* in texture: the compact break with a smooth surface, often conchoidal; the granular have a crystalline granular surface, and the fine varieties resemble loaf sugar.

Granular limestone. The finest and purest white crystalline limestones are used for statuary and the best carving, and are called *statuary marble.* A variety less fine in texture is employed as a building material. Its colors are white, and clouded of various shades. It often contains scales of mica disseminated, and occasionally other impurities, from which the cloudings arise.

The finest statuary marble comes from the Italian quarry at Carrara; from the Island of Paros, whence the name Parian; from Athens, Greece; from Ornofrio, Corsica, of a quality equal to that of Carrara. The Medicean Venus and most of the fine Grecian statues are made of the Parian marble. These quarries, and also those of the Islands of Scio, Samos and Lesbos, afforded marble for the ancient temples of Greece and Rome. The Parthenon at Athens was constructed of marble from Pentelicus.

Statuary marble has been obtained in the United Sates, but not of a quality equal to the foreign. Good building material is abundant along the Western part of Vermont, and south through Massachusetts to Western Connecticut and Eastern New York. In Berkshire county, Mass., marble is quarried annually to the value of $200,000; the principal quarries are at Sheffield, West Stockbridge, New Ashford, New Marlborough, Great Barrington, and Lanesborough.* The columns of the Girard College are from Sheffield, where blocks 50 feet long are sometimes blasted out; the material of the City Hall, New York, came from West Stockbridge; that of the Capitol at Albany, from Lanesboro'. At Stoneham is a fine statuary marble; but it is difficult to obtain large blocks. The variety from Great Barrington is a handsome clouded marble. Some of the West Stockbridge marble is flexible in thin pieces when first taken out. There are Vermont localities at Dorset, Rutland, Brandon, and Pittsford. In New York extensive quarries are opened not far from New York, at Sing Sing; also at Patterson, Putnam county; at Dover in Dutchess county, N. Y.; in Connecticut there are marble quarries at New Preston; in Maine at Thomaston: in Rhode Island at Smithfield, a fine statuary; in Maryland, a few miles east of Hagerstown; in Pennsylvania, a fine clouded variety, 20 miles from Philadelphia. A fine *dun colored* marble is obtained at New Ashford and Sheffield, Mass., and at Pittsford, Vt.

The granular limestone when coarse usually crumbles easily, and is not a good material for building. But the best varieties are not exceeded in durability by any other architectural rock, not even by granite. The impurities are sometimes so abundant as to render it useless. For statuary, it is essential that it should be uniform in tint and without seams or fissures; the liability of finding cloudings within the large blocks makes them useless for statuary. The presence of pyrites or manganese unfits the stone for buildings.

The common minerals in this rock are tremolite, asbestus, scapolite, chondrodite, pyroxene, apatite, besides sphene, spinel, graphite, idocrase, mica.

Verd antique marble—verde antico—is a clouded green marble, consisting of a mixture of serpentine and limestone, as mentioned under Serpentine, page 147. It occurs at Milford, near New Haven, Connecticut, of fine quality; and also in Essex county, N. Y., at Moria and near Port Henry on Lake Champlain. A marble of this kind occurs at Genoa and in Tuscany, and is much valued for its beauty A variety is called *polzivera di Genoa* and *vert d'Egypte.*

* Hitchcock's Geol. Rep., p. 162.

The *Cipolin* marbles of Italy are white, or nearly so, with shadings or zones of green talc. The *bardiglio* is a gray variety from Corsica.

Compact limestone usually breaks out easily into thick slabs, and are a convenient and durable stone for building and all kinds of stone work. It is not possessed of much beauty in the rough state. When polished it constitutes a variety of marbles according to the color; the shades are very numerous, from white, cream and yellow shades, through gray, dove-colored, slate blue or brown, to black.

The *Nero-antico* marble of the Italians is an ancient deep black marble; the *paragone* is a modern one, of a fine black color, from Bergamo; and *panno di morte* is another black marble with a few white fossil shells.

The *rosso-antico* is deep blood-red, sprinkled with minute white dots. The *giallo antico*, or yellow antique marble, is deep yellow with black or yellow rings. A beautiful marble from Sienna, *brocatello di Siena*, has a yellow color, with large irregular spots and veins of bluish-red or purplish. The *mandelato* of the Italians is a light red marble, with yellowish-white spots; it is found at Luggezzana. At Verona, there is a *red marble*, inclining to yellow, and another with large white spots in a reddish and greenish paste.

The *black marble* used in the United States comes mostly from Shoreham, Vt., and other places in that state near Lake Champlain. The *Bristol marble* of England is a black marble containing a few white shells, and the *Kilkenny* is another similar. There are several quarries at Isle La Motte. It is quarried also near Plattsburgh and Glenn's Falls, N. Y.

The *portor* is a Genoese marble very highly esteemed. It is deep black, with elegant veinings of yellow. The most beautiful comes from Porto-Venese, and under Louis XIV a great deal of it was worked up for the decoration of Versailles.

Gray and dove-colored compact marbles are common through New York and the states West.

The *bird's-eye* marble of Western New York is a compact limestone, with crystalline points scattered through it.

Ruin marble is a yellowish marble, with brownish shadings or lines arranged so as to represent castles, towers or cities in ruins. These markings proceed from infiltrated iron. It is an indurated calcareous marl.

Oolitic marble has usually a grayish tint, and is speckled with rounded dots, looking much like the roe of a fish.

Shell marble contains scattered fossils, and may be of different colors. It is abundant through the United States. *Crinoidal* or *encrinital marble* differs only in the fossils being mostly remains of encrinites, resembling thin disks. Large quarries are opened in Onondaga and Madison counties, N. Y., and the polished slabs are much used. *Madreporic marble* consists largely of corals, and the surface consists of delicate stars: it is the *pietra stellaria* of the Italians. It is common in some of the states on the Ohio. *Fire marble*, or *lumachelle*, is a dark brown shell marble, having brilliant *fire* or *chatoyant reflections* from within.

Breccia marbles and *pudding stone marbles* are the polished calcareous breccia or pudding stone, alluded to on page 346.

Stalagmites and stalactites (page 116) are frequently polished, and the variety of banded shades is often highly beautiful. The *Gibraltar stone*, so well known, is of this kind. It comes from a cavern in the Gibraltar rock, where it was deposited from dripping water. It is made into inkstands, letter-holders, and various small articles.

Wood is often petrified by carbonate of lime, and occasionally whole trunks are changed to stone. The specimens show well the grain of the wood, and some are quite handsome when polished.

Marble is sawn by means of a thin iron plate and sand, either by hand or machinery. In polishing, the slabs are first worn down by the sharpest sand, either by rubbing two slabs together or by means of a plate of iron. Finer sand is afterwards used, and then a still finer. Next emery is applied of increasing fineness by means of a plate of lead; and finally the last polish is given with tin-putty, rubbed on with coarse linen cloths or baggings, wedged tight into an iron planing tool. More or less water is used throughout the process.

Quicklime. Limestone when burnt produces quicklime, owing to the expulsion of the carbonic acid by the heat. The purest limestone affords the purest lime, (what is called *fat* lime.) But some impurities are no detriment to it for making mortar, unless they are in excess. *Hydraulic lime*, which is so called because it will set under water, is made from limestone containing some clay, silica, and often magnesia. The French varieties contain 2 or 3 per cent. of magnesia, and 10 to 20 of silica and alumina or clay. The varieties in the United States contain 20 to 40 per cent. of magnesia, and 12 to 30 per cent. of silica and alumina. A variety worked extensively at Rondout, N. Y., afforded Prof. Beck, carbonic acid 34·20, lime 25·50, magnesia 12·35, silica 15·37, alumina 9·13, peroxyd of iron 2·25.* Oxyd of iron is rather prejudicial than otherwise.

In making mortar, the lime is mixed with water and siliceous sand. The final strength of the mortar depends principally on the formation of a compound between water, the silica (or sand) and the lime; of course therefore the finer the sand, the more thorough the combination. In hydraulic lime, there is silica and alumina present in a thoroughly disseminated and finely divided state, which is favorable for the combination alluded to; and to this fact appears to be mainly owing its hydraulic character. Much less sand is added in making mortar from this lime than from that of ordinary limestone.

Pozzuolana (page 347) forms a hydraulic cement when mixed with a little lime and water. Similar cements may be made with tufa, pumice stone, and slate clay, by varying the proportions of lime; these materials consist essentially of silica and alumina or magnesia with alkalies, and often some lime, and therefore produce the same result as with hydraulic limestone.

In the *burning of lime*, the most common mode is to erect a square or circular furnace of stone, with a door for managing the fire below. An arched cavity for the fire is first made of large pieces of limestone, and then the furnace is filled with the stone placed loosely so as to admit of the passage of the flame throughout: the carbonic acid is expelled by the heat, and when the fires are out, the lime now in the state of quicklime, or in other words, pure lime, is taken out. Great economy of fuel is secured by means of what is called a *perpetual kiln*. The cavity within is best made nearly of the shape of an egg with the narrow end uppermost. The inner walls are of quartz rock, mica slate, or some refractory stone or fire brick, and between the inner and outer there is a layer of cinders or ashes, as in the iron furnace, page 233. Below are three or more openings for furnaces which lead into the main cavity, a few feet from the bottom; and alternate with these are other openings at a

---

* Mineralogy of New York, page 78.

lower level for withdrawing the lime. The lime is taken out below and the stone thrown in above, and this may be kept up without intermission as long as the kiln lasts. Beneath the furnaces there are also ash pits. Such a kiln is most convenient for being filled and emptied when situated on a side hill.

The localities of limestone in the United States are too common to need enumeration. Hydraulic limestone is also abundant.

Quicklime is much used for improving lands; also for clarifying the juice of the sugar cane and beet root; for purifying coal gas; for clearing hides of their hair in tanneries, and for various other purposes.

SAND.—CLAY.

The loose or soft material of the surface of the earth consists of sand, clay, gravel or stones, and what we call in general terms, soil or earth. These materials are either in layers or irregular beds. Most clay beds, and many of gravel, when cut through vertically, show indications of horizontal layers, a result of deposition, or distribution, by water.

The ordinary constituents of earth are quartz, feldspar or clay, oxyd of iron and lime; but these vary with the source from whence they are derived When the rock that has afforded the soil is granite, mica slate, or the allied rocks, mica is usually present, as well as feldspar and quartz; so a quartzose rock will furnish siliceous gravel; a magnesian, will give magnesia to the soil; calcareous, lime; trap, the ingredients of decomposed feldspar or hornblende. The material will be coarse or gravelly, or fine earthy, according to the nature of the rock, or the condition under which it is worn down, or its subsequent distribution by flowing waters. Besides the prominent constituents mentioned, there are small proportions of phosphates, nitrates, chlorids, etc., together with the results of vegetable decomposition; and these comparatively rare ingredients are of great importance to growing vegetation. The pebbles of a soil are commonly siliceous, as this kind resists wear most effectually.

Sand is usually pulverized quartz, often with some feldspar. *Clay* is a plastic earth, consisting mainly of pulverized or altered aluminous minerals (largely feldspar) and quartz, the latter about two-thirds the whole. The alumina is often partly in the state of a hydrous silicate like Raolin or Halloysite. It owes its plasticity to the alumina, and ceases to be called clay when the proportion of silica is too great for plasticity. It is afforded by the decomposition of feldspar and all argillaceous rocks. Oxyd of iron, carbonate of lime, and magnesia, are often present in clays.

Sand for glass manufacture should be pure silica, free from a taint of iron. This purity is apparent in the clearness of the grains, under a lens, or their white color. The sand of Cheshire and Lanesboro', in Massachusetts, is a beautiful material.

In the *manufacture of glass*, the object is to form a transparent fusible compound, and not an opaque infusible one as in pottery. This result is secured by heating together to fusion, silica (quartz sand or flint powder) and the alkali potash or soda. The ingredients combine and produce a silicate of potash or soda—in other words, glass.

Besides these ingredients, lime or oxyd of lead are added for glass of different kinds. A small proportion of lime increases the density, hardness, and luster of glass, producing a specific gravity between 2·5 and 2·6; while with lead a still denser material is formed—called *crystal* or *flint* glass—whose specific gravity is from 3 to 3·6.

From 7 to 20 parts of lime are added for 100 of silica, and 25 to 50 of calcined sulphate or carbonate of soda; common salt (chlorid of sodium) may also be employed. A good colorless glass has been found by analysis to consist of silica 76·0, potash 13·6, and lime 10·4 parts, in a hundred. For coarse bottle-glass, wood-ashes and coarse sea-weed soda, called *kelp*, or else pearlashes, are used along with siliceous sand and broken glass. For a hard glass, the proportion of alkali is small.

The best English crystal glass analyzed by Berthier, afforded 59 parts of silica, 9 of potash, 28 of oxyd of lead, and 1·4 of oxyd of manganese. Crown glass contains, in general, less alkali than crystal glass, and is superior in hardness. The alkali, moreover, in England, is soda instead of potash. *Plate glass* also contains soda, and this soda (the carbonate) is prepared with great care. The proportions are 7 parts of sand, 1 of quicklime, 2½ of dry carbonate of soda, besides cullet or broken plate.

The materials are first well pounded and sifted, and mixed into a fine paste; they are then heated together in pots made of a pure refractory clay, until fusion has taken place and the material has settled. The glass is afterwards worked by blowing, or moulded, into the various forms it has in market; and it is finally *annealed*—or in other words, is very slowly cooled—to render it tough. A little oxyd of manganese is usually employed to correct the green color which glass is apt to derive from any oxyd of iron present. But if the manganese is in excess, it gives a violet tinge to it.

The following chemical distribution of glasses has been proposed:

Soluble glass. A simple silicate of potash or soda, or of both of these alkalies.

Bohemian or crown glass. Silicate of potash and lime.

Common window and mirror glass. Silicate of soda and lime; sometimes also of potash.

Bottle glass. Silicate of soda, lime, alumina, and iron.

Ordinary crystal glass. Silicate of potash and lead.

Flint glass. Silicate of potash and lead; more lead than in the preceding.

Strass. Silicate of potash and lead—still more lead.

Enamel. Silicate and stannate, or antimonate of potash or soda and lead.

Glass was manufactured by the Phœnicians, and the later Egyptians. According to Pliny and Strabo, the glass works of Sidon and Alexandria were famous in their times, and produced beautiful articles. The Romans employed glass to some extent in their windows, and remains of this glass are found in Herculaneum. Window glass manufacture was first commenced in England in 1557.

Sand for casting is a fine siliceous sand, containing a little clay to make it adhere somewhat and retain the forms into which it may be moulded. It must be quite free from lime.

Tripoli is a fine grained earthy deposit, having a dry, harsh feel and a white or grayish color. It contains 80 per cent. of silica, mostly derived from the casts of animalcules. It is valuable as a polishing material.

Marl. Marl is a clay containing carbonate of lime. The material is valuable as manure. The term is also improperly applied to any clayey earth used in fertilizing land. The green sand in New Jersey is sometimes called marl.

Fuller's earth is a white, grayish, or greenish-white earth, having a soapy feel, which was formerly used for removing oil or grease from woolen cloth. It falls to pieces in water, and forms a paste which is not plastic. A variety consists of silica 44·0, alumina 23·1, lime 4·1, magnesia 2·0, protoxyd of iron 2·0. Gr=2·45.

Lithomarge is a compact clay of a fine smooth texture, and very sectile. Its colors are white, grayish, bluish-white, reddish-white, or ocher-yellow, with a shining streak. Gr=2·4—2·5. The *tuesite* of Thomson, a white lithomarge from the banks of the Tweed, is said to make good slate pencils.

Clay for bricks is the most ordinary kind; it should have slight plasticity when moistened, and a fine even character without pebbles. It ordinarily contains some hydrated oxyd of iron, which when heated turns red by the escape of the water in its composition, which reduces it to the red oxyd of iron, and gives the usual red color to the brick. It also frequently contains lime; but much lime is injurious, as it renders the brick fusible. A clay is extensively employed at Milwaukie, in Michigan, which contains no iron, and produces a very handsome cream-colored brick. About 9,000,000 of this kind of brick were made at that place in 1847.

In *making bricks*, the clay is first well worked by the treading of cattle or by machinery: after this, it is moulded in moulds of the requisite size, (9⅝ inches, by 4¾ and 2¾,) and then taken out and laid on the ground. A good workman will make by hand 5000 in a day, and the best 10,000. After drying till stiff enough to bear handling, the bricks are trimmed off with a knife when requiring it, and piled up in long walls for farther drying. They are then made into a kiln by piling them in an open manner, (so that the flame and heated draft may have passage among them,) and leaving places beneath for the fires. The heat is continued 48 hours or more.

The best brick are pressed in moulds. They have a smooth, hard surface. Near Baltimore, Md., bricks are thus made by a machine, worked by a single horse, which will mould 30,000 bricks in 12 hours: the bricks are dry enough when first taken from the mould for immediate burning.

Burnt bricks were not used in England before the eleventh century, when they were employed in the construction of the abbey of St. Albans. But they date historically as far back as the city of Babylon. Unburnt bricks have also been used in all ages. Those of Egyptian and Babylonish times were made of worked clay mixed with chopped straw, to prevent it from falling to pieces. The *adobies* of Peru, are large sun-baked bricks or blocks of clay; and in that dry climate they are very durable.

Clay for Fire-bricks should contain no lime, magnesia, or iron, as its value depends on its being very refractory. There is a large manufactory in the United States, at Baltimore, from the tertiary clays of eastern Maryland. In England a slate clay from the coal series is employed.

Potter's clay and *pipe clay* are pure plastic clays, free from iron, and consequently burning white. The clay of Milwaukie, from which the cream-colored bricks are made, is much used also for pottery.

In the *manufacture of coarse pottery*, the clay is worked with water and tempered; and then the required form of a pot or pan is given on a wheel. The ware is dried under cover for a while, and next receives the glaze in a cream-like state. The glaze for the most common ware consists of very finely pulverized galena, mixed with clay and water. The ware after drying again is next placed in the kiln, which is very gradually heated; the heat causes the baking of the clay, and drives off the sulphur of the galena, thus producing an oxyd of lead, which forms a kind of glass (or glaze,) with the alumina. For a better stone ware, common salt is used, and it is put on after the baking has begun.

For the finer earthenware, a mixture of red and white lead, feldspar, silica and flint-glass, is used for a glaze, the proportions differing according to the ware. The clay for this ware is mixed with flint powder (ground flints or sand,) to render it less liable to contract or break, and it is worked with great care, and through various processes to prepare it for moulding. The ware is usually baked to a biscuit, before the glazing is put on, as in the manufacture of porcelain.

Kaolin or *porcelain clay*, is derived from the decomposition of feldspar, as stated on page 117. The foreign kaolin occurs in Saxony; in France at St. Yrieux-la-Perche, near Limoges; in Cornwall, England; also in China and Japan. The kaolin used at the Philadelphia porcelain works comes mostly from the neighborhood of Wilmington, Delaware.

The name kaolin is a corruption of the Chinese *Kauling*, meaning *high-ridge*, the name of a hill near Jauchau Fu, where this material is obtained.

In the *manufacture of porcelain*, the kaolin, and also the other ingredients, are first ground up separately to an impalpable powder The kaolin is mixed with a certain proportion of feldspar, flint and lime. The whole are worked up together in water, by mallets and spades, and well kneaded by the hands and sometimes the feet of the workmen. The plastic material is then laid aside in masses of the size of a man's head, and kept damp till required; the *dough*, as it is called, is now ready for the potter's lathe, (or other means,) by which it is moulded into the various forms of china ware. After moulding, they are slowly and thoroughly dried, and then taken to the kiln, for a preliminary baking. They come out in the state of *biscuit*, and are ready for painting and glazing. The colors are metallic oxyds, which are put on either from a wet copper-plate impression on bibulous paper, or by means of a brush. The former is used for flat surfaces; the paper is rubbed on carefully to transfer the impression to the porcelain, and is then wet and washed off. It is then carefully heated to evaporate any oil or grease employed in the printing. The glaze is made of a quartzose feldspar; it is ground to a very fine powder and worked into a paste with water, and a little vinegar. The articles are dipped for an instant into this milky fluid, and as they absorb the water they come out with a delicate layer of feldspar in a dry state. They are touched with a brush wherever not well covered. They are then ready to be finally baked in the kiln, for which purpose each vessel is placed in a separate baked clay case or receptacle, called a *sagger*. In this process the material undergoes a softening, amounting almost to a partial fusion, and thus receives the translucency which distinguishes porcelain from earthen or stone ware.

The blue color of common china is produced by means of oxyd of cobalt; carmine, purple and violet, by means of chlorid of gold; red of all shades by oxyd of iron; yellow by oxyd of lead, or white oxyd of antimony and sand; green by oxyd of copper or carbonate of lead; brown by oxyd of iron, manganese, or copper. A steel luster is produced from chlorid of platinum.

The best Sèvres ware is made from 63 to 70 parts of kaolin, 22 to 15 of feldspar, nearly 10 of flint, and 5 or 6 of chalk. In China the kaolin is mixed with a quartzose feldspar rock, consisting mainly of quartz, called *peh-tun-tsz*.

Soapstone is sometimes used in this manufacture; and as it substitutes magnesia for a part of the potash, it makes a harder ware; but it is also more brittle.

CHAPTER IX.

CATALOGUE OF AMERICAN LOCALITIES OF MINERALS.

The following catalogue may aid the mineralogical tourist in selecting his routes and arranging the plan of a journey. Only important localities, affording cabinet specimens, are in general included. The names of those minerals which are obtained in good specimens at the several localities, are distinguished by italics. When the specimens are remarkably good, an exclamation mark (!) has been added, or two of these marks (!!) when the specimens are quite unique.

MAINE.

MT. ABRAHAM.—*Andalusite*, staurotide.

ALBANY.—*Beryl ! green and black tourmaline, feldspar, rose quartz.*

ALBION.—Iron pyrites.

AROOSTOOK.—Red Hematite.

BINGHAM.—*Massive pyrites*, galena, blende, andalusite.

BLUE HILL BAY.—*Arsenical iron, molybdenite ! galena, apatite ! fluor spar !* black tourmaline, (Long Cove,) black oxyd of manganese, (Osgood's Farm,) rhodonite, bog manganese, wolfram.

BOWDOINHAM.—*Beryl*, molybdenite.

BRUNSWICK.—*Green mica, garnet ! black tourmaline !* molybdenite.

BUCKFIELD.—*Garnet*, (estates of Waterman and Lowe,) iron ore.

CAMDAGE FARM.—(Near the tide mills,) molybdenite, (wolfram.)

CAMDEN.—*Macle.*

CARMEL, (Penobscot Co.)—Gray Antimony.

CORINNA.—*Iron pyrites, arsenical pyrites.*

DEER ISLE.—*Serpentine, verd antique*, abestus, diallage.

DEXTER.—Galena, pyrites, blende, copper pyrites, green talc.

DIXFIELD.—Native copperas, graphite.

FARMINGTON.—(Norton's ledge,) *pyrites*, graphite, bog ore.

GEORGETOWN.—(Parker's island,) *beryl !* black tourmaline.

GREENWOOD.—Graphite, black manganese.

HARTWELL.—Staurotide.

LENOX.—Galena, pyromorphite.

LEWISTON.—Garnet.

LITCHFIELD.—*Sodalite, cancrinite, nepheline*, zircon.

LUBEC LEAD MINES.—*Galena, copper pyrites, blende*, pyromorphite, an ore of bismuth.

MADRID.—Gold.

NEWFIELD, (Bond's Mt.)—Mispickel, olive phosphate of iron in botryoidal masses.

PARIS.—*Green ! red ! black* and *blue tourmaline ! mica ! lepidolite !* feldspar, *albite, quartz crystals ! rose quartz, blende.*

PARSONSFIELD.—*Idocrase! yellow garnet, pargasite, adularia, scapolite,* galena, blende, copper pyrites.

PERRY.—Prehnite and calc spar, (above Loring's cove,) quartz crystal, calc spar, analcime, apophyllite, agate, (Gin Cove.)

PERU.—*Crystallized pyrites.*

PHIPSBURG.—*Yellow garnet! manganesian garnet, idocrase, pargasite, axinite, laumonite!* chabazite.

POLAND.—Idocrase.

RAYMOND.—*Magnetic iron, scapolite, pyroxene, lepidolite, tremolite,* hornblende.

RUMFORD.—*Yellow garnet, idocrase, pyroxene,* apatite, scapolite, graphite.

SANFORD, York Co.—*Idocrase!* albite, calcite, molybdenite, epidote.

SEARSMONT.—*Andalusite.*

STREAKED MOUNTAIN.—*Beryl! black tourmaline, mica, garnet.*

THOMASTON.—*Calcite, tremolite, hornblende,* sphene, arsenical iron, (Owl's head,) black manganese, (Dodge's mountain.)

WARREN.—Galena, blende.

WATERVILLE.—*Crystallized pyrites.*

WINDHAM, (near the bridge.)—*Staurotide, spodumene, garnet.*

WOODSTOCK, (New Brunswick.)—*Graphite, specular iron.*

NEW HAMPSHIRE.

ACWORTH.—*Beryl!! mica! tourmaline, feldspar, albite, rose quartz, columbite!*

ALSTEAD.—*Mica!! albite, black tourmaline.*

AMHERST.—*Idocrase! yellow garnet,* pargasite, calc spar.

BARTLETT.—Magnetic iron, *specular iron,* brown iron ore in large veins near Jackson, (on "Bald face mountain,") *quartz crystals, smoky quartz.*

BATH.—Galena, copper pyrites.

BELLOWS FALLS.—Kyanite, wavellite, near Saxton's river.

BENTON.—*Quart crystals.*

CAMPTON.—*Beryl!*

CANAAN.—Gold in pyrites.

CHARLESTOWN.—*Staurotide macle, andalusite macle,* bog iron ore.

CORNISH.—Gray antimony, antimonial argentiferous gray copper, *rutile in quartz!* (rare.)

EATON, (3 m. S. of.)—*Galena, blende!* copper pyrites, limonite, (Six Mile Pond.)

FRANCESTON.—*Soapstone,* arsenical pyrites.

FRANCONIA.—*Hornblende, staurotide! epidote! zoisite, specular iron, magnetic iron, black* and *red manganesian garnets: mispickel! (Danaite,)* copper pyrites, molybdenite, prehnite; specimens now hardly obtainable.

GILFORD.—(Gunstock Mt.)—Magnetic iron ore, native "lodestone."

GOSHEN.—*Graphite,* black tourmaline.

GRAFTON.—*Mica!* (extensively quarried at Glass Hill, 2 m. S. of Orange Summit,) *albite!* asparagus stone, blue, green and yellow *beryls!* (1 m. S. of O. Summit,) *tourmaline, garnets.*

GRANTHAM.—*Gray staurotide!*

HANOVER.—*Garnet*, a boulder of quartz containing *rutile! black tourmaline, quartz.*

HAVERHILL.—*Garnet! arsenical pyrites, native arsenic*, galena, blende, iron and copper pyrites, magnetic and white iron pyrites.

HILLSBORO', (Campbell's Mountain.)—*Graphite.*

HILLSDALE.—*Rhodonite*, black oxyd of manganese.

JACKSON.—Drusy quartz, tin ore, *arsenical pyrites*, native arsenic, fluor spar, apatite, *magnetic iron ore, molybdenite*, wolfram, copper pyrites, arseniate of iron.

JAFFREY.—(Monadnock Mt.)—*Kyanite.*

KEENE.—*Graphite, soapstone*, milky quartz.

LANDAFF.—*Molybdenite*, lead and iron ores.

LEBANON.—*Bog iron ore.*

LISBON.—*Staurotide*, black and red *garnets, granular magnetic iron ore, hornblende, epidote, zoisite, specular iron.*

LYME.—Kyanite, (N. W. part,) *black tourmaline*, rutile, iron pyrites, copper pyrites, (E. of E. village,) *sulphuret of antimony.*

MERRIMACK.—*Rutile!* (in gneiss nodules in granite vein.)

MOULTONBOROUGH, (Red Hill.)—*Hornblende*, bog ore, pyrites, tourmaline.

NEWPORT.—Molybdenite.

ORANGE.—*Blue beryls!* Orange Summit, chrysoberyl, *mica*, (w. side of mountain.)

ORFORD.—*Brown tourmaline*, (now obtained with difficulty,) *steatite, rutile*, kyanite, brown iron ore, native copper, green malachite, galena.

PELHAM.—*Steatite.*

PIERMONT.—*Micaceous iron, heavy spar*, green, white and brown mica, apatite.

PLYMOUTH.—Columbite, beryl.

RICHMOND.—*Iolite!* rutile, steatite, iron pyrites.

SADDLEBACK MT.—Black tourmaline, garnet, spinel.

SHELBURNE.—*Argentiferous galena, black blende, copper pyrites, iron pyrites*, manganese.

SPRINGFIELD.—Beryls, (very large, eight inches diameter,) *manganesian garnets!* in mica slate, *albite mica.*

SULLIVAN.—*Tourmalines*, (black,) in quartz, beryl?

SURREY.—Amethyst, calcite.

SWANZEY, (near Keene.)—*Magnetic iron*, (in masses in granite.)

TAMWORTH, (near White Pond.)—Galena.

UNITY, (estate of James Neal.)—*Copper and iron pyrites, chlorophyllite, green mica, magnetic iron, radiated actinolite*, garnet, *titaniferous iron ore, magnetic iron ore.*

WALPOLE, (near Bellows Falls.)—*Macle.*

WARREN.—*Copper pyrites, blende, epidote*, quartz, *iron pyrites, tremolite, galena, rutile, talc*, molybdenite.

WESTMORELAND, (South part.)—*Molybdenite! apatite! blue feldspar, bog manganese*, (north village,) quartz, *fluor spar*, copper pyrites, oxyd of molybdenum and uranium.

WHITE MTS., (notch behind "old Crawford's house.")—Green octahedral fluor, quartz crystals, black tourmaline, chiastolite.

WILMOT.—*Beryl.*

WINCHESTER.—Pyrolusite, diallogite, psilomelane, magnetic iron ore, granular quartz.

VERMONT.

ADDISON.—*Iron sand.*

ALBURGH.—Quartz crystals on calc spar, iron pyrites.

ATHENS.—*Steatite, rhomb spar,* actinolite.

BARNET.—Graphite.

BELVIDERE.—Steatite, chlorite.

BENNINGTON.—*Pyrolusite,* brown iron ore, pipe clay, yellow ochre.

BETHEL.—*Actinolite!* talc, chlorite, octahedral iron, *rutile, brown spar in steatite.*

BRANDON.—Braunite, pyrolusite, *psilomelane,* limonite, lignite, white clay, statuary marble; fossil fruits in the lignite.

BRATTLEBOROUGH.—Black tourmaline in quartz.

BRIDGEWATER.—*Talc, dolomite, magnetic iron,* steatite, chlorite, gold, native copper, blende, galena, blue spinel, copper pyrites.

BRISTOL.—*Rutile,* brown hematite, manganese ores.

BROOKFIELD.—*Mispickel, iron pyrites.*

CABOT.—Garnets, staurotide, hornblende, *albite.*

CASTLETON.—*Roofing slate.*

CAVENDISH.—Garnet, *serpentine.*

CHESTER.—*Asbestus.*

CHITTENDEN.—Psilomelane, pyrolusite, brown iron ore, *specular* and *magnetic iron,* galena.

COLCHESTER.—Brown iron ore, iron sand, jasper, alum.

CORINTH.—*Copper pyrites,* (has been mined;) magnetic iron pyrites.

COVENTRY.—Manganese spar.

CRAFTSBURY.—Mica in concentric balls.

DUMMERSTON.—Rutile.

FLETCHER.—Pyrites, octahedral iron, acicular tourmaline.

GRAFTON.—The *steatite* quarry referred to Grafton is properly in Athens.

GUILFORD.—Scapolite.

HARTFORD.—Calcite, *pyrites! kyanite* in mica slate.

IRASBURGH.—Rhodonite, *psilomelane.*

JAY.—*Chromic iron, serpentine,* picrosmine, amianthus.

LOWELL.—Picrosmine, amianthus.

MARLBORO'.—*Rhomb spar, steatite, garnet, magnetic iron.*

MENDON.—Octahedral iron ore.

MIDDLEBURY.—Zircon.

MIDDLESEX.—Rutile! (exhausted.)

MONKTON.—*Pyrolusite,* brown iron ore.

MORETOWN.—*Smoky quartz! steatite,* talc, wad, rutile.

MORRISTOWN.—Argentiferous galena.

MOUNT HOLLY.—*Asbestus,* chlorite.

NEW FANE.—*Glassy* and *abestiform actinolite, steatite, green quartz,* (called chrysoprase at the locality,) chalcedony, drusy quartz, *garnet, chromic iron, rhomb spar.*

NORWICH.—*Actinolite, feldspar, brown spar* in talc.

PITTSFORD.—*Brown iron ore,* manganese ores.

PLYMOUTH.—Spathic iron, magnetic and specular iron, both in octahedral crystals.

PLYMPTON.—Massive hornblende.

PUTNEY.—Fluor, *brown iron ore, rutile,* and *zoisite* in boulders.

READING.—*Glassy actinolite* in talc.

READSBORO'.—*Glassy actinolite, steatite.*

RIPTON.—*Brown iron ore,* augite in boulders, octahedral iron pyrites.

ROCHESTER.—Rutile, specular iron cryst., *magnetic iron,* in chlorite state.

ROXBURY.—*Dolomite, talc,* serpentine, asbestus.

SALISBURY.—Brown iron ore.

SHARON.—*Quartz crystals,* kyanite.

SHOREHAM.—*Iron pyrites.*

SHREWSBURY.—Magnetic iron and copper pyrites.

SOMERSET.—Magnetic iron, native gold.

STAFFORD.—Magnetic iron and *copper pyrites,* (has been worked,) native copper, hornblende.

STARKSBORO'.—Brown iron ore.

STIRLING.—Copper pyrites, talc, serpentine.

STOCKBRIDGE.—Mispickel, magnetic iron ore.

THETFORD.—Blende, *galena, kyanite ;* chrysolite in basalt.

TROY.—*Magnetic iron,* talc, *serpentine,* picrosmine, amianthus, *steatite,* one mile southeast of village of South Troy, on the farm of Mr. Pierce, east side of Missisco, *chromic iron.*

WARREN.—Actinolite, magnetic iron ore, wad.

WATERBURY.—Mispickel, copper pyrites, *rutile, quartz.*

WATERVILLE.—*Steatite,* actinolite, talc.

WELLS RIVER.—Graphite.

WESTFIELD.—*Steatite,* chromic iron, serpentine.

WESTMINSTER.—Zoisite in boulders.

WARDSBORO'.—*Zoisite.*

WINDHAM.—*Glassy actinolite, steatite.*

WOODBURY.—Massive pyrites.

WOODSTOCK.—*Quartz crystals.*

MASSACHUSETTS

ALFORD.—Galena, iron pyrites.

ATHOL.—*Allanite,* fibrolite (?) *epidote !* babingtonite?

AUBURN.—*Masonite.*

BARRE.—*Rutile ! mica, pyrites, beryl, feldspar, garnet.*

GREAT BARRINGTON.—*Tremolite.*

BEDFORD.—*Garnet.*

BELCHERTOWN.—Allanite.

BERNARDSTON.—Magnetic oxyd of iron.

BEVERLY.—Polymignite, columbite, *green feldspar,* tin ore.

BLANFORD.—*Marmolite, schiller spar, serpentine, anthophyllite, actinolite ! chromic iron,* kyanite, rose quartz in boulders.

BOLTON.—*Scapolite ! petalite, sphene, pyroxene, nuttalite, diopside, boltonite* (chrysolite), *petalite,* apatite, magnesite, rhomb spar, *allanite, yttrocerite,* cerium ochre (on the scapolite), spinel.

BOXBOROUGH.—*Scapolite*, spinel, *garnet*, augite, actinolite, apatite.

BRIGHTON.—Asbestus.

BRIMFIELD, (road leading to Warren.)—*Iolite*, adularia, molybdenite, mica, garnet.

CARLISLE.—*Tourmaline*, *garnet !* *scapolite*, actinolite.

CHARLESTOWN.—*Prehnite*, *laumonite*, stilbite, chabazite, quartz crystals.

CHELMSFORD.—*Scapolite*, *chondrodite*, *blue spinel*, *amianthus !* rose quartz.

CHESTER.—*Hornblende*, *scapolite*, *zoisite*, *spodumene*, *indicolite*, apatite—magnetic iron and *chromic iron*, (west part,)—stilbite, heulandite, analcime and chabazite.

CHESTERFIELD.—*Blue*, *green*, *and red tourmaline*, *cleavelandite* (albite), *lithia mica*, *smoky quartz*, *microlite*, *spodumene*, *kyanite*, apatite, *rose beryl*, *garnet*, *quartz crystals*, *staurotide*, tin ore, *columbite*, erubescite, zoisite, uranite, brookite (eumanite).

CONWAY.—Pyrolusite, fluor spar, zoisite, *rutile ! !* native alum, galena.

CUMMINGTON.—*Rhodonite !* cummingtonite (hornblende), white iron pyrites, *garnet*.

DEDHAM.—Asbestus, galena.

DEERFIELD.—Chabazite, heulandite, stilbite, amethyst, carnelian, chalcedony, *agate*.

FITCHBURG, (Pearl Hill.)—*Beryl*, *staurotide !* garnets, molybdenite.

FOXBOROUGH.—*Iron pyrites*, *anthracite*.

FRANKLIN.—Amethyst.

GOSHEN.—*Mica*, *albite*, *spodumene !* *blue and green tourmaline*, *beryl*, *zoisite*, smoky quartz, columbite, tin ore, galena.

GREENFIELD, (in sandstone quarry, half mile east of village.)—Allophane, white and greenish.

HATFIELD.—*Heavy Spar*, yellow quartz crystals, galena, blende, copper pyrites.

HAWLEY.—*Micaceous iron*, massive pyrites, magnetic iron, zoisite.

HEATH.—*Pyrites*, *zoisite*.

HINSDALE.—Brown iron ore, apatite, zoisite.

HUBBARDSTON.—*Massive pyrites*.

LANCASTER.—*Kyanite*, *chiastolite !* apatite, staurotide, pinite, andalusite.

LEE.—*Tremolite !* *sphene !* (east part.)

LENOX.—Brown hematite, gibbsite, (?)

LEVERETT.—Heavy spar, galena, blende, copper pyrites

LEYDEN.—*Zoisite*, *rutile*.

LITTLETON.—Spinel, scapolite, apatite.

LYNNFIELD.—Magnesite on serpentine.

MARTHA'S VINEYARD.—Brown iron ore, amber, selenite, radiated pyrites.

MENDON.—*Mica !* chlorite.

MIDDLEFIELD.—*Glassy actinolite*, *rhomb spar*, *steatite*, *serpentine*, *feldspar*, drusy quartz, apatite, zoisite, nacrite, chalcedony, *talc !*

MILBURY.—*Vermiculite*.

MONTAGUE.—Specular iron.

NEWBURY.—*Serpentine,* chrysolite, *epidote, massive garnet,* carbonate of iron.

NEWBURYPORT.—*Serpentine, nemalite,* uranite.

NEW BRAINTREE.—*Black tourmaline.*

NORWICH.—*Apatite ! black tourmaline, beryl, spodumene ! triphyline* (altered), blende, quartz crystals.

PALMER, (Three Rivers.)—*Feldspar,* prehnite, calc spar.

PELHAM.—*Asbestus,* serpentine, *quartz crystals,* beryl, *molybdenite, green hornstone.*

PLAINFIELD.—*Cummingtonite, pyrolusite, rhodonite.*

RICHMOND.—*Brown iron ore, gibbsite !*

ROWE.—Epidote, talc.

SOUTH ROYALSTON.—*Beryl ! !* (now obtained with great difficulty,) *mica ! ! feldspar ! ilmenite,* allanite. Four miles beyond old loc., on farm of Solomon Heywood, *mica ! beryl ! feldspar !*

RUSSEL.—Schiller spar, (diallage?) *mica,* serpentine, beryl, galena, copper pyrites.

SAUGUS.—Porphyry.

SHEFFIELD.—*Asbestus,* pyrites, native alum, pyrolusite.

SHELBURNE.—Rutile.

SHUTESBURY, (east of Locke's Pond.)—*Molybdenite.*

SOUTHAMPTON.—*Galena,* white lead ore, anglesite, molybdate of lead, fluor, heavy spar, copper and iron pyrites, blende, corneous lead, pyromorphite.

STERLING.—*Spodumene, chiastolite, spathic iron, mispickel, blende,* galena, iron and copper pyrites.

STONEHAM.—*Nephrite.*

STURBRIDGE.—*Graphite,* garnet, apatite, bog ore.

TAUNTON, (one mile south.)—Paracolumbite.

TURNER'S FALLS, (Conn. River.)—Copper pyrites, prehnite, chlorite, *chlorophæite,* spathic iron, green malachite, magnetic iron sand, anthracite.

TYRINGHAM.—Pyroxene, scapolite.

UXBRIDGE.—Argentiferous galena.

WARWICK.—*Massive garnet, black tourmaline, magnetic iron,* beryl, epidote.

WASHINGTON.—*Graphite.*

WESTFIELD.—*Schiller spar,* (diallage,) *serpentine, steatite,* kyanite, scapolite, actinolite.

WESTFORD.—*Andalusite !*

WEST HAMPTON.—Galena, *argentine, pseudomorphous quartz.*

WEST SPRINGFIELD.—*Prehnite,* ankerite, satin spar, celestine, bituminous coal.

WEST STOCKBRIDGE.—*Hematite,* fibrous pyrolusite, spathic iron.

WHATELY.—*Native copper,* galena.

WILLIAMSBURG.—Zoisite, pseudomorphous quartz, apatite, rose and smoky quartz, galena, pyrolusite, copper pyrites.

WILLIAMSTOWN.—*Cryst. quartz.*

WINDSOR.—*Zoisite,* actinolite, *rutile !*

WORCESTER.—*Mispickel,* idocrase, pyroxene, garnet, amianthus, bucholzite, spathic iron, galena.

WORTHINGTON.—*Kyanite.*

ZOAR.—Bitter spar, *talc.*

RHODE ISLAND.

BRISTOL.—*Amethyst.*

CRANSTON.—Actinolite in talc.

CUMBERLAND.—*Manganese, epidote, actinolite,* garnet, titaniferous iron, magnetic iron, red hematite, copper pyrites.

FOSTER.—*Kyanite.*

JOHNSON.—Talc, brown spar.

NEWPORT.—*Serpentine.*

PORTSMOUTH.—*Anthracite,* graphite, asbestus, iron pyrites.

SMITHFIELD.—*Dolomite, calc spar, bitter spar, nacrite,* serpentine (bowenite), tremolite, asbestus, quartz, magnetic iron in chlorite slate, *talc ! !*

WARWICK, (Natic village.)—*Masonite,* garnets, graphite.

WESTERLY.—*Ilmenite.*

CONNECTICUT.

BERLIN.—Heavy spar, datholite, blende, quartz crystals.

BOLTON.—Staurotide, copper pyrites.

BRADLEYVILLE, (Litchfield.)—Laumonite.

BRISTOL.—*Copper glance, copper pyrites,* heavy spar, *erubescite,* talc, *allophane,* pyromorphite.

BROOKFIELD.—Galena, calamine, *blende,* spodumene, magnetic pyrites.

CANAAN.—*Tremolite* and *augite !* in dolomite.

CHATHAM.—Mispickel, smaltine, chloanthite (chathamite), scorodite, copper nickel, *beryl.*

CHESHIRE.—*Heavy spar ! copper glance cryst.,* erubescite, *green malachite,* kaolin, natrolite, prehnite, chabazite, datholite.

CHESTER.—*Sillimanite !* zircon, epidote.

CORNWALL, near the Housatonic.—*Graphite, pyroxene.*

DANBURY.—*Danburite, oligoclase, moonstone,* brown tourmaline.

FARMINGTON.—*Prehnite, chabazite !* agate, native copper.

GRANBY.—Green malachite.

GREENWICH.—*Black tourmaline.*

HADDAM.—*Chrysoberyl ! beryl ! epidote ! tourmaline ! feldspar, anthophyllite, garnet ! iolite ! oligoclase, chlorophyllite ! automolite, magnetic iron, adularia,* apatite, *columbite !* zircon (calyptolite), *mica, white* and yellow iron pyrites, *molybdenite,* allanite, bismuth, bismuth ochre.

HADLYME.—Chabazite and stilbite in gneiss, with epidote and garnet.

HARTFORD.—*Datholite,* (Rocky Hill quarry.)

KENT.—*Brown iron ore,* pyrolusite, ochrey iron ore.

LITCHFIELD.—*Kyanite* with corundum, apatite and andalusite, *ilmenite,* (washingtonite,) copper pyrites.

LYME.—Garnet, sunstone.

MERIDEN.—Datholite.

MIDDLEFIELD FALLS.—Datholite, chlorite, &c., in amygdaloid.

MIDDLETOWN.—*Mica, lepidolite* with green and red tourmaline, *albite, feldspar, columbite ! prehnite, garnet,* beryl, topaz, uranite, apatite, pitchblende; at lead mine, *galena, copper pyrites,* blende, quartz, *calcite,* fluor, *iron pyrites,* sometimes capillary.

MILFORD.—Sahlite, *pyroxene, asbestus,* zoisite, verd-antique marble, pyrites.

NEW HAVEN.—Serpentine, asbestus, chromic iron, sahlite, stilbite, prehnite.

NORWICH.—*Sillimanite, monazite!* zircon, *iolite,* corundum, feldspar.

OXFORD, near Humphreysville.—Kyanite, copper pyrites.

PLYMOUTH.—Galena, *heulandite, fluor.*

ROARING BROOK, (Cheshire.)—*Datholite!* calc spar, prehnite, saponite.

READING, (near the line of Danbury.)—Pyroxene, *garnet.*

ROXBURY.—Massive *spathic iron,* blende.

SALISBURY.—*Brown iron ore,* ochery iron, *pyrolusite,* triplite.

SAYBROOK.—*Molybdenite,* stilbite, plumbago.

SIMSBURY.—*Copper glance,* green malachite.

SOUTHBURY.—Rose quartz, laumontite, prehnite, calc spar, heavy spar.

SOUTHINGTON.—Heavy spar, datholite, asteriated quartz cystals.

STAFFORD.—Massive pyrites.

STONINGTON.—Stilbite and *chabazite* on gneiss.

THATCHERSVILLE, (near Bridgeport.)—Stilbite on gneiss, babingtonite?

TOLLAND.—Staurotide, massive pyrites.

TRUMBULL and MONROE.—*Chlorophane, topaz, beryl,* diaspore, magnetic pyrites, iron pyrites, *tungstate of lime, wolfram,* (pseudomorph of tungsten,) rutile, native bismuth, tungstic acid, spathic iron, mispickel, argentiferous galena, blende, scapolite, *tourmaline, garnet,* albite, augite, graphic tellurium, (?) *margarodite.*

WASHINGTON.—*Triplite, ilmenite!* (washingtonite of Shepard,) diallogite, natrolite, *andalusite,* (New Preston,) kyanite.

WATERTOWN, near the Naugatuck.—White sahlite, monazite.

WEST FARMS.—Asbestus.

WINCHESTER and WILTON.—Asbestus, garnet.

NEW YORK.

ALBANY CO.—COEYMAN'S LANDING.—Epsom salt.

GUILDERLAND.—*Petroleum.*

WATERVLIET.—*Quartz crystals.*

ALLEGANY CO.—CUBA.—*Petroleum.*

CATTARAUGUS CO.—FREEDOM.—*Petroleum.*

CAYUGA CO.—AUBURN.—Fluor, epsom salt.

CAYUGA LAKE.—Sulphur.

LUDLOWVILLE.—Epsom salt.

SPRINGVILLE.—Nitrogen springs.

CHATAUQUE CO.—FREDONIA.—*Petroleum, carburetted hydrogen.*

LAONA.—Petroleum.

COLUMBIA CO.—ANCRAM LEAD MINE.—Galena, *blende, copper pyrites*, heavy spar.
AUSTERLITZ.—*Earthy manganese*, molybdate of lead, copper mica.
HUDSON.—*Selenite!*
LEBANON.—Nitrogen spring.

DUCHESS CO.—DOVER.—*Garnet* (Foss ore bed.)
FISHKILL.—*Graphite, green actinolite! talc*, hydrous anthophyllite.
RHINEBECK.—Granular epidote.
UNION VALE.—*Gibbsite*, (at Clove mine.)
AMENIA.—Brown hematite.

ESSEX CO.—ALEXANDRIA.—Kirby's graphite mine, *graphite, pyroxene, scapolite*, sphene.
CROWN POINT.—*Apatite*, (eupyrchroite of *Emmons*,) *brown tourmaline!* in the apatite, chlorite, quartz crystals, pink and blue calcite, pyrites; a short distance south of J. C. Hammond's house, *garnet, scapolite*, copper pyrites, *aventurine feldspar*, zircon; magnetic iron (Peru).
LEWIS.—*Tabular spar, colophonite*, garnet, *labradorite.*
LONG POND.—Apatite, *garnet, pyroxene*, idocrase, *coccolite!! scapolite*, magnetic iron ore, *blue calc spar.*
McINTYRE.—*Labradorite*, garnet, *magnetic iron ore.*
MORIAH, at Sandford Ore Bed.—*Magnetic iron*, apatite, *allanite!* actinolite, and feldspar; at Fisher Ore Bed, *magnetic iron*, feldspar, quartz; at Hall Ore Bed, or "New Ore Bed," *magnetite, zircons.*
NEWCOMB.—*Labradorite*, feldspar.
PORT HENRY.—*Brown tourmaline, mica, rose quartz, serpentine, green* and *black pyroxene*, hornblende, *cryst. pyrites, magnetic pyrites, adularia. Phlogopite!* at Cheever Ore Bed, with magnetite and serpentine.
ROGER'S ROCK.—*Graphite, tabular spar, garnet, colophonite, feldspar*, adularia, *pyroxene, sphene*, coccolite.
SCHROON.—*Calc spar*, pyroxene, *chondrodite.*
TICONDEROGA.—*Graphite, pyroxene, sahlite, sphene*, black tourmaline.
WESTPORT.—Labradorite, prehnite.
WILLSBORO'.—*Tabular spar, colophonite*, garnet, *green coccolite*, hornblende.

FRANKLIN CO.—CHATEAUGAY.—Nitrogen springs.
MALONE.—*Massive pyrites*, magnetic iron ore.

GENESEE CO.—*Acid springs* containing sulphuric acid.

GREENE CO.—CATSKILL.—*Calc spar.*
DIAMOND HILL.—Quartz crystals.

HERKIMER CO.—LITTLE FALLS.—*Quartz crystals*, heavy spar, *calc spar*, anthracite.
MIDDLEVILLE.—*Quartz crystals! calc spar*, brown and pearl spar.
SALISBURY.—*Quartz crystals!* blende, galena, iron and copper pyrites.
STARK.—Fibrous celestine, *gypsum.*

JEFFERSON CO.—ALEXANDRIA.—Hornblende, *orthoclase, tourmaline*, celestine.

ADAMS,—Fluor, calc tufa, barytes.

ANTWERP.—Stirling iron mine, *specular iron, chalcodite, spathic iron, millerite*, nickeliferous iron pyrites, *quartz crystals*, pyrites; at Oxbow, *calc spar!* porous coralloidal heavy spar; near Vrooman's lake, *calc spar!* idocrase, *phlogopite! pyroxene, sphene*, fluor, calcite, pyrites, copper pyrites; also *feldspar, bog iron ore*, scapolite, (farm of David Eggleson,) *serpentine*, tourmaline (yellow, rare).

HIGH ISLAND, (in the St. Lawrence.)—*Tourmaline*.

PAMELIA.—*Agaric mineral*, calc tufa.

PILLAR POINT.—*Massive heavy spar* (exhausted).

THERESA.—*Fluor, calcite*, specular iron ore, hornblende, *quartz crystals*, serpentine, (associated with the specular iron,) celestine, strontianite: the Muscolonge lake locality of fluor is exhausted.

WATERTOWN.—*Tremolite, agaric mineral*, calc tufa, celestine.

[This county adjoins St. Lawrence Co., and the localities of Rossie, Hammond and Gouverneur, near Oxbow, are in the latter county.]

LEWIS CO.—DIANA, (localities mostly near junction of crystalline and sedimentary rocks, and within two miles of Natural Bridge.) *Scapolite! tabular spar, green coccolite, feldspar, tremolite*, black pyroxene, sphene, mica, *quartz crystals*, drusy quartz, cryst. pyrites, magnetic pyrites, blue calc spar, serpentine, *rensselaerite*, zircon, graphite, chlorite, specular iron, bog iron ore, iron sand.

GREIG.—*Magnetic iron ore*, pyrites.

LOWVILLE.—*Calc spar*, fluor spar, pyrites, galena, blende, calc tufa.

MARTINSBURGH.—Wad, galena, etc., but mine not now opened.

WATSON, BREMEN.—Bog iron ore.

MONROE CO.—ROCHESTER.—*Pearl spar*, calc spar, snowy gypsum, fluor, celestine, galena, blende.

MONTGOMERY CO.—ROOT.—Pearl spar, *drusy quartz, blende*.

PALATINE.—*Quartz crystals*, drusy quartz.

NEW YORK CO.—CORLAER'S HOOK.—Apatite.

KINGSBRIDGE.—*Tremolite, pyroxene, mica, tourmaline*, pyrites, rutile.

HARLEM.—Epidote, apophyllite, stilbite, tourmaline, vivianite, lamellar feldspar, mica.

NEW YORK.—*Serpentine, amianthus*, actinolite, talc, *pyroxene*, hydrous anthophyllite, garnet, staurotide, molybdenite, graphite.

NIAGARA CO.—LEWISTON.—*Epsom salt*.

LOCKPORT.—*Celestine, calc spar, selenite, anhydrite, fluor, pearl spar, blende*.

NIAGARA FALLS.—*Calc spar*, fluor, blende.

ONEIDA CO.—BOONVILLE.—*Calc spar, tabular spar, coccolite*.

CLINTON.—Blende, *lenticular argillaceous iron ore;* in rocks of the Clinton Group, strontianite, celestine, the former covering the latter.

ONONDAGA CO.—CAMILLUS.—*Selenite* and *fibrous gypsum*.

MANLIUS.—*Gypsum* and fluor.

SYRACUSE.—*Serpentine*, celestine.

ORANGE CO.—CORNWALL.—*Zircon, chondrodite, hornblende, spinel, massive feldspar, fibrous epidote,* hudsonite, ilmenite, *serpentine,* boltonite.

DEER PARK.—*Cryst. pyrites,* galena.

MONROE.—*Mica ! sphene ! garnet,* colophonite, *epidote, chondrodite, allanite,* bucholzite, brown spar, boltonite, *spinel,* hornblende, talc, *ilmenite, magnetic pyrites,* common pyrites, chromic iron, *graphite.*

At WILKS and O'NEIL Mine in Monroe.—Aragonite, *magnetite,* dimagnetite (pseud?), jenkinsite.

At TWO PONDS in Monroe.—*Pyroxene ! chondrodite, hornblende, scapolite ! zircon, sphene,* apatite.

At GREENWOOD FURNACE in Monroe.—*Chondrodite, pyroxene ! mica, hornblende, spinel, scapolite, biotite !* ilmenite.

At FOREST OF DEAN.—*Pyroxene, spinel,* zircon, scapolite, hornblende, boltonite.

Town of WARWICK.

WARWICK VILLAGE.—*Spinel, zircon, serpentine ! brown spar, pyroxene ! hornblende ! pseudomorphous steatite, feldspar !* (Rock Hill,) *ilmenite, clintonite,* tourmaline, (R. H.,) *rutile,* sphene, molybdenite, mispickel, *white iron pyrites,* common pyrites, yellow iron sinter.

AMITY.—*Spinel, garnet, scapolite, hornblende, idocrase, epidote ! clintonite ! magnetic iron ! tourmaline,* warwickite, *apatite, chondrodite,* ilmenite, *talc ! pyroxene !* rutile, *zircon, corundum, feldspar,* sphene, calc spar, serpentine, schiller spar. (?)

EDENVILLE.—*Apatite, chondrodite ! hair brown hornblende !* tremolite, *spinel, tourmaline, warwickite, pyroxene, sphene, mica, feldspar, mispickel,* orpiment, *rutile, ilmenite,* scorodite, copper pyrites.

WEST POINT.—*Feldspar, mica,* scapolite, *sphene, hornblende,* allanite.

PUTNAM CO.—CARMEL, (Brown's quarry.)—Anthophyllite, schiller spar, (?) orpiment, mispickel.

COLD SPRING.—Chabazite, mica, sphene.

PATTERSON.—*White pyroxene ! calc spar, asbestus, tremolite,* dolomite, massive pyrites.

PHILLIPSTOWN.—*Tremolite, amianthus, serpentine, sphene, diopside, green crocolite,* hornblende, *scapolite,* stilbite, mica, laumontite, gurhofite, calc spar, magnetic iron, chromic iron.

PHILLIPS Ore Bed.—Hyalite, *actinolite, massive pyrites.*

RENSSELAER CO.—HOOSIC.—Nitrogen springs.

LANSINGBURGH.—Epsom salt, quartz crystals, *iron pyrites.*

TROY.—*Quartz crystals, iron pyrites, selenite.*

RICHMOND CO.—ROSSVILLE.—Lignite, *cryst. pyrites.*

QUARANTINE.—*Asbestus, amianthus,* aragonite, *dolomite, gurhofite,* brucite, serpentine, *talc.*

ROCKLAND CO.—CALDWELL.—*Calc spar !*

GRASSY POINT.—Serpentine, actinolite.

HAVERSTRAW.—*Hornblende.*

LADENTOWN.—Zircon, red copper ore, green malachite.

PIERMONT.—Datholite, stilbite, apophyllite, stellite, prehnite, thomsonite, calc spar.

STONY POINT.—Kerolite, lamellar hornblende, asbestus.

ST. LAWRENCE CO.—CANTON.—*Massive pyrites, calc spar*, brown tourmaline, *sphene, serpentine*, talc, *rensselaerite*, pyroxene, specular iron, copper pyrites.

DEKALB.—*Hornblende*, heavy spar, *fluor, tremolite, tourmaline*, blende, graphite, pyroxene, quartz (spongy), serpentine.

EDWARDS.—*Brown and silvery mica!* scapolite, apatite, *quartz crystals*, actinolite, tremolite, specular iron, serpentine, magnetite.

FINE.—*Black mica*, hornblende.

FOWLER.—*Heavy spar, quartz crystals! specular iron, blende*, galena, tremolite, chalcedony, bog ore, satin spar, (assoc. with serpentine,) iron and copper pyrites, actinolite, *rensselaerite*, (near Somerville.)

GOUVERNEUR.—*Calc spar! serpentine! hornblende! scapolite! orthoclase, tourmaline!* idocrase, (one mile south of G.,) pyroxene, apatite, *rensselaerite*, serpentine, *sphene*, fluor, *heavy spar* (farm of Judge Dodge,) black *mica*, phlogopite, *tremolite!* asbestus, *specular iron*, graphite, idocrase; (near Somerville in serpentine), *spinel*, houghite, scapolite, *phlogopite*, dolomite; three quarters of a mile west of Somerville, *chondrodite*, spinel; two miles north of Somerville, *apatite*, pyrites.

HAMMOND.—*Apatite! zircon!* (farm of Mr. Hardy), *orthoclase, pargasite*, heavy spar, pyrites, purple fluor, dolomite.

HERMON.—*Quartz crystals, specular iron, spathic iron*, pargasite, pyroxene, serpentine, tourmaline, bog iron ore.

MACOMB.—Blende, mica, *galena* (on land of James Averil), sphene.

MINERAL POINT, Morristown.—Fluor, blende, galena, *phlogopite* (Pope's Mills,) heavy spar.

OGDENSBURG.—Labradorite.

PITCAIRN.—Satin spar, associated with serpentine.

POTSDAM.—*Hornblende!*—eight miles from Potsdam on road to Pierrepont, *feldspar, tourmaline, black mica*, hornblende.

ROSSIE, (Iron Mines.)—*Heavy spar, specular iron*, corralloidal aragonite in mines near Somerville, limonite, *quartz*, (sometimes stalactitic at Parish iron mine,) *pyrites, pearl spar*.

ROSSIE Lead Mine.—*Calc spar, galena, pyrites, celestine*, copper pyrites, *spathic iron!* white lead ore, anglesite.

Elsewhere in ROSSIE.—*Calc spar, heavy spar*, quartz crystals, chondrodite (near Yellow Lake), *feldspar! pargasite! apatite, pyroxene* hornblende, sphene, zircon, *mica*, fluor, serpentine. automolite, pearl spar, graphite.

RUSSEL.—*Pargasite, specular iron. quartz* (dodec.), calcite, serpentine, rensselaerite, magnetite.

SARATOGA CO.—GREENFIELD.—*Chrysoberyl! garnet, tourmaline mica, feldspar*, apatite, graphite, aragonite, (in iron mines.)

SCHOHARIE CO.—BALL'S CAVE, and others.—Calc spar, stalactites.

CARLISLE.—***Fibrous sulphate of baryta, cryst. and fib. carbonate of lime.***

SCHOHARIE.—**Fibrous celestine, *strontianite! cryst. pyrites!***

SENECA CO.—CANOGA.—*Nitrogen springs.*

SULLIVAN CO.—WURTZBORO'.—*Galena, blende, pyrites, copper pyrites.*

ULSTER CO.—ELLENVILLE.—*Galena*, blende, *copper pyrites! quartz, brookite.*
MARBLETOWN.—Pyrites.

WARREN CO.—CALDWELL.—*Massive feldspar.*
CHESTER.—*Pyrites*, tourmaline, rutile, copper pyrites.
DIAMOND ISLE, (Lake George.)—*Calc spar, quartz crystals.*
GLENN'S FALLS.—Rhomb spar.
JOHNSBURG.—*Fluor! zircon!! graphite, serpentine, pyrites.*

WASHINGTON CO.—FORT ANN.—*Graphite.*
GRANVILLE.—*Lamellar pyroxene*, massive feldspar, epidote.

WAYNE Co.—WOLCOTT.—*Heavy spar.*

WESTCHESTER CO.—ANTHONY'S NOSE.—*Apatite*, pyrites, *calcite!* in very large tabular crystals, grouped and sometimes incrusted with drusy quartz.
DAVENPORT'S NECK.—*Serpentine*, garnet, sphene.
EASTCHESTER.—Blende, copper and iron pyrites, dolomite.
HASTINGS.—*Tremolite, white pyroxene.*
NEW ROCHELLE.—*Serpentine*, brucite, quartz, *mica*, tremolite, garnet.
PEEKSKILL.—Mica, feldspar, hornblende, stilbite.
RYE.—*Serpentine, chlorite, black tourmaline*, tremolite. kerolite.
SINGSING.—*Pyroxene, tremolite, iron pyrites*, copper pyrites, beryl, azurite, green malachite, white lead ore, pyromorphite, anglesite, vauquelinite, galena, native silver.
WEST FARMS.—Apatite, tremolite, garnet, stilbite, heulandite, chabazite, epidote, sphene.
YONKERS.—*Tremolite*, apatite, calc spar, analcime, *pyrites*, tourmaline.
YORKTOWN.—*Sillimanite, monazite*, magnetic iron.

NEW JERSEY.

ABBOTTSVILLE.—Serpentine, chrysotil.
ANDOVER IRON MINE, (Sussex Co.)—Willemite, brown garnet, *magnetite. calcite*, blende, fluor, galena, copper pyrites, talc.
ALLENTOWN, (Monmouth Co.)—Vivianite.
BELVILLE.—Copper mines.
BERGEN.—*Calc spar, datholite, thomsonite, pectolite* (called stellite), *analcime, apophyllite, prehnite*, sphene, stilbite, natrolite, heulandite, laumontite, chabazite, pyrites, pseudomorphous steatite imitative of apophyllite.
BRUNSWICK.—Copper mines; *native copper, malachite, mountain leather.*
BRYAM.—Chondrodite.
CANTWELL'S BRIDGE, Newcastle Co., three miles west.—Vivianite.
DANVILLE.—(Jemmy Jump Ridge.)—*Graphite*, chondrodite, augite, mica.

FLEMINGTON.—*Copper Mines.*

FRANKFORT.—*Serpentine.*

FRANKLIN and STERLING.—*Spinel! garnet! rhodonite! willemite! franklinite! red zinc ore! dysluite! hornblende, tremolite, chondrodite, white scapolite, black tourmaline, epidote, pink calc spar, mica,* actinolite, augite, sahlite, coccolite, asbestus, *jeffersonite* (augite), calamine, graphite, fluor, beryl, galena, serpentine, honey-colored sphene, quartz, chalcedony, amethyst, zircon, molybdenite, vivianite. Also *algerite* in gran. limestone. The zinc ores and franklinite, especially at Sterling Hill in Sterling, the jeffersonite at Mine Hill, in Franklin.

FRANKLIN and WARWICK MTS.—*Pyrites.*

GREENBROOK.—Copper mines.

GRIGGSTOWN.—Copper mines.

HAMBURGH.—One mile north, *spinel! tourmaline! phlogopite, hornblende, &c., limonite,* specular iron.

HOBOKEN.—Serpentine, *brucite! nemalite* (or fibrous brucite), aragonite, dolomite.

HURDTOWN.—*Apatite, magnetic pyrites,* magnetite, chalcedony, feldspar, hornblende.

IMLEYTOWN.—Vivianite.

LOCKWOOD.—*Graphite, chondrodite, talc, augite, quartz, green spinel.*

MONTVILLE., Morris Co.—Serpentine, *chrysotile.*

MOUNT HOPE.—Three miles N. W. of Rockaway, iron mines, magnetite, pyrites, *hornblende, apatite;* Mt. Tabo mines, *spathic iron,* pyrites; at Mt. Pleasant, *apatite, hornblende,* feldspar.

MULLICA HILL, Gloucester Co.—*Vivianite* lining belemnites and others fossils.

NEWTON.—*Spinel,* blue and white corundum, (exhausted,) *mica, idocrase, hornblende, tourmaline, scapolite,* rutile, pyrites, talc, calc spar, heavy spar, *pseudomorphous steatite.*

PATTERSON.—*Datholite.*

ROSEVILLE, (Bryan Township, Sussex Co.)—Magnetite, calcite, *epidote,* garnet, mica.

SCHUYLER'S MINES.—Green malachite, red copper ore, native copper, *chrysocolla.*

SOMERVILLE.—*Red copper ore, native copper, chrysocolla,* green malachite, bitumen, (two miles to the northeast.)

SPARTA.—*Chondrodite! spinel,* sapphire, green talc, graphite, epidote, augite.

STANHOPE.—Few miles south, several iron mines.

SUCKASUNNY, on the Morris canal.—*Brown apatite* in magnetic pyrites.

TRENTON.—Zircon, amber, lignite.

VERNON.—*Green spinel, chondrodite, red sapphire, hornblende, pyrites, phlogopite, graphite, limonite,* rutile, sphene, ilmenite, zircon, fluor, margarite.

WOODBRIDGE.—Copper mine.

NOTE.—From Amity, N. Y., to Andover, N. J., a distance of about thirty miles, the outcropping limestone, at different points, affords more or less of the minerals enumerated as occurring at Franklin. (See Geol. Rep. on N. J., by H. D. Rogers.)

PENNSYLVANIA.

ADAMS CO.—Reading.—Molybdenite in quartz, zircon, magnetic iron ore.

BERKS CO.—At Jones's Mines, near Morgantown, *green malachite! chrysocolla!* oct. and dodec. magnetic iron, iron pyrites, copper pyrites;—two miles to the northeast, graphite, sphene; at Steel's mines, *octahedral and micaceous iron ore*, coccolite; Eckhardt's furnace, allanite.

BUCK'S CO.—Opposite New Hope, *tourmaline!* near Attleboro', at Vanarsdale's limestone quarry, *sahlite*, scapolite, sphene, green coccolite, graphite, green mica.

CARBON CO.—At Mauch Chunk, *cryst. iron pyrites*, selenite.

CHESTER CO.—Birmingham.—Kerolite, amethyst, quartz cryst., serpentine.

E. Bradford.—On Minorcus Hill, green, blue and gray *kyanite*, apatite, allanite; on A. Tayfor's farm, sphene, cryst. smoky quartz; on the farms of B. Jones, B. Price, L. Sharpless, and S. Entrikin, amethyst; near Strode's mill, asbestus, magnesite, marmolite, garnet; near T. Hoope's saw mill, *epidote, asbestus;* on Osborn's Hill, sphene, manganesian garnet, wad, tourmaline, actinolite, anthophyllite, feldspar, fetid calcite; near the Black Horse Inn, rutile.

W. Bradford.—Near A. Jackson's limestone quarry, green *kyanite*, rutile, scapolite, iron pyrites; near Marshall's mill, chromic iron, serpentine; at Poor House (limestone) quarry, (called also Baldwin's,) four miles north of Unionville, and six west of Westchester, *rutile!* in brilliant acicular crystals; cryst. calc spar, cryst. *dolomite*, zoisite in quartz, *talc* in implanted crystals on dolomite, orthoclase! (in fine crystals implanted on dolomite,) quartz crystals.

Chester Springs.—*Gibbsite*, in an iron mine; near Coventryville, in Chrisman's limestone quarry, augite, sphene, graphite, *zircon!* in iron ore about half a mile from the village on French Creek.

West Goshen.—Amianthus, asbestus, precious serpentine, cellular quartz, jasper, chalcedony, drusy quartz, chlorite, marmolite, dolomite, cryst. *carb. magnesia! chromic iron! magnetic iron;* near Westchester Water Works, zoisite, (rare, not found now.)

Keim's Iron Mine near Knauertown.—*Flos-ferri*, pyroxene, *metaxite, micaceous iron ore, aplome! actinolite, yellow octahedral pyrites*, copper pyrites in tetrahedrons, *red garnet!* malachite, hornblende (var. byssolite.)

Kennet Township.—*Actinolite!* (rare on Gregg's farm,) brown tourmaline, brown mica, *epidote*, tremolite, scapolite, aragonite; at Pearce's paper mill, zoisite, epidote, *sunstone;* on R. Lamborne's farm, chabazite in small brownish yellow crystals, (rare,) zeolite; at Gause's corner, *epidote.*

Knauertown.—North of Pughtown, graphite, sphene, cryst. magnetic iron; in Chrismard's Iron Mine, zircon.

London Grove.—In Jackson's limestone quarry, *yellow tourmaline!* (rare,) fib. tremolite; at Pusey's quarry, rutile, tremolite.

New Garden Township.—At Nevin's limestone quarry, *brown tourmaline!* scapolite, brown and green mica, rutile, *aragonite, kaolin.*

Newlin.—See Unionville, below.

East Marlboro.—*Epidote,* and nearly white *tourmaline,* (rare).

Oxford.—Iron pyrites, garnets.

Nottingham.—At Scott's chrome mine, *chromic iron, foliated talc,* marmolite, serpentine, chalcedony; at the Magnesian Quarry, magnesite, marmolite, serpentine.

Parksburg, (in township of Sadsbury.)—In the soil for seven miles along the valley, *rutile!;* northeast of the village, amethyst, tourmaline, epidote, (in a boulder.)

Penn.—Garnets, figure stone.

Pennsbury Township.—On Cephas Cloud's farm, *brown garnets!;* J. Dilsworth's farm, near Pennsville, *mica!!* (in six-sided prisms from one quarter to seven inches across); at Harvey's lime quarry, on the Brandywine, *chondrodite;* quarter of a mile above the last, at Wm. Burnett's lime quarry, sphene, diopside, augite, coccolite.

Phenixville.—In Railroad Tunnel, *pearl spar* (exhausted), *dolomite, yellow blende,* iron pyrites; at Wheatley's Mine, *pyromorphite! cerusite!* cryst. quartz, *galena, anglesite! copper pyrites,* heavy spar, *fluor, wulfenite! calamine,* cerasine? vanadinite? phosphate of copper, chromate of lead, *calcite?*

Pottstown, near French Cr.—(Elizabeth Mine.)—*Iron pyrites!* (in octahedrons), *copper pyrites, magnetite, dark brown garnet,* molybdenite.

Unionville.—One and a half miles northeast, on Serpentine Barrens, *corundum!* massive and cryst. (often in loose crystals and also in albite, the loose crystals mostly covered with a thin coating of steatite, sometimes with gibbsite), talc, *green tourmaline* (with flat or pyramidal terminations), ligniform asbestus, *yellow beryl* (rare), *serpentine,* brucite, *chromic iron,* quartz crystals, *green quartz,* actinolite, *clinochlore* in cryst., diallage, granular albite (H=7), adularia, *oligoclase,* halloysite, *margarite, euphyllite,* allanite, hematite, chalcedony; half a mile southwest, on T. Webb's farm, *serpentine, chromic iron,* (mas.); two and a half miles southwest, in R. Bailey's lime quarry, fib. tremolite, mussite; *kyanite, margarodite;* two miles southwest, at Pusey's saw mill, *zircon* (cryst. small, loose in the soil, rare), rutile; one mile south, on the farm of Baily and Brothers, bright *yellow* and nearly *white tourmaline!* (rare), *orthoclase* (chesterlite), *albite!* (inaccessible); two miles east, near Marlborough meeting house, *epidote!* (rare), serpentine, acicular black tourmaline in white quartz; one mile west, near Logan's quarry, staurotide, kyanite, yellow tourmaline (rare); at Edward's lime quarry, near the last, purple fluor, *rutile;* four miles west, in limestone quarries of West Marlborough, near Doe River Village, scapolite, rutile, tremolite. At Steamboat, Wavellite with limonite.

Westchester.—One and a half mile north, *hydromagnesite, clinochlore, brucite,* in serpentine, zircon, two miles west; one and a half mile northwest, pitch-black allanite; B.B. intumesces very readily (G. 2·5); three miles south, *chlinoclore, phlogopite.*

Willistown.—Magnetic iron, chromic iron, actinolite.

COLUMBIA CO.—At Webb's mine, yellow blende in calc spar; near Bloomburg, cryst. magnetic iron.

DAUPHIN CO.—Near Hummerstown, green garnets, cryst. *smoky quartz,* cryst. feldspar.

DELAWARE CO.—Aston.—Near Village Green, *amethyst, corundum, emerylite,* staurotide, sillimanite, black tourmaline, pearl mica, asbestus, anthophyllite; near Tyson's Mill, garnet, staurotide; at head of Peter's Mill Dam, in a brook, garnet resembling pyrope.

Birmingham.—At Bullock's quarry, zircon, bucholzite, fibrolite, nacrite.

Chester.—*Amethyst,* black tourmaline; in Burk's quarry, *beryl! ! black tourmaline! ! feldspar!* manganesian garnet, cryst. pyrites; on Chester Creek, at Carter's, *molybdenite, molybdic ochre,* copper pyrites, *tourmaline, kaolin;* at Little's quarry, brown garnets, tourmaline; near Henvi's quarries, *amethyst* in geodes; six miles northwest of Chester, chromic iron, in sand, consisting of crystals.

Chichester.—Near Trainer's Mill Dam, beryl, tourmaline, cryst. feldspar, kaolin; on W. Eyre's farm, *tourmaline! !*

Concord.—On Green's Creek, *garnets* resembling pyrope, *bucholzite, mica!* in hexagonal prisms, beryl, actinolite, anthophyllite, fibrolite, *rutile!* in capillary crystals in the cavities of cellular rose quartz.

Darby.—Kyanite, zoisite, (in a boulder); near Gibbon's, garnets, staurotide.

Edgemont.—One mile east of Edgemont Hall, near the road, rutile in quartz, amethyst, oxyd of manganese, cryst. feldspar.

Leiperville.—*Beryl!* in granite; in Judge Leiper's Quarries, *beryl, tourmaline, apatite,* garnet, cryst. feldspar, mica; at Morris's Ferry, *kyanite, sillimanite, apatite, red garnet,* mica; at Hill's Quarries, chabazite, stilbite, zeolite, epidote, sphene, albite, calcite, cryst. pyrites; near Leiper's Church, on the edge of a wood, *andalusite,* apatite, tourmaline, mica, gray *kyanite.*

Marple.—*Tourmaline!;* on A. Worrall's farm, *andalusite,* tourmaline; near C. Palmer's Mills, beryl, tourmaline, actinolite, amethyst.

Mineral Hill.—*Corundum! aventurine feldspar* (sunstone), chatoyant feldspar (moonstone), *actinolite, green coccolite, green feldspar!* chromic iron, cryst. green quartz, ferruginous quartz, asbestus, hydrous anthophyllite, *brown garnet!* magnesite, marmolite, bronzite, chalcedony, limonite, labradorite, float stone, red garnet, beryl, serpentine.

Providence.—At Blue Hill, serpentine, cryst. green quartz in green talc, asbestus, talc, anthophyllite, *actinolite,* hydrous anthophyllite; on M. Hunter's farm, *amethyst!* (one finely colored crystal found weighing over 7 lbs.), *andalusite.*

Radner.—Garnets, marmolite, deweylite, serpentine, chromic iron, asbestus, magnesite.

Springfield.—*Andalusite;* on Abby Worral's farm, *tourmaline,* beryl, ilmenite? garnets; on Fell's Laurel Hill, *beryl,* garnet; near

Beattie's Mill, staurotide, apatite; near Lewis's Paper Mill, tourmaline, mica.

HUNTINGTON CO.—Near Frankstown, in the bed of a stream, and on the side of a hill, *fibrous celestine*, abundant.

LANCASTER CO.—Near Texas, in the south part of the county, at Wood's Chrome Mine, *emerald nickel, pennite, kæmmererite*, millerite, *baltimorite, chromic iron, marmolite*, picrolite, *hydromagnesite, brucite, dolomite*, cryst. magnesite, calcite, serpentine; at Lowe's Mine, *hydromagnesite*, brucite, *picrolite! magnesite*, chromic iron, talc, emerald nickel, serpentine, baltimorite; on M. Boice's farm, N. of the village in the soil, *cryst. pyrites!* anthophyllite, marmolite, magnesite; near the Rock Spring, chalcedony, carnelian, moss agate, green tourmaline in talc, titanic iron, cryst. magnetic iron in chlorite; at Reynold's Mine, calcite, *talc*, picrolite; at Gap Mine, *magnetic* pyrites (containing nickel), copper pyrites, actinolite; at Safe Harbor, iron ores; Pequea Valley, 8 m. S. of Lancaster, argentiferous galena (250 to 300 oz. of silver to the ton); 4 m. N. W. of Lancaster, on L. and H. Railroad, *calamine*, galena, blende, buratite.
LITTLE BRITAIN.—*Anthophyllite.*

LEBANON CO.—CORNWALL, adjoining Lancaster Co.—*Pyrites!* in cubo-octahedrons, brilliant steel tarnish, magnetite, native copper, *red copper*, azurite, chrysocolla.

LEHIGH CO.—Near Friedensville in the Saucon Valley, *calamine!* (valuable mine), lanthanite, cryst. quartz, malachite, pyrolusite, wad; near Allentown, magnetic iron, pipe iron ore; near Bethlehem, in S. Mountain, *allanite* in syenite, zircon.

MONROE CO.—In Cherry Valley, calc spar, chalcedony, cryst. quartz; in Poconoe Valley, near Judge Mervine's, cryst. quartz.

MONTGOMERY CO.—At Perkiomen Copper Mine, azurite, blende, galena, *pyromorphite, cerusite, molybdate of lead, anglesite, heavy spar*, calamine, copper pyrites, green malachite, chrysocolla; at Henderson's Marble Quarry, *calc spar;* about one mile N. of Henderson's, in the bank of railroad, cryst. quartz in geodes; at Spring Mills, cacoxene, lepidokrokite, spathic iron: near the Gulf Mills, *limonite*, garnets, chromic iron; in Franconia Township, (?) gold.

NORTHUMBERLAND CO.—Opposite Selim's grove, calamine.

NORTHAMPTON CO.—Near Easton, *zircon!!* (exhausted), nephrite, serpentine in pseudomorphs, coccolite, tremolite, calamite, pyroxene, sahlite, limonite, magnetic iron, purple calc spar; near Bethlehem, at the South Mountain, on Mr. Weaver's farm, *allanite*, magnetite, epidote, *zircon*, sphene, brown garnet, black spinel and tourmaline in syenitic gneiss.

PHILADELPHIA CO.—On the Schuylkill, near foot of inclined plane, garnet, tourmaline, *mica;* on the Schuylkill, a fourth of a mile from the Suspension Bridge, yellow uranite; one hundred yards above bridge, on east side, *laumontite* in hornblende slate.

CHESNUT HILL.—*Mica, serpentine, dolomite, asbestus,* nephrite, talc, *tourmaline*, sphene, apatite, tremolite.

GERMANTOWN.—Mica, apatite, feldspar, beryl, garnet.

BANKS OF WISSAHICCON.—Actinolite garnet, staurotide.

FRANKFORD.—Garnet, staurotide, iron pyrites.

CONCHIKOCEN.—Staurotide, garnet, argillaceous iron ore; near Manyunk Tunnell, stilbite, chabazite (rare in small brownish-yellow crystals).

YORK CO.—*Calc spar* (transparent), *cryst. smoky quartz, cryst. pyrites;* in Slate Quarries near the Susquehannah, *wavellite.*

DELAWARE.

NEWCASTLE CO.—Brandywine Springs, *bucholzite, fibrolite* abundant, sahlite, pyroxene; near Middletown, *vivianite* in green sand.

Dixon's Feldspar Quarries, 6 miles N. W. of Wilmington, (these quarries have been worked for the manufacture of porcelain), *adularia, albite, beryl, apatite, cinnamon stone!!* (both granular, like that from Ceylon, and crystallized, rare), magnesite, serpentine, asbestus, black *tourmaline!* (rare), *indicolite!* (rare), sphene in pyroxene, kyanite.

Dupont's Powder Mills, "hypersthene."

Eastburn's Limestone Quarries, near the Pennsylvania line, *tremolite, bronzite.*

QUARRYVILLE.—Garnet, spodumene, fibrolite, sillimanite.

Near Newark on the railroad, sphærosiderite on drusy quartz, jasper (ferruginous opal), cryst. spathic iron in the cavities of cellular quartz.

WILMINGTON.—In Christiana quarries, *metalloidal diallage.*

Kennett turnpike, near Centreville, kyanite and garnet.

KENT CO.—Near Middletown, in Wm. Polk's marl pits, *vivianite!*

On Chesapeake and Delaware Canal, retinasphalt, iron pyrites, amber.

SUSSEX CO.—Near Cape Henlopen, vivianite.

MARYLAND.

BALTIMORE, (Jones's Falls, 1¾ miles from B.)—Chabazite (haydenite), heulandite (beaumontite of Levy), pyrites, lenticular carbonate of iron, *mica, stilbite.*

Sixteen miles from Baltimore, on the Gunpowder.—*Graphite.*

Twenty-three miles from B., on the Gunpowder.—*Talc.*

Twenty-five miles from B., on the Gunpowder.—*Magnetic iron, sphene,* pycnite.

Thirty miles from B., in Montgomery Co., on farm of S. Eliot.—Gold in quartz.

Eight to twenty miles north of B., in limestone.—*Tremolite, augite, pyrites,* brown and yellow tourmaline.

Fifteen miles north of B.—*Sky-blue chalcedony* in granular limestone.

Eighteen miles north of B., at Scott's Mills.—*Magnetic iron*, kyanite.

BARE HILLS.—*Chromic iron, asbestus, tremolite, talc*, hornblende, serpentine, chalcedony, meerschaum, baltimorite, *copper pyrites*, magnetite.

CAPE SABLE, near Magothy R.—Amber, pyrites, alum slate.

CARROLL Co.—Near Sykesville, Liberty Mines, gold, magnetic iron, *pyrites* (*octahedrons*), *copper pyrites*, linnæite (carrollite), an ore of nickel (not analyzed); at Patapsco Mines, near Finksburg, *erubescite, malachite, linnæite, remingtonite*, magnetic iron, *copper pyrites;* at Mineral Hill Mine, *erubescite*, copper pyrites, ore of *nickel* (sieginite), gold, magnetic iron.

CECIL Co., north part.—*Chromic iron* in serpentine.

COOPTOWN, Harford Co.—Olive colored *tourmaline, diallage, talc* of green, blue and rose colors, *ligniform asbestus, chromic iron, serpentine.*

DEER CREEK.—*Magnetic iron!* in chlorite slate.

FREDERICK Co.—Old Liberty Mine, near Liberty Town, black copper, malachite, copper glance, specular iron; at Dollyhyde Mine, *erubescite*, copper pyrites, iron pyrites, argentiferous galena in dolomite.

MONTGOMERY Co.—*Oxyd of manganese.*

SOMERSET AND WORCESTER Cos., north part.—*Bog iron ore, vivianite.*

ST. MARY'S RIVER.—*Gypsum!* in clay.

VIRGINIA AND DISTRICT OF COLUMBIA.

ALBEMARLE Co., a little west of the Green Mts.—*Steatite, graphite;* galena.

AMHERST Co., along the west base of Buffalo ridge.—*Copper ores*, etc.

AUGUSTA Co.—At Weyer's (or Weir's) cave, sixteen miles northeast of Staunton, and eighty-one miles northwest of Richmond, calc spar and stalactites.

BUCKINGHAM Co.—*Gold* at Garnett and Moseley Mines, largely worked, also pyrites, pyrrhotine, calcite, garnet; at the Eldridge Mine (now London and Virginia Mines) near by, and the Buckingham Mines near Maysville, gold, auriferous pyrites, copper pyrites, tennantite, *heavy spar ; kyanite, tourmaline, actinolite.*

CHESTERFIELD Co.—Near this and Richmond Co., bituminous coal, native coke.

CULPEPPER Co., on Rapidan river.—Gold, pyrites.

FRANKLIN Co.—Grayish steatite.

FAUQUIER Co., Barnet's Mills.—Asbestus ; gold mines, *barytes, calcite.*

FLUVANNA Co.—Gold at Stockton's Mine ; also tetradymite at "Tellurium Mine."

PHENIX Copper Mines.—*Copper pyrites*, etc.

GEORGETOWN, D. C.—Rutile.

GOOCHLAND Co.—Gold Mines, (Moss and Busby's.)

HARPER'S FERRY, on both sides of the Potomac.—Thuringite (owenite), with quartz.

JEFFERSON Co., at Sheperdstown.—Fluor.

KENAWHA Co.—At Kenawha, *petroleum*, brine springs, cannel coal.

LOUDON Co.—*Tabular quartz, prase, pyrites, talc, chlorite, soapstone,* asbestus, *chromic iron, actinolite, quartz crystals; micaceous iron,* erubescite, malachite, epidote, near Leesburg, (Potomac Mine.)

LOUISA Co.—Walton gold mine, gold, pyrites, copper pyrites, argentiferous galena, spathic iron, blende, anglesite; boulangerite, blende (at Tinder's Mine).

NELSON Co.—Galena, copper pyrites, malachite.

ORANGE Co.—Western part, Blue Ridge, specular iron; gold at the Orange Grove† and Vaucluse gold mines, worked by the "Freehold" and "Liberty" Mining Companies.

ROCKBRIDGE Co., three miles southwest of Lexington.—Heavy spar

SHENANDOAH Co., near Woodstock.—Fluor spar.

MT. ALTO, Blue Ridge.—Argillaceous iron ore.

SPOTSYLVANIA Co., two miles northeast of Chancellorville.—*Kyanite;* Gold mines at the junction of the Rappahannock and Rapidan ("Gardiner" Co.); on the Rappahannock (Marshall Mine); Whitehall Mine, affording also tetradymite.

STAFFORD Co., eight or ten miles from Falmouth.—Micaceous iron, gold, tetradymite, silver, galena, vivianite.

WASHINGTON Co., eighteen miles from Abingdon.—*Rock salt* with *gypsum.*

WYTHE Co., (Austin's mines).—*Cerusite, minium, plumbic ochre,* blende, *calamine, galena.*

On the Potomac, twenty-five miles north of Washington City.—*Native sulphur* in gray compact limestone.

NORTH CAROLINA.

ASHE Co.—Malachite, copper pyrites.

BUNCOMBE Co.—Corundum (from a boulder), *margarite,* corundophilite, *garnet,* chromic iron, barytes, *fluor,* rutile, iron ores, oxyd of manganese.

BURKE Co.—Gold, monazite, zircon, beryl, *corundum, garnet,* sphene, *graphite,* iron ores.

CABARRAS Co.—Phenix mine, gold, barytes, *copper pyrites,* auriferous pyrites, quartz pseudomorph after barytes, tetradymite; Pioneer Mines, *gold,* limonite, pyrolusite, *barnhardite, wolfram, scheelite,* tungstate of copper, *wolframine,* diamond, chrysocolla, copper glance, molybdenite, *copper pyrites, iron pyrites;* White Mine, needle ore, copper pyrites, barytes; Long and Muse's Mine, argentiferous galena, iron pyrites, copper pyrites, limonite; Boger Mine, tetradymite; Fink Mine, valuable copper ores; Mt. Makins, tetrahedrite? magnetite, talc, blende, pyrites, galena; Geo. Luderick's farm, *scorodite,* limonite, gray copper, copper pyrites, iron pyrites.

CALDWELL Co.—Chromic iron.

CHATHAM Co.—Mineral coal, pyrites.

CHEROKEE Co —Iron ores, gold, galena, corundum, rutile.

DAVIDSON Co.—King's, now Washington Mine, native silver, cerusite, anglesite, scheelite, pyromorphite, galena, blende, malachite, black copper, *wavellite,* garnet, stilbite. Five miles from Washington Mine, on Faust's Farm, gold, *tetradymite,* oxyd of bismuth and tellurium, copper pyrites, limonite, spathic iron, epidote; near Squire Ward's, gold in crystals, electrum.

Gaston Co.—Iron ores, corundum, margarite. Near Crowder's Mountain (in what was formerly Lincoln Co.), *lazulite, kyanite, garnet*, graphite; also twenty miles northeast, near south end of Clubb's Mountain, lazulite, kyanite, talc.

Guilford Co.—McCullock copper and gold mine, twelve miles from Greensboro', *gold, pyrites, copper pyrites* (worked for copper), *quartz*, spathic iron. The North Carolina Copper Co. are working the copper ore at the old Fentress mine.

Henderson Co.—*Zircon*.

Jackson Co.—Smoky Mountain, alunogen.

Lincoln Co.—Diamond; at Randleman's, *amethyst!* rose quartz.

Macon Co.—Chromic iron.

McDowell Co.—Brookite, monazite, corundum in small crystals red and white, *zircons*, garnet, beryl, sphene, xenotime, rutile, elastic sandstone, iron ores.

Mecklenburg Co.—Near Charlotte (Rhea and Cathay Mines) and elsewhere, *copper pyrites, gold;* chalcotrichite at McGinn's Mine; barnhardite near Charlotte; pyrophyllite in Cotton Stone Mountain, diamond.

Rowan Co.—Gold Hill Mines, thirty eight miles northeast of Charlotte, and fourteen from Salisbury, gold, auriferous pyrites; ten miles from Salisbury, *feldspar* in crystals.

Rutherford Co.—*Gold, graphite*, bismuthic gold, diamond, euclase, *pseudomorphous quartz*, chalcedony, corundum in small crystals, *epidote, pyrope*, brookite, zircon, monazite, rutherfordite, samarskite, *quartz crystals*, itacolumite; on the road to Cooper's gap, kyanite.

Stokes and Surrey Cos.—Iron ores, graphite.

Union Co.—Lemmond Gold Mine, eighteen miles from Concord, (at Stewart's and Moore's Mine), gold, quartz, blende, argentiferous galena (containing 29·4 oz. of gold, and 86·5 oz. silver to the ton, Genth), pyrites, some copper pyrites.

Yancey Co.—*Iron ores*, amianthus, *chromic iron*.

SOUTH CAROLINA.

Abbeville Dist.—Oakland Grove, *Gold* (Dorn Mine), galena, pyromorphite, amethyst, garnet.

Anderson Dist.—At Pendleton, *actinolite*, galena, kaolin, *tourmaline*.

Charleston.—*Selenite*.

Cheowee Valley.—Galena, tourmaline, gold.

Chesterfield Dist.—Gold (Brewer's mine), talc, chlorite, pyrophyllite, pyrites, native bismuth, carbonate of bismuth, red and yellow ochre, whetstone.

Darlington.—Kaolin.

Edgefield Dist.—Psilomelane.

Greenvile Dist.—Galena, phosphate of lead, kaolin, chalcedony in burhstone, beryl, plumbago, epidote, *tourmaline*.

Kershaw Dist.—*Rutile*.

Lancaster Dist.—Gold (Hale's mine), talc, chlorite, kyanite, elastic sandstone, pyrites; gold also at Blackman's mine, Massey's mine, Ezell's mine.

NEWBERRY DIST.—Leadhillite (?).

PICKEN'S DIST.—Gold, manganese ores, kaolin.

RICHLAND DIST.—Chiastolite, novaculite.

SPARTANBURG DIST.—*Magnetic iron ore,* chalcedony, *hematite;* at the Cowpens, *brown hematite, graphite,* limestone copperas.

SUMTER DIST.—Agate.

UNION DIST.—Fairforest gold mines, pyrites, copper pyrites.

YORK DIST.—Limestones, whetstones, witherite, heavy spar.

GEORGIA.

BURKE AND SCRIVEN COS.—Hyalite.

CLARK CO., near Clarksville.—Gold, xenotime, zircon, rutile, kyanite, specular iron, garnet, quartz.

HABERSHAM CO.—*Gold,* iron and copper pyrites, *galena,* hornblende, garnet, quartz, kaolin, soapstone, chlorite, *rutile,* iron ores, galena, tourmaline, staurotide, zircon.

HALL CO.—*Gold,* quartz, kaolin, diamond.

HANCOCK CO.—Agate, chalcedony.

HEARD CO.—Molybdote of iron.

LUMPKIN CO.—Gold, quartz crystals.

RABUN CO.—Gold, *copper pyrites.*

WASHINGTON CO., near Saundersville.—*Wavellite, fire opal.*

CANTON Mine.—*Harrisite,* copper pyrites, melaconite, galena, pyromorphite, pyrites, marcasite, erubescite, blende, native copper, automolite, staurotide, kyanite, ilmenite, Hitchcockite, covelline.

ALABAMA.

BIBB CO., CENTREVILLE.—*Iron ores,* marble, *heavy spar,* coal, cobalt.

TUSCALOOSA CO.—*Coal,* galena, pyrites, vivianite, limonite, calcite, dolomite, kyanite, steatite, quartz crystals, manganese ores.

FLORIDA.

NEAR TAMPA BAY.—Limestone, sulphur springs, chalcedony, carnelian, agate, silicified shells and corals.

KENTUCKY.

MAMMOTH CAVE.—*Gypsum,* in imitative forms, stalactites, nitre, epsom salt.

Near the line between Livingston and Union Cos., galena, copper pyrites.

TENNESSEE.

BROWN'S CREEK.—Galena, blende, heavy spar, celestine.

CARTER'S CO., foot of Roan Mt.—*Sahlite,* magnetic iron.

CLAIBORNE CO.—*Calamine,* galena, smithsonite, chlorite, steatite, and magnetic iron.

Cocke Co., near Brush Creek.—Cacoxene, kraurite, iron sinter, stilpnosiderite, brown hematite.

Davidson Co.—Selenite, with granular and snowy *gypsum*, or alabaster, crystallized and compact *anhydrite, fluor* in crystals? *calc spar* in crystals. Near Nashville, blue *celestine* (crystallized, fibrous and radiated), with *heavy spar* in limestone. Haysboro', galena, blende, with heavy spar as the gangue of the ore.

Dickson Co.—Manganite.

Jefferson Co.—*Calamine*, galena, fetid heavy spar.

Knox Co.—Magnesian limestone.

Maury Co.—Wavellite in limestone.

Morgan Co.—Epsom salt, nitrate of lime.

Polk Co., Hiwassee mine, southeast corner of state, near Ocoee river. *Black copper!* copper pyrites, iron pyrites (mines valuable), allophane.

Roan Co., eastern declivity of Cumberland Mts.—Wavellite in limestone.

Severn Co., in caverns.—Epsom salt, soda alum, saltpetre, nitrate of lime.

Smith Co.—Fluor.

Smoky Mt., on declivity.—Hornblende, garnet, staurotide

Unaka Mts., Eastern Tennessee, at Sevier, etc., in caverns.—Alum.

OHIO.

Brainbridge, (Copperas Mt., a few miles east of B.)—Calc spar, heavy spar, iron pyrites, copperas, alum.

Canfield.—*Gypsum!*

Duck Creek, Monroe Co.—Petroleum.

Liverpool.—Petroleum.

Marietta.—Argillaceous iron ore; iron ore abundant also in Scioto and Lawrence Counties.

Poland.—*Gypsum!*

MICHIGAN.

Lake Superior Mining Region.—The four principal regions are Keweenaw Point, Isle Royale, the Ontonagon, and Portage Lake. The mines of Keweenaw Point are along two ranges of elevation, one known as the Greenstone Range and the other as the Southern or Bohemian Range, (Whitney.) The copper occurs in the trap or amygdaloid, and in the associated conglomerate. *Native copper! native silver!* copper pyrites, horn silver, gray copper, manganese ores, epidote, *prehnite, laumontite, datholite,* heulandite, *stilbite, analcime,* chabazite, *mesotype,* (Copper Falls mine), *leonhardite,* (ib.), *analcime,* (ib.), *apophyllite,* (at Cliff Mine), *wollastonite,* (ib.), *calc spar, quartz* (in crystals, at Minesota mine), *saponite, black oxyd* of copper, (near Copper Harbor, but exhausted), chrysocolla; on Chocolate river, galena and sulphuret of copper; copper pyrites and native copper at Presq' Isle. At Albion mine, *domeykite;* at Prince Vein, *amethyst;* at Michipicoten Ids., copper nickel, stilbite, analcime.

Isle Royale, 48° N., 89° W.—*Native copper, epidote,* harmotome (?) *datholite,* wollastonite (exhausted), *pectolite,* chlorastrolite.

ILLINOIS.

Gallatin Co., on a branch of Grand Pierre Creek, sixteen to thirty miles from Shawneetown, down the Ohio, and from half to eight miles from this river.—*Violet fluor spar!* in carboniferous limestone, heavy spar, *galena,* blende, brown iron ore; near Rosiclare, *calcite,* galena, blende; five miles back from Elizabethtown, bog iron; one mile north of the river, between Elizabethtown and Rosiclare, *nitre.*

In Northern Illinois, townships 27, 28, 29, several important mines of *galena.*

Pope Co.—Pyromorphite.

INDIANA.

Limestone Caverns; Corydon Caves, &c.—*Epsom salt.*

In most of the southwest counties, *pyrites, sulphate of iron,* and *feather alum;* on Sugar Creek, pyrites and *sulphate of iron;* in sandstone of Lloyd Co., near the Ohio, *gypsum;* at the top of the blue limestone formation, *brown spar, calc spar.*

MINESOTA.

North Shore of Lake Superior, (range of hills running nearly northeast and southwest, extending from Fond du Lac Superieure to the Kamanistiqueia river in Upper Canada.)—*Scolecite, apophyllite, prehnite, stilbite, laumontite, heulandite, harmotome,* thomsonite, *fluor spar, sulphate of baryta, tourmaline, epidote,* hornblende, calcareous spar, quartz crystals, iron pyrites, magnetic iron ore, steatite, blende, black oxyd of copper, malachite, native copper, copper pyrites, amethystine quartz, ferruginous quartz, *chalcedony, carnelian, agate,* drusy quartz, hyalite? fibrous quartz, jasper, prase (in the debris of the lake shore), dogtooth spar, augite, native silver, spodumene? arsenite? of cobalt, chlorite; between Pigeon Point and Fond du Lac, near Baptism river, saponite (thalite), in amygdaloid.

Kettle River Trap Range.—Epidote, nail-head calc spar, amethystine quartz, calcareous spar, undetermined zeolites, saponite.

Stillwater.—Blende.

Falls of the St. Croix.—Green carbonate of copper, native copper, epidote, nail-head spar.

Rainy Lake.—Actinolite, tremolite, fibrous hornblende, garnet, iron pyrites, magnetic iron, steatite.

WISCONSIN.

At Mineral Point and elsewhere, copper and lead ores, principally silicate and carbonate of copper, copper pyrites and *galena,* (only the last abundant.) Also *pyrites, capillary pyrites, blende, white lead ore,* leadhillite (?), *smithsonite* (carbonate of zinc), anglesite, *heavy spar,* and calc spar.

SANK CO.—Specular iron! malachite, copper pyrites.

MONTREAL RIVER PORTAGE.—Galena in gneissoid granite.

LAC DU FLAMBEAU R.—Garnet, kyanite.

BIG BULL FALLS, (near.)—Bog iron.

LEFT HAND R., (near small tributary.)—Malachite, copper glance, native copper, red copper ore, earthy malachite, epidote, chlorite? quartz crystals.

IOWA.

DU BUQUE LEAD MINES, and elsewhere.—*Galena! calc spar*, black oxyd of manganese; at Ewing's and Sherard's diggings, *calamine!* or smithsonite; at Des Moines, quartz crystals, selenite; Mahoqueta R., *brown iron ore.*

CEDAR RIVER, a branch of the Des Moines.—*Selenite* in crystals, in the bituminous shale of the coal measures; also elsewhere on the Des Moines, gypsum abundant; argillaceous iron ore, spathic iron; copperas in crystals on the Des Moines, above the mouth of Saap and elsewhere, *iron pyrites*, blende.

MISSOURI.

BIRMINGHAM.—Limonite.

JEFFERSON CO., at Valle's Diggings.—*Galena, white lead ore*, anglesite, calamine, pyritous copper, blue and green malachite, carbonate of baryta.

MINE A BURTON.—*Galena, white lead ore, anglesite, heavy spar*, calc spar.

DEEP DIGGINGS.—Carbonate of copper, *white lead ore* in crystals, and manganese ore.

MINE LA MOTTE.—*Galena!* malachite, *earthy cobalt* and *nickel*, bog manganese, sulphuret of iron and nickel, *white lead ore* in crystals, caledonite, plumboresinite, wolfram.

PERRY'S DIGGINGS, and elsewhere.—Galena, &c.

Forty miles west of the Mississippi and ninety south of St. Louis, the iron mountains, specular iron, limonite.

ARKANSAS.

BATESVILLE.—In bed of white R., some miles above Batesville, Gold.

OUACHITA SPRINGS.—*Quartz!* whetstones.

MAGNET COVE.—*Brookite! schorlomite, elæolite*, magnetic iron, quartz, green coccolite, garnet, apatite.

CALIFORNIA.

Along the Sierra Nevada, *gold*, platinum (rare), iridosmine, molybdenite, molybdine, zircon, magnetic iron; near bay of San Francisco, actinolite, talc, serpentine, jasper, salt, gypsum (island in the Caquines Straits); ridges of Sierra Azul, south of San José, *cinnabar.* Gold also found in the Umpqua region, Oregon and the Shasty Mountains. Pt. Orford, gold, platinum, iridosmine.

CANADA.

CANADA EAST.

Abercrombie.—Labradorite.

Aubert, Gallion.—Gold, iridosmine, platinum.

Bay St. Paul.—*Ilmenite!* apatite, allanite, rutile, (or brookite?)

Bolton.—*Chromic iron, magnesite,* serpentine, picrolite, steatite, bitter spar, wad.

Boucherville Mountain.—*Augite* in trap.

Brome.—*Magnetic iron,* copper pyrites, *sphene,* ilmenite, phyllite, sodalite, cancrinite, galena.

Chambly.—Analcime, chabazite and calcite in trachite.

Chateau Richer.—*Labradorite, ilmenite, hypersthene.*

Daillebout.—Blue spinel, with clintonite.

Grenville.—*Tabular spar, sphene,* idocrase, calcite, pyroxene, *garnet* (cinnamon stone), *zircon, graphite, scapolite.*

Ham.—Chromic iron in serpentine.

Inverness.—*Variegated copper.*

Lake St. Francis.—*Andalusite* in mica slate.

Landsdowne.—*Barytes.*

Mille Isles.—*Labradorite!* ilmenite, hypersthene, andesine, *zircon.*

Montreal.—*Calcite, augite,* sphene in trap

Morin.—*Sphene, apatite, labradorite.*

Polton.—Chromic iron, *steatite,* serpentine, *amianthus.*

Rougemont Mts.—Augite in trap.

St. Armand.—Micaceous iron ore with quartz, epidote.

St. François Beauce.—Gold, platinum, iridosmine, ilmenite, magnetite, serpentine, chromic iron, soapstone, magnetite, heavy spar.

St. Jerome.—*Sphene, apatite, chondrodite, phlogopite, tourmaline, zircon,* molybdenite, *magnetic pyrites.*

St. Norbert.—Apatite in greenstone.

St. Roch.—On the Achigan, two miles below St. Roch, *apatite* in trappean rocks.

Stukely.—Serpentine, *verde antique!* schiller spar.

Sutton.—*Magnetic iron* in fine crystals, *specular iron, rutile,* dolomite, *magnesite,* chromiferous *talc,* bitter spar, steatite.

Upton.—Copper pyrites, malachite, calcite.

Vaudreuil.—Limonite, vivianite.

Yamaska.—Sphene in trap.

CANADA WEST.

Balsam Lake.—*Molybdenite,* scapolite, quartz.

Brantford.—Sulphuric acid spring, (4·2 parts of pure sulphuric acid in 1000).

Bathurst.—Heavy spar, *black tourmaline, perthite* (orthoclase), *peristerite* (albite), *bytownite.*

Brome.—Magnetite.

Burgess.—*Pyroxene,* albite, *mica, sapphire,* sphene, copper pyrites, *apatite, black spinel!* spodumene (in a boulder).

BYTOWN.—*Calcite, bytownite,* chondrodite, spinel.

CAPE IPPERWASH, Lake Huron.—Oxalite in shales.

CLARENDON.—*Idocrase.*

DALHOUSIE.—Hornblende, dolomite.

DRUMMOND.—Labradorite.

ELMSLEY.—Pyroxene, sphene, feldspar, *tourmaline.*

FITZROY.—Amber, brown *tourmaline,* in quartz.

GŒTINEAU RIVER, Blasdell's Mills.—Calcite, apatite, tourmaline, hornblende, pyroxene.

GRAND CALUMET ISLAND.—*Apatite, phlogopite! pyroxene!* sphene, *idocrase!!* serpentine, tremolite, *scapolite,* brown and black *tourmaline!* pyrites, loganite.

HIGH FALLS OF THE MADAWASKA.—*Pyroxene!* hornblende.

HULL.—*Magnetite,* garnet, graphite.

HUNTERSTOWN.—*Scapolite, sphene,* idocrase, garnet, *brown tourmaline!*

INNISKILLEN.—Petroleum.

LAC DES CHATS, Island Portage.—*Brown tourmaline!* pyrites, calcite, quartz.

LANARK.—Raphilite (hornblende), serpentine, asbestus.

LANDSDOWN.—*Barytes!* vein 27 in. wide, and fine crystals.

MADOC.—Magnetite.

MARMORA.—Magnetite, chalcolite, garnet, epsomite, specular iron.

McNAB.—Specular iron.

SOUTH CROSBY.—Chondrodite in limestone, magnetite.

ST. ADELE.—Chondrodite in limestone.

SYDENHAM.—Celestine.

TERRACE COVE, Lake Superior.—Molybdenite.

WALLACE MINE, Lake Huron.—*Specular iron,* arsenical nickel, sulphuret of nickel, nickel vitriol.

BRUCE MINES.—Copper pyrites, copper glance, erubescite.

NEW BRUNSWICK, St. John.—*Graphite.*

NOTE.—The rock of the Mississippi valley containing the remarkable deposits of galena, (sometimes regarded as the equivalent of the "Cliff" or "Upper Magnesian" Limestone,) is considered by James Hall as between the Hudson River and Trenton Groups of New York in age, and as having no representative in the eastern part of the United States. The sandstones and conglomerates of the Lake Superior copper region in Michigan are referred by Foster and Whitney, Hall, Owen and Logan, to the age of the Potsdam sandstone, or the *lowest Silurian;* while the copper bearing red sandstone of Connecticut and New Jersey, is shown by Redfield, Rogers and Hall, to be as recent as the *Liassic* Period.

BRIEF NOTICE OF FOREIGN MINING REGIONS.

The geographical positions of the different mining regions are learned with difficulty from the scattered notices in the course of a mineralogical treatise. A general review of the more important is therefore here given, to be used in connection with a good map.

A course across Europe from southeast to northwest, passes over a large part of the mining regions, and it will be found most convenient to the memory to mention them in this order, commencing with the borders of Turkey.

1. The mines of the Bannat in southern Hungary, near the borders of Turkey, (about latitude 45°) situated principally at Orawitza, Saszka, Dognaszka, and Moldawa. *Ores.* Argentiferous copper ores, vitreous copper, malachite, copper pyrites, red copper ore, galena, ores of zinc, cobalt, native gold, yielding silver, gold, copper, and lead. *Rock.* Syenite, and granular limestone.

2. The mines of western Transylvania, about latitude 46°, situated between the rivers Maros and Aranyos, at Nagyag, Offenbanya, Salathna, and Vöröspatak. *Ores.* Native gold, telluric gold, telluric silver, white tellurium, with galena, blende, orpiment, realgar, gray antimony, fahlerz, carbonate of manganese, manganblende; especially valuable in gold and silver.

3. In the mountain range, bounding Transylvania on the north, about latitude 47° 40′, at Nagy-banya, Felso-banya, and Kapnik. *Ores.* Native gold, red silver, argentiferous gray copper, pyritous copper, blende, realgar, gray antimony. *Rock.* Porphyry.

4. In the Königsberg mountains, northern Hungary, about latitude 48° 45′, at Schemnitz and Kremnitz. *Ores.* Argentiferous galena and copper pyrites, native gold, red silver ore, gray antimony, some cobalt ores and bismuth, mispickel; particularly valuable for gold, silver, and antimony. *Rock.* Diorite and porphyry.

5. To the east of the Königsberg mountains, at Schmolnitz and Retzbanya. *Ores.* Pyritous copper, gray copper ore, blende, gray antimony, particularly valuable for copper. *Rock.* Clay slate.

6. Illyria, west of Hungary, at Bleiberg and Raibel, (in Carinthia.) *Ores.* Argentiferous galena, calamine, with

some copper pyrites and other ores, affording silver and zinc abundantly. *Rock.* Mountain limestone.—Also at Idria, native mercury and cinnabar, in argillaceous schist.

7. In Western Styria, at Schladming. *Ores.* Arsenical nickel, copper nickel, native arsenic, arsenical iron, largely worked for nickel. *Rock.* Argillaceous slate. Illyria and Styria are noted also for their iron ores, especially spathic iron.

8. In the Tyrol, at Zell. *Ores.* Argentiferous copper and iron ores, auriferous pyrites, native gold. *Rock.* Argillaceous slate.

9. In the Erzgebirge separating Bohemia from Saxony, and consisting principally of gneiss.

A. Bohemian or southern slope, at Joachimstahl, Mies, Schlackenwald, Zinnwald, Bleistadt, Przibram, Katherinenberg. *Ores.* Tin ores, argentiferous galena, (worked principally for silver,) arsenical cobalt ores, copper nickel, affording tin, silver, cobalt, nickel, and arsenic.

B. Saxon or northern slope, at Altenberg, Geyer, Marienberg, Annaberg, Schneeberg, Ehrenfriedersdorf, Johanngeorgenstadt, Freiberg. *Ores.* Argentiferous galena, (worked only for silver,) tin ore, various cobalt and nickel ores, vitreous and pyritous copper, affording silver, tin, cobalt, nickel, bismuth, and copper.

10. In Silesia, in the Riesen-gebirge, an eastern extension of the Erz-gebirge, at Kupferberg, Jauer, Reichenstein. *Ores* of copper, cobalt, affording copper, cobalt, arsenic and sulphur.

11. In Silesia, in the low country east of the Riesen-gebirge, near the boundary of Poland, at Tarnowitz. *Ores.* Calamine, electric calamine, blende, argentiferous galena, affording zinc, silver and lead. *Rock.* Mountain limestone.

12. Northwest of Saxony, near latitude 51° 30′, at Eisleben, Gerlstadt, Sangerhausen, and Mansfeld. *Ores.* Gray copper, somewhat argentiferous, variegated copper ore, affording copper. *Rock.* A marly bituminous schist (kupferschiefer) more recent than the coal strata.

13. In the Harz-gebirge, (Hartz mountains,) north of west from Eisleben, about latitude 51° 50′, at Clausthal, Zellerfeld, Lauthenthal, Wildemann, Grund, Andreasberg, Goslar, Lauterberg. *Ores.* Vitreous copper, gray copper, pyritous copper, cobalt ores, copper nickel, ruby silver ore, argentiferous galena, blende, antimony ores, affording silver, lead, copper, and some gold.

14. In Hesse-Cassel to the southwest of the Hartz, at Riechelsdorf. *Ores.* Arsenical cobalt, arsenical nickel, nickel ocher, native bismuth, bismuth glance, galena, affording cobalt. *Rock.* Red sandstone. Also at Bieber, cobalt ores in mica slate.

15. In the Bavarian or Upper Rhine, (Palatinate,) near latitude 49° 45′, at Landsberg near Moschel, Wolfstein, and Morsfeld. *Ores.* Cinnabar, native mercury, amalgam, horn quicksilver, pyrites, brown iron ore, some gray copper ore, and copper pyrites. *Rocks.* Coal formation.

16. Province of the Lower Rhine, at Altenberg, near Aix la Chapelle (or Aachen.) *Ores.* Calamine, electric calamine, galena, affording zinc. *Rock.* Limestone. The same, just south in Netherlands, at Limburg, and also to the west at Vedrin, near Namur.

17. There are also copper mines at Saalfeld, west of Saxony, in Saxon-Meiningen, in Southern Westphalia near Siegen, in Nassau at Dillenberg, and elsewhere.

18. In Switzerland, Canton du Valais. *Ores.* Argentiferous lead, and valuable nickel and cobalt ores.

19. The range of the Vosges, in France, parallel with the Rhine, about St. Marie-aux-Mines. *Ores.* Argentiferous galena, (affording 1-1000 of silver,) with phosphate of lead, gray copper, antimonial sulphuret of silver, native silver, arsenical cobalt, native arsenic, and pyrites, occasionally auriferous; affording silver and lead. *Rocks.* Argillaceous schist, syenite, and porphyry.

20. In France there are also the mining districts of the Alps, Auvergne or the Plateau of Central France, Brittany, and the Pyrenees, but none are very productive, except in iron ores. Brittany resembles Cornwall, and formerly yielded some tin and copper. The valley of Oisans in the Alps, at Allemont, contains argentiferous galena, arsenical cobalt and nickel, gray copper, native mercury, and other ores, in talcose, micaceous, and syenitic schists, but they are not now explored. The region of Central France is worked at this time only at Pont-Gibaud, in the department of Puy-de-Dome, and at Vialas and Villefort in the Gard. The former is a region of schistose and granite rocks, intersected by porphyry, affording some copper, antimony, lead, and silver; the latter of gneiss, affording lead and silver, from argentiferous galena. The French Pyrenees are worked at the present time only for iron.

21. In England there are two great metalliferous districts.

A. On the southwest, in Cornwall, and the adjoining county of Devonshire. *Ores.* Pyritous copper and various other copper ores, tin ore, galena, with some bismuth, cobalt, nickel, and antimony ores, affording principally copper, tin, and lead. *Rocks.* Granite, gneiss, micaceous and argillaceous schist.

B. On the North, in Cumberland, the adjoining parts of Durham, with Yorkshire and Derbyshire, just south. *Ores.* Galena, and other lead ores, blende, copper ores, calamine, (the last especially at Alstonmoor in Cumberland, and Castleton and Matlock, in Derbyshire,) affording largely of zinc, and three-fifths of the lead of Great Britain, and some copper. *Rock.* Carboniferous limestone.

C. There is also a rich vein of calamine, blende, and galena, in the same limestone at Holywell, in Flintshire, on the north of Wales; another of calamine at Mendip Hills, in Southern England, south of the Bristol channel, in Somersetshire, occurring in magnesian limestone; mines of copper on the isle of Anglesey, in North Wales, in Westmoreland and the adjacent parts of Cumberland and Lancashire, in the southwest of Scotland, the Isle of Man, and at Ecton in Staffordshire, &c.

22. In Spain, there are mines—

A. On the south, in the mountains near the Mediterranean coast, in New Grenada, and east to Carthagena, in Murcia; situated in New Grenada, in the Sierra Nevada, or the mountains of Alpujarras, the Sierra Almagrera, the Sierra de Gador, just back of Almeria, and at Almazarron near Carthagena. *Ore.* Galena, which is argentiferous at the Sierra Almagrera, and at Almazarron, affording full 1 per cent. of silver. *Rock.* Limestone, associated with schist and crystalline rocks.

B. The vicinity of the range of mountains running westward from Alcaraz, (in the district of La Mancha,) to Portugal. 1. On the south, near the center of the district of Jaen, at Linares, latitude 38° 5′, longitude 3° 40′. *Ores.* Galena, carbonate of lead, red copper ore, malachite, in granite and schists; affording lead and copper. 2. In La Mancha, at Alcaraz, northeast of Linares, latitude 38° 45′. *Ores.* Calamine affording abundantly zinc. 3. In the west extremity of La Mancha, near latitude 38° 38′, at

Almaden. *Ores.* Cinnabar, native mercury, horn quicksilver, pyrites, in clay slate. 4. Southwest of Almaden, in Southern Estremadura, and Northwestern Sevilla, at Guadalcanal, Cazalla, Rio Tinto. *Ores.* Gray copper, copper vitriol, malachite, with some red silver ore, and native silver, in ancient schists or limestones.

There are also mines of lead and copper at Falsete in Catalonia; in Galicia, a little tin ore; in the Asturias at Cabrales, copper ores.

23. In Sweden:—1. At Fahlun, in Dalecarlia. *Ores.* Copper pyrites, variegated copper. *Rock.* Syenite and schists.—At Finbo and Broddbo. *Ores.* Columbium ores, tin ore.—At Sala. *Ore.* Argentiferous galena, affording lead and silver. *Rock.* Crystalline limestone.—At Vena, (or Wehna,) and at Tunaberg. *Ores.* Arsenical cobalt, arsenate of cobalt. *Rock.* Mica slate and gneiss.—At Dannemora and elsewhere. *Ore.* Magnetic iron.

24. In Norway, at Kongsberg, vitreous silver, native silver, horn silver, native gold, galena, native arsenic, blende. *Rock.* Mica slate.—At Modum and Skutterud. *Ores.* Cobalt ores, native silver. *Rock.* Mica slate.—At Arendal, magnetic iron.

25. In Russia:—1. In the Urals, (mostly on the Asiatic side,) at Ekatherinenberg, Beresof, Nischne Tagilsk, &c. *Ores.* Native gold, platinum, iridium, native copper, red oxyd of copper, malachite. 2. The Altai, (southern Siberia,) at Kolyvan and Zmeof. *Ores.* Native gold, native silver, argentiferous galena, carbonate of lead, native copper, oxyds of copper, malachite, pyritous copper, calamine. *Rocks.* Metamorphic beds, and porphyry. 3. In the Daouria mountains, east of Lake Baikal, at Nertchinsk. *Ores.* Argentiferous galena, carbonate of lead, arsenate of lead, gray antimony, arsenical iron, electric calamine, cinnabar. *Rocks.* Ancient compact limestone and schists.

Other important foreign mines, are the copper mines of Cuba, South America, Southern Australia; the silver mines of South America and Mexico; the gold mines of South America, Africa, and the East Indies; the quicksilver mines of Huanca Velica, Peru, and those of China; the tin of Malacca, (principally on the island of Junck Ceylon,) of Banca; of zinc, in China; of platinum, in Brazil, Columbia, St. Domingo, and Borneo; of palladium, in Brazil; of arsenic in Khoordistan, China. Copper mines are also reported from New Zealand.

MINERALOGICAL IMPLEMENTS.

For the examination and collection of minerals, the mineralogist should be provided with a few simple implements.

1. A three-cornered or small flat file, for testing hardness.

2. A knife with a pointed blade, of good steel, for trying hardness. Berzelius suggests that it may be magnetized, to be used as a magnet.

3. The series of crystallized minerals, constituting the scale of hardness (see page 64.) The diamond and talc are least essential.

4. Small glass-stoppered bottles (one-ounce) of each of the acids muriatic, sulphuric, and nitric, in a dilute state, (page 66.)

5. A blowpipe, (page 67.)

6. The common fluxes, (page 69.)

7. Pieces of charcoal for blowpipe purposes, (page 69.) Also strips of mica for holding the assay when platinum is not at hand.

8. A candle or lamp for blowpipe trials, (page 68.)

9. Platinum foil, wire, and forceps, (page 69.)

10. Also a pair of small steel spring forceps, for holding fragments of minerals in the blowpipe flame, and for managing the assay.

11. A piece of glass tube, $\frac{1}{6}$ inch bore; and two or three test tubes (of hard glass,) or small mattresses, for trying the action of acids, and testing the presence of water by the blowpipe.

12. A pair of cutting pliers, for removing chips of a mineral for blowpipe or chemical assay.

13. A common goniometer; or a pair of arms pivoted together to use with a scale, as explained on pages 47, 48. The reflecting goniometer (page 50) is also a desirable instrument.

14. Models of the common crystalline forms; they may be made by the student, out of chalk, or wood; and when finished, a coat of varnish or gum will give great hardness to the chalk.

15. A pair of balances for specific gravity, (page 63.)

16. A hammer weighing about two pounds, resembling a

stone cutter's hammer, having a slightly rounded face, and at the opposite end, an edge having the same direction as the handle. The handle should be made of the best hickory, and the mortice to receive it should be as large as the handle. A similar hammer, having the upper part prolonged to a blunt point, to be used like a pick.

17. Another hammer of half a pound weight, similar to the preceding, except that the face should be flat; to be used in trimming specimens.

18. A small jeweller's hammer, for trying the malleability of globules obtained by the blowpipe, and for other purposes.

19. A piece of steel, say ½ inch thick, 1 or 2 wide, and 2 or 3 long, to be used as an anvil. A fragment may be broken or pulverized upon it, by first folding it in a piece of thin paper, to prevent its flying off when struck. A half inch circular cavity on one side, and a pestle to correspond, will be found very convenient.

20. Two steel chisels of the form of a wedge, as in the annexed figure; one 6 inches long, and the other 3. When it is desired to pry open seams in rocks with the larger chisel, two pieces of steel plate should be provided to place on opposite sides of the chisel, after an opening is obtained; this protects the chisel and diminishes friction while driving it.

21. Bone ashes, to be used upon mica, or in a small cavity in charcoal, in cupelling for silver, with the blowpipe. A rounded cavity should be made in the charcoal, as large as the end of the little finger, and the bone ashes (slightly moistened, and mixed with a little soda,) should be pressed into it firmly with the head of a small pestle; after thoroughly drying, it is in a condition to receive the assay.

22. A pocket microscope.

23. A small agate mortar and pestle.

24. A magnetic needle.

25. A pair of scissors.

26. A box of matches.

For blasting and other heavy work, the following tools and appliances are necessary:—

1. Three hand-drills, 18, 24, and 36 inches long, an inch in diameter. The best form is a square bar of steel, with a diagonal edge at one end. The three are designed to follow one another.

2. A sledge hammer of 6 or 8 pounds weight, to use in driving the drill.

3. A sledge hammer of 10 or 12 pounds weight, for breaking up the blasted rock.

4. A round iron spoon, at the end of a wire 15 or 18 inches long, for removing the pulverized rock from the drill-hole.

5. A crowbar, a pickaxe, and a hoe, for removing stones and earth before or after blasting.

6. Cartridges of blasting powder, to use in wet holes. They should one-third fill the drill-hole. After the charge is put in, the hole should be filled with sand and gravel alone without ramming. If any ramming material is used, plaster of Paris is the best, which has been wet and afterwards scraped to a powder.

7. Patent fuse for slow match, to be inserted in the cartridge, and to lead out of the drill-hole.

WEIGHTS, MEASURES, AND COINS.

For the convenience of the student, the following information is here inserted, of such weights, measures, and coins, of different countries, as are likely to be met with in the course of his ordinary reading on minerals and mining.

24 grains, Troy,	= 1 pennyweight (dwt.)
20 dwt. "	= 1 ounce (oz.)
12 oz. "	= 1 pound (lb.)
16 drams Avoirdupois,	= 1 oz.
16 oz. "	= 1 pound.
112 lbs. "	= 1 hundred (cwt.)
20 cwt. "	= 1 ton.

1 lb. troy = 5760 grs. troy = 13 oz. 2·65143 drams av.
1 lb. av. = 7000 grs. troy = 1 lb. 2 oz. 1 dwt. 16 gr. troy.

To reduce *pounds troy*, to pounds avoirdupois, multiply by the decimal .822857 ; or, approximately, diminish by 3-17.

To reduce *pounds avoirdupois*, to pounds troy, multiply by 1·215.

100 lbs. av. is now the usual 1 cwt., and 25 lbs. the quarter cwt.

112 pounds, formerly	= 1 quintal.
100 pounds, now usually	= 1 quintal.
1 French *gramme*	= 15·433159 grs. troy.

1 French *kilogramme* = 1000 grammes = 2·21 lbs. av. nearly = 2·68 lbs. troy = 2·0429 French livres.

To reduce			Approximately.
Fr. kilograms to Eng. av. pounds,	mult. by	2·2055	or add 6-5.
Prussian, (including Hanoverian, Brunswick, and Hessian,) pounds, to Eng. avoir. pounds,	" "	1·031114	" " 1-32.
Fr. livre, (poids de marc) to Eng. av. lbs.	" "	1·079642	" " 2-25.
Eng. av. lb. to French *kilogram,*	" "	0·453414	" sb. 11-20.
Eng. av. lb. to French *livre,*	" "	0·9262	" " 1-13.
Eng. cwt. (112 lbs.) to a *metric quintal,* (= 100 kilog. French,)	" "	0·5078	
Eng. cwt. to a Prus. *centner,* (=110lbs.)	" "	0·9875	" " 1-80.
Eng. cwt. to a quintal, (old measure= 100 livres,)	" "	1·0385	" add 2-53.
A metric quintal to an English cwt.	" "	1·971	
A quintal, old meas. to an Eng. cwt.	" "	0·963	" sub. 1-27.
A Prussian centner to an Eng. cwt.	" "	1.0127	" add 1-80.

•The old French livre contained 2 marcs, or 16 ounces; a marc = 3778 Eng. grs. A marc at Cologne, (Hamburgh, etc.,) = 8 oz. = 3608 Eng. grs.

The Russian *pood* (or *pud*) = 40 Russian pounds = 36 English pounds avoirdupois.

12 inches English,	1 foot.
3 feet,	1 yard.
40 rods,	1 furlong.
8 furlongs,	1 mile.
3 miles,	1 league.
6 feet,	1 fathom.
60 geographical miles,	1 degree.
69½ statute miles (nearly,)	1 degree.

A French meter=3 feet, 3·371 inches English, or more correctly, 39·37079 inches English=3 feet, 0 inches, 11·296 lines French. A kilometer=3280·9 English feet, or $\frac{80}{129}$ths of a statute mile.

A French toise=6·3946 English feet=6 old French feet.

English.		French.		Prussian, Danish and Rhenish.
Foot :—	=	·9382928	=	·9711361.

To reduce			Approximately.
French feet to English,	multiply by	1·065765	or add 1-15.
English feet to French,	" "	0·9382928	or subt. 1-16.
French meters to English feet,	" "	3·280899	or add 23-7.
French meters to English yards,	" "	1·093633	or add 1-11.
English feet to French meters,	" "	0·3047945	or subt. 7-10

The French foot according to an act in 1812, is a ⅓ of a

meter, but this measure has not been adopted, the old French foot, (=1·066 English feet) continuing to be used.

A German geographical mile=4 English geographical miles, or about 4·633 Eng. statute miles = 7407·40 meters.

French stere, (cubic measure) = 35·34384 cubic ft. U. S.

French litre (liquid and dry measure,) = 61·07416 cubic inches, or 1·05756 quarts wine measure.

Value of different weights, in English avoirdupois pounds, of measures in English feet and inches, and of coins in American dollars.

Amsterdam.—1 centner (100lbs.) = 108·923 av. lbs.

Batavia.—1 picul = nearly 136 av. lbs.

Bremen.—1 centner = 116 av. lbs. ; 1 lb. = 1·1 av. lbs.; 1 foot = 11⅜ in ; 1 rix dollar, (silver) = $0·787 ; 72 grotes = 1 rix dollar.

Calcutta.—1 rupee, (gold) =$6.75 ; 1 rupee (silver,)= $0.45,6 ; 1 candy = 20 maunds, = 500 lbs. av.

Canton.—1 picul = 133⅓ av. lbs. ; 1 catty = 1⅓ av. lbs.; 1 tael = 1⅓ oz. ; 1 tael = $1·48 ; 10 mace = 1 tael.

Denmark.—1 centner (100 lbs.) = 110¼ av. lbs. ; 1 foot =12⅓ inches ; 1 rix dollar, (silver) $0·52 ; 6 marcs = 1 rix dollar ; 16 skillings = 1 marc.

Florence and Leghorn.—1 cantaro, (100 lbs.) = 74·86 av. lbs. ; 1 palmo = 9¾ inches.

France.—1 franc = $0·186 ; 10 decimes = 1 franc ; 10 centimes = 1 decime.

Genoa.—1 peso grosso (100 lbs.) = 76⅝ av. lbs ; 1 peso sottile = 69·89 av. lbs ; 1 palmo = 9⅕ in.

Great Britain.—£1 = 20 shillings sterling = $4·84 ; 1 guinea = 21 shillings sterling = $5·08⅕.

Hamburg.—1 foot = 11·3 inches ; 1 mile = 4·68 miles ; 1 marc banco = $0·35 ; current marc = $0·28 ; 3 marcs = 1 rix dollar.

Malta.—1 foot, 10⅙ inches ; 1 cantaro, (100 lbs.) = 174·5 av. lbs. ; 1 pezza = $1.

Manilla.—1 arroba = 26 av. lbs. ; 1 picul = 143 av. lbs.; 1 palmo = 10·38 in.; 8 rials = $1 ; 34 maravedis = 1 rial.

Naples.—1 cantaro grosso = 196·5 av. lbs. ; 1 cantaro piccolo = 106 av. lbs. ; 1 palmo = 10⅜ in. ; 1 ducat, (silver) = $0·80 ; 10 carlini = 1 ducat ; 10 grani = 1 carlino.

Portugal.—100 lbs. = 101·19 av. lbs. ; 1 arroba = 22·26

av. lbs.; 1 quinta. = 89·05 av. lbs.; 1 pe or foot, = $12\frac{1}{3}$ in.; 1 mile = $1\frac{1}{4}$ mile; 1 milree, or crown = $1·12 = 1000 rees; 400 rees = 1 cruzado.

Prussia.—100 lbs. = 103·11 av. lbs.; 1 quintal, (110 lbs.) = 113·42 av. lbs.; 1 foot = 1·03 feet; 1 mile = 4·68 miles; 1 thaler, $0·69 = 30 groschen; 12 pfennigs = 1 grosch.

Rome.—100 libras = 74·77 av. lbs.; 1 foot = $11\frac{3}{4}$ in.; 1 canna = $6\frac{1}{2}$ feet; 1 mile = $7\frac{2}{5}$ fur.

Russia.—100 lbs. = 90·26 av. lbs.; 1 pood, (40 lbs.) = 36 lbs.; 1 Russian pound = 32 loths = 96 zolotniks; 1 verst, (mile) = 3500 Eng. feet = 5·3 fur.; 1 inch = 1 English inch; 1 foot (in general) = 1 Eng. foot; 1 ruble, (silver) = $0·78 = 100 copecks. Bank ruble = $0·223, or nearly $22\frac{1}{3}$ cents.

Sicily.—100 libras = 70 av. lbs; 1 cantaro grosso = 192·5 av. lbs.; 1 cantaro sottile = 175 av. lbs.; 1 palmo = $9\frac{1}{2}$ in.; 1 canna = $6\frac{1}{3}$ feet; 1 oncia, (gold) = $2·40 = 30 tari; 20 grani = 1 taro.

Spain.—1 quintal = 101·44 av. lbs.; 1 arroba = 25·36 av. lbs.; 1 fanega = 1·6 bu.; 1 foot = 11·128 in.; 1 league = 4·3 m. nearly; 1 vara = 2·78 feet; 20 rials = $1; 16 quintos = 1 rial; 2 maravedis = 1 quinto.

Sweden.—100 lbs. (victualie) = 73·76 av. lbs.; 1 foot = 11·69 in.; 1 mile = 6·64 m.; 1 ell = 1·95 feet.

Smyrna.—100 lbs. (1 quintal) = 129·48 av. lbs.

Trieste.—100 lbs. = 123·6 av. lbs.; 1 foot Austrian = 1·037 feet; 1 mile Austrian = 4·6 miles; 1 florin, (silver) = $0·485; 60 kreutzers = 1 florin.

Venice.—1 peso grosso, (100 lbs.) = 105·18 av. lbs.; 1 peso sottile = 64·42 av. lbs.; 1 foot = 1·14 feet; 1 lira = 1 franc French = $0·186; 100 centesimi = 1 lira.

A troy pound of fine silver is worth at the mint, $15·51,515.
A troy pound of standard silver, (American) $13·86,615.
A troy pound of fine gold, $248·27,586.
A troy pound of standard gold, (American) $223·25,581.
1 dwt. of fine gold, $1·034.
1 dwt. of American native gold, usually, $0·95 to 1·01.
A troy pound of platinum in bars, $90 to $100.
A pound av. of copper, about $0·25 to 0·27.
A pound av. of tin, about $0·20.
A carat, see page 82.

TABLES FOR THE DETERMINATION OF MINERALS.

In the following tables, the more common mineral species (comprising all the American) are arranged in subdivisions, to afford aid in ascertaining the names of species. These tables will be found valuble as a means of instruction ; the use of them fixes the attention on distinctive characters, and thereby impresses the peculiarities of species on the mind.

A general view of the arrangement in Table I. is here annexed.

I.—Soluble Minerals.

A. No effervescence with muriatic acid.
  a. No deflagration on burning coals.
  b. Deflagration on burning coals.
B. Effervesce with muriatic acid when heated, if not without.

II.—Insoluble Minerals.

Luster unmetallic.
  A. Streak uncolored.
    a. No odorous or colored fumes before the blowpipe, on charcoal.
      1. Wholly soluble in one or more of the three acids.
        * Infusible.*
        † Fusible with more or less difficulty.
      2. Soluble, except the silica which separates as a jelly.
        * Infusible.
        † Fusible with more or less difficulty.
      3. Not acted on by acids, or partially soluble without forming a jelly.
        * Infusible.
        † Fusible with more or less difficulty.
    b. Colored or odorous fumes before the blowpipe, alone or on charcoal.
  B. Streak colored.
    a. No fumes before the blowpipe.

* By infusible is meant, not capable of being melted alone or on charcoal by the flame of the common blowpipe.

* Fusible.
† Infusible.
b. Fumes before the blowpipe.

II. Luster metallic.
A. Streak unmetallic.
* No fumes before the blowpipe on charcoal.
† Fumes before the blowpipe.
B. Streak metallic.
* Malleable.
† Not malleable; no fumes when heated.
‡ Not malleable; fumes when heated.

The abbreviations used in these tables are as follows:

Ad.	Adamantine.	Limest.	Limestone.
Amyg.	Amygdaloidal.	Mag.	Magnetic.
Antim.	Antimony.	Mam.	Mammillary.
Arsen.	Arsenical.	Mas.	Massive.
B, bh.	Blue, bluish.	Met.	Metallic.
Bl.	Blowpipe.	*Mur.*	Muriatic acid.
Bn, bnh.	Brown, brownish.	*Nit.*	Nitric acid.
Bk, bkh.	Black, blackish.	Op.	Opaque.
Bor.	Borax.*	*Phos.*	Salt of phosphorus.*
Bot.	Botryoidal.	P'ly.	Pearly.
Cleav.	Cleavable.	Pms.	Prisms.
Char.	Charcoal.	*Prim.*	Primary rocks.†
Col.	Columnar.	R, rdh.	Red, reddish.
Cryst.	Crystals, crystalline.	Rad.	Radiated.
Decrep.	Decrepitate.	Ren.	Reniform.
Deliq.	Deliquescent.	Res.	Resinous.
Dif.	Difficult, difficultly.	*Soda,*	Carbonate of soda.*
Div.	Divergent.	Sol.	Soluble.
Efferv.	Effervescence.	St.	Streak.
Exfol.	Exfoliate.	Stalact.	Stalactitic.
Fib.	Fibrous.	Stel.	Stellate.
Flex.	Flexible.	Strl.	Translucent on edges only
Fol.	Foliated.	Strp.	Semitransparent.
Fus.	Fusible.	Sulph.	Sulphureous
Gelat.	Gelatinize.	Submet.	Submetallic.
Glob.	Globule.	*Sul.*	Sulphuric acid.
Gn, gnh.	Green, greenish.	Trl.	Translucent.
Gran.	Granular.	Trp.	Transparent.
Gy, gyh.	Gray, grayish.	Vit.	Vitreous
Infus.	Infusible.	Vol.	Volatile.
Insol.	Insoluble.	Volc.	Volcanic rocks.
Intum.	Intumesce.	W, wh.	White, whitish.
Lam.	Laminæ.	Yw, ywh.	Yellow, yellowish.

* Blowpipe flux.

† This term as here used means simply, granite and the allied crystalline rocks, syenite, gneiss, mica slate, talcose slate, hornblende rock, without reference to age.

The Roman numerals refer to the systems of crystallization, (page 32.)

I. Monometric. IV. Monoclinic.
II. Dimetric. V. Triclinic.
III. Trimetric. VI. Hexagonal or Rhombohedral.

The page on which each species is described is mentioned, that the student may conveniently turn to the fuller descriptions for a farther examination of a mineral.

The kinds of rock in which the species occur is often added after the description.

I.—SOLUBLE MINERALS.

A. No Effervescence with muriatic acid, even if heated.

a. Not deflagrating on burning coals.

Sal ammoniac, 100. I; crusts; G 1·5—1·6; wh, ywh; taste acute and pungent; not deliquescent; *Sul*, effervesce; mixed in powder with quicklime ammoniacal odor; volatile.

Alum, 127. I; wh; very soluble, sweetish astringent: *Bl*, fus! intumesces.

Common salt, 104. I; G 2·2—2·3; w, rdh, gyh; saline; crystals cubic: *Bl*, decrepitates.

Epsom salt, 124. III; G 1·7—1·8; w; bitter saline: *Bl*, deliq.

White vitriol, 271. III; G 2—2·1; wh; astringent-met: *Bl*, w coating on charcoal.

Borax, 107. IV; G 1·7—1·8; wh; slow efflor; sweetish alkaline: *Bl*, swells up and becomes w and opaque.

Glauber salt, 102. IV; G 1·4—1·5; wh, gyh; cooling and bitter: *Bl*, watery fusion.

Copperas, 246. IV; G 2; gn, ywh, wh; astringent-met: *Bl*, red; *Bor* gn glass.

Blue vitriol, 297. V; G 2·2—2·3; sky-blue; nauseous met: *Bl*, copper reaction.

White arsenic, 226. Capillary cryst; bot, mas; Gr 3·7; w; taste astringent, sweetish: *Bl*, volatile, alliaceous fumes.

b. Deflagrate on burning coals.

Niter, 101. III; G 1·9—2; w, not deliquescent or efflorescent.

Nit. of soda, 103. VI; G 2—3; wh; deliq; burns with a deep yellow light.

Nitrate of lime, 123. Cryst efflorescences; G 1·62; w, gy; very deliquescent: *Bl*, watery fusion, scarcely detonates.

B. Effervescing with muriatic acid.

Natron, 103. IV; G 1·4—1·5; w, gyh; efflorescent.

II.—INSOLUBLE MINERALS.

I. LUSTER UNMETALLIC.

A. Streak Uncolored.

a. No fumes before the blowpipe on charcoal.

1. *Wholly soluble in one or more of the acids, (cold or hot), usually with effervescence.*

* Infusible.

Hardness.

Hydromagnesite, 125. 1·0—2·0 Whitish crusts; G 2·8; adheres to the tongue. *Serpentine.*

Brucite, 125. 1·5—2·0 VI; fol, laminæ flexible; G 2·3—2·4; w, gnh; p'ly trl; no effervescence. *Serpentine.*

		Hardness.	
Websterite,	129.	1·5—2·0	Ren, mas; G 1·6—1·7; dull; w, op; adheres to the tongue: *sul*, sol, no effervescence.
Nemalite,	125.	2	Silky fib; G 2·3—2·5; gyh, bh-w; fibres separable, brittle on exposure. *Serpentine.*
Calc spar,	115.	3;0—2·0	VI; cleav! fib, mas; G 2·3—2·5; vit, p'ly, w, gy, bnh, bk; trp—op.; sometimes soft and earthy: *Bl*, intense light.
Aragonite,	118.	3·5	III; mas, fib; G 2·8—3; vit; w, gyh, bnh; trp—op; effervesce; *Bl.* intense light, crumbles.
Diallogite,	261.	3·5	VI; cleav; mas; G 3·5—3·6; vit, p'ly; rdh; trl op; *nit*, effervesces: *Bl*, *bor*, violet glass.
Magnesite,	124.	3·0—4·0	VI; cleav! fib, mas; G 2·9—3; vit, silky; w, ywh, bn; trp, op; little effervescence.
Blende,	269.	3·5—4·0	I; dodec cleav; mas; G 4—4·1; resin-yw, rdh, w; trp, strl; *nit* sol, emitting sul. hydrogen: *Bl*, *bor* infus.
Dolomite,	118.	3·5—4·0	VI; cleav! mas; G 2·8—2·9; vit, p'ly; w, gy, bn; no effervescence unless heated.
Mesitine spar,	249.	4·6	VI; cleav! mas; G 3·3—3·65; vit; ywh, bn on exposure; *mur* slow solution.
Oligon spar,	248.	"	VI; cleav; mas; G 3·7—3·8; vit; bn on exposure: *Bl*, *bor* amethystine glob.
Yttrocerite,	206.	4·5—5·0	III; cleav; mas; violet b, gyh, rdh-bn; vit, p'ly; strl, op; hot *mur*, sol; *Bl.* whitens. *Prim.*

† Fusible with more or less difficulty.

Witherite,	109.	3·0—3·5	III; mas, fib; G 4·2—4.4 w, ywh, gyh; trl—op; *nit* efferv: *Bl* fus! op. glob.
White lead ore,	281.	3·0—3·5	III; mas; G 6·1—6·5; w, gyh, bnh; ad, res; trp, trl; brittle; *mur* eff: *Bl*, fus! on *char*, lead.
Strontianite,	111	3·5	III; cleav; fib, mas; G 3·6—3·7; res, vit; gnh, ywh, gyh; effervescence: *Bl*, fus dif! colors flame reddish.
Pyromorphite,	283.	3·5—4·0	VI; hexag pms; bot, fib; G 6·5—7·1; bright gn, yw, bn; res; strl, strp; brittle; hot *nit* sol: *Bl*, fus! *With lead ores.*
Spathic iron,	247.	3—4·0	VI, cl, mas; G 3·7—3·9; ply; ywh, bnh, gyh; darkens on exposure; trl, op; pulverized, some eff: *Bl*, fus dif!; blackens; iron reaction.
Wavellite,	130.	"	III; fib, glob; G 2·3—2·4; p'ly, vit; w, ywh, bnh, gyh; trl; hot *nit*, sol, vapors corrode glass: *Bl*, fus, intum, colorless glass.
Cacoxene,	249.	"	Div, rad, fib, silky; G 3·3—3·4; ywh-bn, ywh; bn on exposure: *Bl*, infus.
Fluor spar,	121.	4·0	I; cl! mas; G 3·1—3·2; vit; w, yw, b, violet, gn, r, often lively; trp, trl; *sul*, affords fumes that corrode glass: *Bl*, fus, decrep; phosphoresces when heated.
Apatite,	120.	4·5—5·0	VI; hexag; mas; G 3—3·3; vit, res; gn, bh, w, rh, bn; trp, op; brittle; *nit* sol slowly in powder,

		Hardness.	
			without efferv: *Bl,* fus dif! *bor* fus! *Prim. Gran. limestone, volc.*
Triplite,	260.	5·0	Lam, mas; G 3·4—3·8; bkh-bn; res, ad; *nit* sol, no ef: *Bl,* fus! bk scoria; *bor* violet glass.
Triphyline,	249.	5·5	IV; mas; G 3—3.6; gnh, yw, gy, rdh-br; vit. res; trp, trl; *mur,* sol; *Bl,* fus *bor,* green glass, *soda* manganese reaction.
Boracite,	126.	7·0	I; hemihed cubes; G 2·9—3; w, gyh; vit, ad; strp, trl; pyro-electric; *mur,* sol: *Bl,* fus. *Gypsum.*

2. *Soluble, excepting the silica, which separates as a jelly.*

* Infusible.

Halloylite,	161.	1·0—2·0	Mas, earthy or waxy; G 1·8—2·1; w, bh; adheres to the tongue; *sul,* gelat! *Bl,* infus.
Allophane,	162.	3·0	Mas, ren; G 1·8—1·9; vit, res; bh, gnh, ywh, trl; very brittle; gelat! *Bl,* intum.

† Fusible.

Mesole,	167.	3·5	III; fib rad; G 2·3—2·4; p'ly; gyh-w, ywh; trl: *Bl,* fus! *Amyg.*
Laumontite,	166.	"	IV; mas; G 2·2—2·4; vit, p'ly; w, gyh; trl; w and friable on exposure; gelat! *Bl,* fus w, frothy. *Amyg. prim.*
Phillipsite,	168.	4·0—4·5	III; rad, cryst often crossed; G 2—2·2; w, rdh; vit; trp. op; *mur* gelat: *Bl,* fus. *Amyg.*
Tabular spar,	141.	4·0—5·0	V; cl, subfib; G 2·7—2·9; p'ly, vit; w, gyh; trl: *mur* gelat: *Bl,* fus dif, pearl semiop. *Prim. amyg.*
Thomsonite,	167.	4·5—5·0	III; cl, fib, rad; G 2·3—2·4; w, bnh; trp—trl; brittle; gelat: *Bl,* fus! intum, w, op. *Amyg. prim.*
Dysclasite,	142.	4·5—5·0	fib, div; G 2·2—2·4; p'ly, vit; w, bh; trl, strp; very tough under the hammer; *mur* gelat; *Bl,* fus, op. *Amyg.*
Pectolite,	142.	"	fib, div; G 2·69; vit, p'ly; w, gyh; after heating gelat in *mur: Bl,* fus trp glass. *Amyg.*
Calamine,	272.	"	III; cl; mas, bot, fib; G 3·2—3·5; w, b, gn, yw, bn; trp—trl; hot *nit* gelat: *Bl,* fus dif!! intum; phosphoresces. *Stratified rocks.*
Natrolite,	166.	4·5—5·5	III; acic, cryst; div fib; G 2·1—2·3; vit; w, ywh; trp, trl; gelat! *Bl,* fus! op glass. *Amyg. volc.*
Analcime,	168.	5·0—5·5	I; trapezohed; mas; G 2—2·3; vit; w, rhd, gyh; trp—op; brittle; *mur* gelat: *Bl* fus! intum, glassy glob. *Amyg. volc.*
Scolecite,	167.	"	III; div, fib, rad; G 2·2—2·3; vit, p'ly; w; trp, trl; *nit* and *mur;* gelat! *Bl* fus! op, curls up in outer flame. *Amyg. volc.*
Datholite,	142.	"	IV; glassy crystals; fib, bot, mas; G 2—2·3; w, gnh, rdh; trp, trl; *nit* gelat! *Bl,* fus! *Amyg. prim.*
Sodalite,	197.	5·5—6·0	I; dodec cryst; mas; G 2·2—2·45; vit; gyh, bn, b; trp—strl; *nit* gelat: *Bl,* fus, colorless glass.
Nepheline,	180.	5·5—6·0	VI; hexag; coarse massive, subfib; G 2·4—2·6; vit greasy; w, ywh, gnh, bnh, rdh; trp—op; gelat. *Bl,* fus dif, blebby glass. *Volc. prim.*

3. *Not acted on by acids, or partially soluble without forming a jelly.*

† Infusible.

Mineral	No.	Hardness.	Characters
Talc,	143.	1·0—1·5	VI; fol! mas; G 2·7—2·9; light gn, gnh-w, gyh, p'ly, unctuous; laminæ flexible, inelastic. *Prim.*
Pyrophyllite,		"	Fol! gran; apple-gn, w, bnh-gn, ywh; p'ly; strp, strl: *Bl*, swells up! *Prim.*
Mica,	191.	"	Fol!! lam. thin elastic, tough; G 2·8—3; colors various, often bright; p'ly; trp, strl: *Bl*, fus dif! *Prim*, etc.
Chrysocolla,	300.	2·0—3·0	mas, bot; G 2—2·3; bluish-gn; smooth vit, or earthy; strl, op; *nit* sol, except silica.
Gibbsite,	131.	3·0—3·5	Stalact, crusts; G 2·3—2·4; gyh-w, gnh-w; dull.
Emerald nickel,	264.	3·0—3·5	minute globular, crust; G 3·05; emerald-gn; St paler; vit; trp, trl; *Bl*, loses its color.
Blende,	269.	3·5—4·0	I; dodec cleav, mas; G 4—4·1; resin-yw, bn, bh, w, rdh; trp—op: *Bl*, *bor* inf.
Plumbo-resinite,	285.	4·0—4·5	reniform; G 6·3—6·4; ywh, bnh, rdh; resinous, or like gum arabic; til; *Bl*, decrep; enam on char.
Clintonite,	148.	4·0—5·0	IV; fol! lam brittle; G 3—3·1; rdh-bn; met-p'ly; strl: *Bl*, *bor* trp pearl.
Alum stone,	129.	5·0	III; acic, stel, mas; G 2·6—2·8; vit, p'ly, earthy; w, rh, gyh; *sul*, mostly sol: *Bl*, decrep; *soda* infus.
Monazite,	206.	5·0	IV; imbedded, cryst, cleav! in one direction; G 4·8—5·1; bn, bnh-r; vit, res; strp, op; brittle; *mur*, decomposed. *Prim.*
Leucite,	175.	5·5—6·0	I; trapezohedrons; G 2·4—2·5; w, gyh; vit; strp, trl: *Bl*, *bor*, fus dif. *Volc.*
Anatase,	211.	"	II; in cryst; G 3·8—3·9; fine bn, b; met-ad, res; strp, trl: *Bl*, loses col; *bor* fus dif. *Prim.*
Turquois,	130.	6·0	Reniform; G 2·8—3; b, bh-gn; waxy, dull; trl, op: *Bl*, flame green; *bor*, fus.
Opal,	139.	5·5—6·5	Massive, uncleav; w, yw, r, bn, gn, gy, pale; in some a play of colors; vit, p'ly; trp, strl: *Bl*, decrep, op.
Kyanite,	173.	5·0—7·0	V; in prisms or bladed cryst; G 3·5—3·7; b, w, bnh; p'ly, vit; trp, strl: *Bl*, *bor* fus dif, trp. *Prim.*
Nephrite,	147.	6·0—7·0	Mas, subgran; G 2·9–3·1; leek-gn, bh, wh; vit; trl, strl: *Bl*, whitens; *bor* clear glass. *Prim.*
Bucholzite,	172.	"	Col, fib; G 3·2—3·6; w, gyh, bnh; p'ly; trl, strl: brittle. *Prim.*
Tin ore,	214.	"	II; mas, fib; G 6·5——7·1; bn, bk, w, gy, r, yw; ad, res, cryst often brilliant; strp, op: *Bl*, *bor* on char with soda affords tin. *Prim.*
Chrysolite,	156.	6·5—7·0	III; imbeded grains or masses of a glassy appearance; G 3·3—3·6; gn, bottle glass gn: *Bl*, darkens *bor* gn glass; [rarely fusible.] *Basalt*, etc.
Sillimanite,	172.	6·5—7·5	V; col, fib; G 3·0—3·4; bn, gyh; p'ly, vit; trl, strl brittle: *Bl*, *bor* infus. *Prim.*
Andalusite,	174.	"	III; stout prisms; mas; G 2·9–3·2; vit, p'ly; gyh,

		Hardness.	
			rdh; tough; structure sometimes tesselated* *Bl, bor* fus dif, trp glass. *Prim.*
Quartz,	132.	7·0	VI; mas; G 2·6—2·8; colors various; vit; trp, op: *Bl, soda* fus! trp glass, efferv.
Staurotide,	174.	7·0—7·5	III; stout prisms; G 3·5—3·8; bn, rdh-bn, bk; vit, res; strp, op. *Prim.*
Zircon,	200.	7·5	II; cryst, seldom mas; G 4·4—4·8; bn, r, yw, gy, gn, w, some bright; subad; trp, trl: *Bl, bor,* clear glass. *Prim; gran limest.*
Topaz,	194.	7·5—8·0	III; prisms with basal cleavage! mas, col; G 3·4—3·6; pale yw, gn, b, w; vit; trl, strl: *Bl, bor* slowly trp glass. *Prim.*
Spinel,	160.	8·0	I; octahedrons, etc; G 3.5—4·6; r, bh, gnh, yh, bn, bk; vit; trp, strl, (some impure crystals soft): *Bl, bor* fus dif. *Prim; gran limest,* etc.
Chrysoberyl,	199.	8·5	III; cryst; G 3·5—3·8; bright gn, ywh, gyh; vit; trp, trl: *Bl, bor* fus dif! *Prim.*
Sapphire,	158.	9·0	VI; mas; in grains; G 3·9—4·2; b, r, yw, bn, gyh-b, gy, w; vit; trp, trl: *Bl, bor* fus dif. *Prim; gran limest.*
Diamond,	80.	10·0	I; G 3·4—3·7; w, b, r, yw, gn, bn, gy, bk; adamantine; trp; strl.
			† Fusible with more or less difficulty.
Talc,	143.	1·0—1·5	III; fol! mas; G 2·7—2·9; light gn, gnh-w, gyh; p'ly, unctuous; laminæ flexible, not elastic: *Bl,* infus, or fus dif!! *Prim; gran limest.*
Chlorite,	145.	1·5	Fol; mas gran; G 2·6—2·9; olive green; p'ly; *sul* decomp: *Bl.* fus dif! sometimes to a black glassy bead. *Prim.*
Gypsum,	112.	1·5—2·0	IV; fol! gran, stel; G 2·2—2·4; w, gyh, bnh, rh, bk; trp, trl; lam flexible, inelastic: *Bl,* fus dif; whitens, exf, and becomes friable; *Strat. prim. volc.*
Mica,	191.	2·0—2·5	Fol!! lam thin elastic, tough; G 2·8—3; colors various, often bright; p'ly; trl, strl: *Bl,* fus dif! *Prim,* etc.
Cryolite,	132.	"	Mas, fol; G 2·9—3; w; vit, p'ly; fusible in a candle. *Prim.*
Serpentine	145.	2·0—3·5	III; mas; sometimes thin fol, fol brittle; fib; G 2·4—2·6; dark or light gn, gnh-w, bh-w; trl—op; feel often greasy: *Bl,* fus dif!!
Chlorophyllite,	162.	2·0—4·0	VI; fol prisms; fol brittle; G 2·7—2·8; dull green, gyh, bnh; p'ly, vit: *Bl,* fus dif!! *Prim with iolite.*
Anglesite,	281.	2·5—3·0	III; mas; lam; Gr 6·2—6·3; w, ywh, gyh; gnh, ad, vit, res; trp, trl: *Bl,* fus!! decrep; on *char,* lead globule.
Anhydrite,	114.	2·5—3·5	III; rectang cleav! mas; G 2·8—3; w, rh, bh, gyh; p'ly, vit: *Bl,* fus dif; whitens; not exf.

* This tesselated variety is often quite soft, owing to impurities.

		Hardness.	
Celestine,	110.	3·0—3·5	III; mas, fib, lam; G 3·8—4; w, bh, rh; vit, res; trp, strl: *Bl*, fus, decrep; phosphoresces.
Heavy spar,	108.	"	III; mas, fib, lam, G 4·3—4·8; w, gyh, ywh, bn; vit, p'ly, res; trp, strl: *Bl*, fus, decrep. *Strat; prim.*
Heulandite,	164.	3·5—4·0	IV; fol! fol brittle; G 2·2; w, rdh, gy, bnh; p'ly, vit; trp, strl; *acids* sol, except silica: *Bl*, fus, intum, phosphorescent. *Amyg, prim.*
Stilbite,	165	3·5—4·0	III; fol! rad, div; G 2·1—2·2; w, ywh, rh, bn; p'ly; trl, strp; *nit*, silica deposited: *Bl*, fus! intum, colorless glass. *Amyg, prim*, &c.
Schiller spar,	148.	"	Mas, fol; fol brittle; G 2·5—2·7.; dark gn, or submet: *Bl*, fus dif!! gives off water.
Chabazite,*	169.	4·0—4·5	VI; in rbdns, nearly cubes, and complex small crystals; G 2—2·2; w, rdh, ywh; vit; strp, trl; *mur*, silica deposited: *Bl*, fus! blebby enamel. *Amyg, volc, prim.*
Harmotome,	168.	"	III; crystals often crossed; G 2·3—2·5; w, rdh; vit; strp, op; *mur*, silica deposited: *Bl*, fus, clear w glass; phosphoresces. *Amyg, prim*, etc.
Tungstate of lime,	219.	"	II; mas; G 6—6·1; vit, res; ywh, w; strp—op; *nit*, becomes yw, but is not dissolved: *Bl*, fus dif!! decrepitates. *Prim.*
Apophyllite,	165.	4·5—5·0	II; glassy cryst; transverse cleav; G 2·3—2·4; w; gnh, ywh, rdh; p'ly, vit; trp, op; *nit*, sol, but hardly gelat: *Bl*, fus, exfoliates. *Amyg.*
Monazite,	206.	5·0	IV; imbedded cryst, one cleavage! G 4·8—5·1; bn, bnh-r; vit, res; strp, op; brittle; *mur*, decomposed: *Bl*, fus dif!! *Prim.*
Pyrochlore,	208.	5·0	IV; imbedded oct cryst; G 3·8—4·3; yw, ywh; res, vit; St slightly colored; trl: *Bl*, fus dif!!
Sphene,	211.	5·0	IV; usually in acute, thin crystals; G 3.2—3·5; bn, yw, gy, bk; res, ad; strl, op: *Bl*, fus dif! *bor* yw glass. *Prim, gran limestone*, etc.
Scapolite,	180.	5·0—6·0	II; mas; subcol; G 2·6—2·8; w, gyh, rh; vit, p'ly; trp—op: *Bl*, fus. *Prim, gran limestone.*
Hornblende,†	152.	"	IV; fib, rad; mas; (some var fib like flax) G 2·9—3·4; gn to bk and w; vit, p'ly; trp, op: *Bl*, fus, or dif! fus. *Prim, trap, trachyte*, etc.
Pyroxene,†	150.	"	IV; fib; mas; cleav; G 3·1—3·5; gn to bk and w, vit, p'ly; trp—op: *Bl*, fus; glassy globule. *Prim, basalt, volc*, etc.
Lazulite.	131.	5·5—6·0	IV; mas; G 3—3·1; pure b, gnh-b; vit; strp—op: *Bl*, fus, *bor* clear glob.

* The var. *Gmelinite* gelatinizes in acids.

† Some fibrous varieties (asbestus) of hornblende and pyroxene are quite soft, and resemble those of serpentine · and others are like flax, or have nearly the texture of felt.

		Hardness.	
Lapis Lazuli,	196.	5·5—6·0	I; dodec; mas; G 2·5—2·9; rich b; vit; trl—op: *Bl*, fus, trl or op glass.
Feldspar,	176.	6·0	IV; cleav, mas; G 2·3—2·6; wh, gyh, rh, bh, gnh; p'ly, vit; trp, strl; *mur*, no action: *Bl*, fus dif; *bor* trp glass.
Albite,	177.	6·0	V; cleav, mas; G 2·6—2·7; w, gyh, gnh, rh, bh; p'ly, vit; trp, strl; *mur*, no action: *Bl*, fus dif; flame yellow; may generally be distinguished from feldspar by its purer white color.
Labradorite,	178.	6·0	V; cleav, mas; G 2·6—2·8; chatoyant, gy, gnh, bn, rdh-bn: p'ly, vit; strl; hot *mur* decomp: *Bl*, fus easily, colorless glass. *Prim.*
Chondrodite,	157.	5·5—6·5	IV; gran mas; G 3·1—3·2; ywh, bnh-yw, rh, gnh; vit, res; trp, strl; brittle: *Bl*, fus dif!! *bor* fus! ywh-gn. *Gran limestone.*
Obsidian,	341.	5·5—6·5	Mas, like glass; G 2·2—2·8; bk, gy, gn; vit, p'ly: *Bl*, fus.
Manganese spar,	239.	5·5—7·0	V; mas; G 3·4—3·7; flesh-r, dark bn on exposure; vit; trp, strl: *Bl*, fus bkh glass; *bor* violet. *Prim.*
Petalite,	182.	6·0—6·5	Cleav mas, gran; G 2·4—2·5; w, bh, rh, gnh; vit, p'ly; trl; phosphoresces. *Bl*, fus; *bor* trp glass. *Prim.*
Idocrase,	184.	6·5	II; mas; G 3·3—3·4; bn, gn, w; vit, res, cryst often brilliant; trp, strl: *Bl*, fus! trl glob. *Prim; volc; gran limest.*
Prehnite,	170.	6·0—6·5	III; bot, mas; G 2·8—3; light gn, w; vit, p'ly; trl, strl; tough; *mur* sol, exc't silica: *Bl*, fus. *Amyg, prim.*
Epidote,	182.	6·0—7·0	IV; mas, gran, col; G 3·2—3·5; ywh-gn, gy, bn, bh, rh; vit, p'ly; trp, op: *Bl*, fus. *Prim*, etc.
Spodumene,	156.	6·5—7·0	Cleav mas, gran; G 3·1—3·2; gyh-w, gnh; p'ly: *Bl*, fus, intum, exf, colorless glass. *Prim.*
Axinite,	190.	"	V; cryst acute-edged; Gr 3·2—3·3; deep bn; vit, brilliant; trp, strl: *Bl*, fus! intum dark gn glass; *Prim*, etc.
Garnet,	184.	6·5—7·5	I; cryst, mas, gran; G 3·5—4·3; r, bn, w, gn, bk, often bright; vit, res; trp, trl: *Bl*, fus, no efferv bk glob. *Prim*, etc.
Boracite,	126.	7·0	I; hemihed cubes; G 2·9—3; w, gyh; vit, ad; strp trl; pyro-electric: *Bl*, fus, intum. *Gypsum.*
Iolite,	190.	7·0	III; mas, glassy; G 2·6—2·8; b, gyh-b, bnh; trp, trl: *Bl*, fus dif! *bor* trp, glass. *Prim.*
Tourmaline,	187.	7·0—8·0	VI; col, mas; G 3—3·1; bk, bn, gn, r, b, w, often bright; vit, res; trp, op; pyro-electric: *Bl*, fus, intum. *Prim*, etc.
Euclase,	199.	7·5	IV; in crystals, cleav; G 2·9—3·1; pale-gn, b, w; vit, brilliant; trp, strl: *Bl*, fus dif! intum. *Prim*
Beryl,	197.	7·5—8·0	VI; hexag pms, mas; G 2·6—2·8; gn, bright or dull, bh, ywh; trp, strl: *Bl*, fus dif; *bor* trp glass *Prim.*

b. Colored or odorous fumes before the blowpipe on charcoal.

		Hardness.	
Horn silver,	327.	1·0—1·5	I; mas, like wax; G 5·5—5·6; gy, bh, gnh; trl, strl; sectile; fus. in candle, yielding odorous fumes. *Silver ores.*
Mimetene,	284.	2·7—3·5	VI; mas; G 6·4—6·5; pale yw, bnh, bnh-r; strp, trl; hot *nit* sol: *Bl.* fus!! on *char* alliaceous fumes.—*Lead ores.*
Scorodite,	249.	3·5—4·0	III; mas; G 3·1—3·3; leek-gn, gnh w, bh, bnh; ad, vit; strp, strl: *Bl* fus! alliaceous fumes.
Blende,	269.	"	I; dodec cleav! mas; G 4—4·1; resin-yw, rdh, wh; trp, strl; *nit* sol, emitting sulph hydrogen *Bl.* on *char* at a high heat fumes of zinc.
Bismuth blende,	291.	3·5—4·5	I; mas, col; G 5·9—6·1; bn, gyh, ywh; res, ad *Bl,* fus, w fumes. *Prim.*
Smithsonite,	272.	5·0	VI; mas, ren, bot; G 4·2—4·5; gyh-w, gnh, bnh vit, p'ly; strp, trl; *nit* eferv: *Bl,* infus; on *char* w fumes. Usually *with lead ore.*

B. STREAK COLORED.

a. No fumes before the blowpipe.

* Fusible.

Minium,	280.	soft	Mas, pulv; G 4·6; bright red: *Bl,* fus; on *char* glob lead. *Lead ores.*
Vivianite	248.	1·5—2·0	IV; fol! lam. flex; mas; G 2·6—2·7; bkh-gn dark b; St, bh-w, b; *nit* or *sul* sol: *Bl,* fus! decrep, dark bn scoria, magnetic.
Uranite,	228.	2·0—2·5	II; fol! mas; G 3—3·6; bright gn, yw; St, paler; p'ly, ad; trp, strp; *nit* sol, no efferv: *Bl,* fus, bk glob. *Prim.*
Cup. anglesite,	281.	2·5—3·0	IV; cleav! G 5·3—5·5; fine azure blue; St, paler; ad, vit; trl, strl: *Bl,* reaction of copper and lead. *Lead ores.*
Chromate of lead,	284.	2·5—3·0	IV; mas, col; G 6; bright r; St orange; ad; trl; sectile; *nit* sol, no efferv: *Bl,* blackens, decrep, shining slag. *Lead ores.*
Green malachite,	298.	3·5—4·0	IV; mam, bot, crust; G 4—4·1; gn; St, paler; vit, silky, earthy; trl, op; *nit* sol, efferv: *Bl,* fus! bk; *bor* gn. *Copper ores.*
Red copper ore,	296.	"	I; mas, fib; G 5·9—6; deep red; St, bnh-r; ad submet; strp, strl; *nit* sol, efferv: *Bl* fus! on *char* metallic copper. *Copper ores.*
Pyromorphite,	283.	"	VI; mas; G 6·8—7·1; gn, bn, gy; St yw; res; strp, strl; hot *nit* sol, no efferv: *Bl* fus! *Lead ores.*
Azurite	300.	"	IV; mas, earthy; G 3·5—3·9; azure b, dark b; St paler; vit, ad; trp, strl; *nit* efferv: *Bl,* fus! copper reaction. *Copper ores.*

		Hardness.	
Pyrochlore,	208.	5·0	I; octahed cryst; G 4·2—4·3; rdh-bn, yw, ywh; St paler; res, vit; strl, op: *Bl* ywh-bn, fus dif! *bor* yw glob in outer flame. *Prim.*
Triplite,	**260.**	**5·0—5·5**	Mas, cleav; G 3·4—3·8; bkh-bn; St ywh-gy; res, ad; strp, op; *nit* sol, no efferv: *Bl* fus! bk scoria; *bor,* violet.
Monazite,	**206.**	"	IV; cryst; G 4·8—5·1; bn, rdh-bn, gyh; St rdh-w, bnh-w; vit, res; strp, op: *Bl* fus dif!! yw, op. *Prim.*
Chondrodite,	**157.**	**6·0—6·5**	IV; gran mas; G 3·1—3·3; light yw, bn, rdh; St paler; res, vit; trp, strl; very brittle: *Bl* fus dif!! loses color. *Gran limest. Prim.*
Allanite,	**207.**	"	V; acic cryst; mas; G 3·2—4·1; bnh-bk, gnh; submet, res; St gnh-gy; op, strl: *Bl* fus, froths, bk scoria. *Prim.*
			† Infusible.
Wad,	**260.**	**1·0**	Mas, often earthy; G 3·7; bn, bk, soils: *Bl,* manganese reaction.
Black copper,	296.	"	Mas, or earthy; bk, bnh-bk; St bk; soils: *Bl,* copper reaction. *Copper ores.*
Earthy cobalt,	267.	"	Earthy, mas; bk: *Bl, bor,* blue from cobalt.
Cacoxene,	249.	3·0—4·0	Fib, rad; G 3.3—3·4; ywh-bn, yw; St ywh; silky: *Bl, bor* dark red bead. *Iron ores.*
Blende,	269.	"	I; dodec cleav; mas; fib; G 4—4·1; resin yw, bn, bk, red; St pale; strp, op; *nit* sol, emitting sulph hydrogen: *Bl, bor* infus; on *char,* at high heat, fumes of zinc.
Warwickite,		3·0—4·0	Prismatic cryst; G 3—3·3; bnh, tarnished bh, or wh; St. bnh; met-p'ly; res. *Gran limest.*
Red zinc ore,	270.	4·0	III; fol! mas; G 5·4—5·6; bright r; St. orange; subad; strl, op; *nit* sol, no efferv: *Bl, bor* yw glass; *soda* a zinc slag.
Dioptase,	301.	5·0	VI; cryst; G 3·2—3·3; emerald-gn; St. gn; vit, res; trp, trl; *mur,* sol, no efferv: *Bl,* decrep, ywh-gn flame; *copper ores.*
Limonite.	239.	5·0—5·5	Mas, mam, stalact, bot; earthy; G 3·9—4·1; dull bn, bk, ywh; res, submet; strp, op: *Bl,* bk, magnetic, iron reaction.
Chromic iron,	241.	5·5	I; mas, uncleav; G 4·3—4·5; iron bk, St bn; nearly dull, submet; op: *Bl, bor* fine gn glob. *Serpentine.*
Pitchblende,	228.	5.5	Mas, bot; G 6·47; bnh-bk, velvet bk; St bk; submetallic or dull; *nit* slow sol; *Bl, bor* a gray scoria. *Prim.*
Psilomelane,	259.	5·0—6·0	Mas, bot; G 4—4·4; bk, dark steel gy; St bkh; submet; op; *mur* sol, odorous fumes: *Bl,* manganese reaction.
Rutile,	210.	6·0—6·5	II; rarely mas; G 4·2—4·3; rdh-bn, ywh, gy; St paler; ad, met-ad; trl, op: *Bl, bor* ywh-r glass; crystals often acicular. *Prim,* etc.

		Hardness.	
Tin ore,	214.	6·0—7·0	II; mas, fib; G 6·5—7·1; bn, bk, yw, r; St paler; ad; strp, op: *Bl*, on *char*, with *soda*, tin glob. *Prim.*

b. *Fumes before the blowpipe.*

Red antimony,	224.	1·0—1·5	IV; capil tufts and div; G 4·4—4·6; cherry-r; St bnh-r; ad, met; strl; *nit* w coating: *Bl*, fus!! or *char*, volat. *Prim.*
Cobalt-bloom,	267.	1·5—2·0	IV; fol! fib, stel, earthy; 2·9—3; crimson and peach-blossom r, gyh, gnh; St paler; dry powder lavender b; lam flex: *Bl*, fus! on *char* alliaceous, *bor* fine blue glob. *Prim, cobalt ores.*
Orpiment,	226.	"	III; fol! lam flex; mas; G 3·4—3·5; lemon yw; St paler; p'ly, res; strp, strl; sectile: *Bl*, sulphur and arsenical fumes. (*Realgar*, p. 305, differs in its red color and orange streak.)
Copper mica,	301.	2·0	VI; fol! mas; G 2·55; emerald gn, grass gn; St paler; p'ly, vit; trp, trl; sectile: *Bl*, alliaceous fumes, rdh-bn scoria.
Sulphur,	98.	1·5—2·5	III; mas; G 2·07; yw; rdh, gnh; res; trp, strl; burns, b flame.
Red silver,	327.	2·0—2·5	VI; mas; G 5·4—5·9, light r, to bk; St r; ad, met; strp, op: *Bl*, fus!! sulph and arsen fumes; *silver ores.*
Cinnabar,	287.	"	VI; cleav; mas; G 8—8·1; bright r, bnh-r, bn; st r, bnh; ad, sub-met; strp, op; *nit*, sol, r fumes: *Bl*, wholly vol. *Strat, prim.*
Atacamite,	302.	2·5—3·0	III; cleav; mas; G 4·4—4·5; bright gn, olive-gn; St gnh; ad, vit; strl: *Bl*, fus! muriatic fumes; copper reaction; *copper ores.*

II. LUSTER METALLIC.

A. Streak unmetallic.

* No fumes before the blowpipe on charcoal.

Wad,	260.	1·0	Mas, often earthy; G 3·7; bn, bh; soils; submet: *Bl*, manganese reaction.
Earthy cobalt,	267.	"	Mas, earthy, bot; G 2·2—2·3; bh-bk, bnh-bk; St bh bk; sectile: *Bl*, arsen fumes; *bor* blue glass.
Pyrolusite,	259.	2·0—2·5	III; col, rad; mas; G 4·8—5; iron-bk, St bk; *mur*, odor of chlorine: *Bl*, infus; *bor* amethyst, glob.
Cinnabar,	287.	"	VI; cleav; mas: G 8—8·1; r, bnh-r, gyh, dark bn; St r; strp, op; *nit* sol, r fumes: *Bl*, volatile. *Strat, prim.*
Blende,	269.	3·5—4·0	I; dodec cl! mas; G 4—4·1; bn, bk; St yw, bnh; op; submet, bright: *Bl*, fus. *Prim, strat*, etc.
Manganite,	261.	4·0—4·5	III; mas; G 4·3—4·4; dark steel-gy, iron-bk; St rdh-bn, bkh: *Bl*, infus; *bor*, amethystine glob.

	Hardness.	
Limonite,	239 5·0—5·5	mam, bot, stalact, mas; G 3·9—4; bn, bkh; St ywh-bn; strp, op; no action on magnet: *Bl*, infus, bk and magnetic.
Wolfram,	244. 5·0—5·5	III; mas; col, lam; G 7·1—7·4; gyh-bk, bnh-bk; St dark rdh-bn; submet: *Bl* fus! decrep, *bor* gn bead. *Prim.*
Chromic iron,	241. "	I; mas; G 4·3—4·5; iron bk, rather dull, brittle; St bn; often slightly magnetic: *Bl*, infus; *bor* fine gn, fus dif. *Serpentine.*
Pitchblende,	228. 5·5	Mas, bot; G 6·47; bnh-bk, velvet-bk; St, bk; submet; *nit* slow sol: *Bl*, *bor* gray scoria. *Prim.*
Psilomelane,	259. 5·0—6·0	Mas, bot; G 4—4·4; bh, gyh to dark steel gy; St bnh-bk, shining; brittle: *Bl*, infus, *bor* violet. *Manganese ores.*
Columbite,	243. "	III; mas; G 5·9—6·1; bnh-bk, bk, often with a steel blue tarnish; St dark rdh-bn, bnh-bk; submet: *Bl*. infus, *bor* fus dif. *Prim.*
Yenite,	245. 5·5—6·0	III; mas, col; G 3·8—4·1; iron bk, bnh; St gnh, bnh; submet; brittle: *Bl*, fus; *bor* bk mag glob. *Prim.*
Specular iron,	218. 5·5—6·5	VI; mas; G 4·5—5·3; iron-bk and cryst brilliant; St r, rdh-bn: *Bl* infus, *bor* iron reaction, glob finally mag. *Prim, strat, volc.*
Magnetic iron,	235. "	I; mas; G 5—5·1; iron-bk; St bk; strongly magnetic: *Bl*, infus, *bor* iron reaction. *Prim, strat.*
Franklinite,	240. "	I; mas; G 4·8—5·1; iron-bk; St dark rdh bn; slightly magnetic: *Bl*, infus; at high heat zinc fumes. *Prim.*

† Fumes before the blowpipe.

Dark red silver,	327. 2·5	VI; mas; G 5·7—5·9; iron-bk, lead-gy; St red; met-ad: *Bl*, fus!! b flame, sulph and antimony fumes. *Silver ores.*
Erubescite,	294. 3·0	I; mas; G 5—5·1; pinchbeck-bn, copper-r, bh tarnish: St pale gyh-bk; brittle: *Bl*, fus; on *char* sulph odor, glob mag. *Prim, strat, with copper ores.*
Copper pyrites,	292. 3·5—4·0	II; mas; G 4—4·2; brass yw; St gnh-bk; brittle. *nit* sol, gn: *Bl*, fus; on char, sulph odor. *Prim, strat, with copper ores.*
Magnetic pyrites,	233. 3·5—4·5	VI; mas; G 4·5—4·7; bronze-yw, copper-r; St gyh-bk; magnetic; brittle; dilute *nit* sol: *Bl*, fus, sulph odor.
Leucopyrite,	235. 5·0—5·5	III; mas; G 7·2—7·4; silver-w, steel-gy; St gyh-bk; brittle: *Bl*, fus; on *char*, arsen fumes.
Copper nickel,	263. "	VI; mas; G 7·3—7·7; copper-r; St pale bnh-bk; brittle: *Bl*, fus! on *char*, arsen fumes. *Prim usual with cobalt ores.*

Hardness.

Nickel glance, 263. 5·0—5·5 I; mas; G 6—6·2; silver w, steel-gy; St gyh-bk: *Bl*, fus! decrep; sulph and arsen fumes in glass tube.

Cobaltine, 266. " I; mas; G 6·2—6·4; silver-w, rdh; St gyh-bk; brittle: *Bl*, fus; on *char*, arsen fumes, bh, glob, mag; *bor* blue. *Prim.*

Smaltine, 266. " I; mas; G 6·4—7·2; tin-w, steel-gy; St gyh-bk; brittle: *Bl*, fus! arsen odor, gyh bk mag pearl; *bor* blue. *Prim.*

White ir'n pyrites, 233. " III; mas; crests; G 4·6—4·9; pale bronze yw; St gyh, bnh-bk; brittle: *Bl*, fus; on *char*, sulph fumes.

Mispickel, 234. 5·5—6·0 III; mas; G 6·1; silver-w; St dark gyh-bk; brittle: *Bl*, on *char*, arsen fumes, and leaves a magnetic globule.

Iron pyrites, 231. 6·0—6.5 I; mas; G 4·8—5·1; light bronze-yw; St bnh-bk: *Bl*, fus; on *char*, sulph odor. *Prim, strat, volc, etc.*

B. STREAK METALLIC.

* Malleable.

Native mercury, 287. fluid G 13—14; tin-w: *Bl*, volatilizes. *Strat, prim.*

Native lead, 277. 1·0—1·5 I; in membranes and glob; G 11—12; lead gray; soils: *Bl*, fus!! volatilizes and colors charcoal yellow.

Native copper, 290. 2·5—3·0 I; mas, in strings; G 8·5—8·6; copper-r; *nit* sol! r fumes; *Bl*, fus, colors flame green.

Native silver, 323. " I; mas, capil; G 10—11; silver-w; *nit* sol: *Bl*, fus.

Native gold, 312. " I; mas, capil; G 12—20, pale to deep yw, according to the proportion of silver present; *nit* not sol: *Bl*, fus.

Native platinum, 308. 4·0—4·5 In grains and lumps; Gr 16—19; pale steel gy; hot *nit-mur* sol: *Bl*, infus.

Native iron, 230. 4·0—5·0 I; mas; G 7·3—7·8; iron-gy, magnetic.

Native palladium, 311. 5·0—5·5 In grains, structure rad; G 10—12; steel-gy, silver-w: *Bl*, infus.

† Not malleable: no fumes when heated.

Graphite, 91. 1·0—2·0 Mas, fol! gran; G 2—2·1; iron-bk, dark steel-gy; sectile; soils; *nit*, no action: *Bl*, infus. *Prim, strat.*

Ilmenite, 241. 5·0—6·0 VI; mas; G 4·4—4·8; dark iron-bk; slightly magnetic; strong *mur* sol: *Bl*, infus. *Prim, volc.*

‡ Not malleable: fumes when heated.

Molybdenite, 217. 1·0—1·5 VI; mas, fol! lam flex; G 4·5—4·8; pure lead-gy; sectile; *nit*, partly sol: *Bl*, infus, on *char* sulph odor. *Prim.*

		Hardness.	
Fol. Tellurium,	280.	1·0—1·5	II; fol! gran; G 7—7·1; bkh lead-gy; lam flex sectile; *nit* sol! *Bl*, on *char*, w fumes, flame b. *Prim.*
Gray antimony,	222.	2·0	III; cleav; col, div; G 4·5—4·7; lead-gy, steel-gy; tarnishes; lam subflex: *Bl*, fus!! on *char* sulph odor and wholly volat. *Prim.*
Vitreous silver,	325.	2·0—2·5	I; mas, retic; G 7·1—7·4; bkh lead-gy; *nit* sol: *Bl*, fus!! intum, glob of silver. *Silver ores.*
Native tellurium,	219.	"	VI; mas; G 5·7—6·1; tin-w, rather brittle: *Bl*, fus!! on *char* gnh flame, w inodorous fumes, wholly volat. *Prim.*
Brittle silver,	326.	"	III; mas; G 6·2—6·3; iron-bk; sectile; hot *nit* sol: *Bl*, fus!! sulph and antim fumes; on *char*, glob of silver. *Silver ores.*
Native bismuth,	220.	"	I; mas, cleav! G 9·7—9·8; silver-w, rdh; *nit* sol, and solution w if diluted: *Bl*, fus!! volat, inod; yw on *char*. *Prim.*
Vitreous copper,	292.	2·5—3·0	III; mas; G 5·5—5·8; bkh, lead-gy; *nit* sol, and polished iron put in the solution covered with copper: *Bl*, fus! on *char* sulph fumes. *Prim, strat.*
Galena,	277.	2·5—3·0	I; cleav! mas; G 7·5—7·7; pure lead-gy; rather sectile: *Bl*, fus! decrep; on *char* sulph fumes and glob of lead. *Prim, strat.*
Amalgam,	287.	2·0—3·5	I; mas; G 10·5—14; silver-w; *nit* sol: *Bl*, fumes of mercury, and silver glob.
Native antimony,	222.	3·0—3·5	VI; cleav; lam, mas; G 6·6—6·8; tin-w: *Bl*, fus!! volat; on *char* w fumes. *Prim.*
Native arsenic,	225.	3·5	VI; mas; G 5·6—5·8; tin-w, lead-gy, darker from tarnish; brittle: *Bl*, wholly volat, garlic odor. *Prim.*
Gray copper,	295.	3·0—4·0	I; tetrahed; mas; G 4·7—5·1; steel-gy to iron-bk: *Bl*, fus!! arsen and antim fumes; copper reaction. *Prim, copper ores.*
White nickel,	263.	5·0—5·5	I; mas; G 7·1—7·2; tin-w: *Bl*, arsen fumes; also nickel reaction. *Prim.*

In determining the name of a mineral by the preceding table, trials should be made of the hardness and of the other characters upon which the arrangement is based, as shown in the general view on page 188. The particular subdivision containing the species is thus arrived at, and also, by means of the hardness, the place of the species in the subdivision. Afterwards, by a comparison of the other characters, (specific gravity, color, etc.,) with the brief descriptions given in the table, the name of the mineral will be ascertained. If any doubt still remains, the fuller descriptions in the body of the work may be referred to, for the convenience of which reference, the page is added for each species.

The following hints may be of service to the beginner in the science, by enabling him to overcome a difficulty in the outset, arising from the various forms and appearance of the minerals quartz and limestone. Quartz occurs of nearly every color, and of various degrees of glassy luster to a dull stone without the slightest glistening. The common grayish cobble stones of the fields are usually quartz, and others are dull red and brown; from these there are gradual transitions to the pellucid quartz crystal that looks like glass itself. Sandstones and freestones are often wholly quartz, and the seashore sands are mostly of the same material. It is therefore probable that this mineral will be often encountered in mineralogical rambles. Let the first trial of specimens obtained be made with a file or the point of a knife, or some other means of trying the hardness; if the file makes no impression, there is reason to suspect the mineral to be quartz; and if on breaking it, no regular structure or cleavage plane is observed, but it breaks in all directions with a similar surface and a more or less vitreous luster, the probability is much strengthened that this conclusion is correct. The blowpipe may next be used; and if there is no fusion produced by it, when carefully used on a thin splinter, there can be little doubt that the specimen is in fact quartz.

Carbonate of lime (calc spar, including limestone,) is another very common species. If the mineral collected is rather easily impressible with a file, it may be of this species: if it effervesces freely when placed in a test-tube containing dilute muriatic acid, and is finally dissolved, the probability of its being carbonate of lime is increased: if the blowpipe produces no trace of fusion, but a brilliant light from the fragment before it, but little doubt remains on this point. Crystalline fragments break with three equal oblique cleavages.

Familiarized with these two Protean minerals by the trials here alluded to, the student has already surmounted the principal difficulties in the way of future progress. Frequently the young beginner, who has devoted some time to collecting all the different colored stones in his neighborhood, on presenting them for names to some practised mineralogist, is a little disappointed to learn that, with two or three exceptions, his large variety includes nothing but limestone and quartz. He is perhaps gratified, however, at being told that he may call this specimen yellow jasper, that red jasper, another flint, and another hornstone, others chert, granular quartz, ferruginous quartz, chalcedony, prase, smoky quartz, greasy quartz, milky quartz, agate, plasma, hyaline quartz, quartz crystal, basanite, radiated quartz, tabular quartz, etc. etc.; and it is often the case, in this state of his knowledge, that he is best pleased with some treatise on the science in which all these various stones are treated of with as much prominence as if actually distinct species; being loth to receive the unwelcome truth, that his whole extensive cabinet contains only one mineral. But the mineralogical student has already made good progress when this truth is freely admitted, and quartz and limestone, in all their varieties, have become known to him.

To facilitate still farther the study of minerals, the following tables are added.

TABLE II. FOR THE DETERMINATION OF MINERALS.

The general arrangement in this table is the same as in the preceding: but the order of the species, instead of being that of their hardness, is that of their *specific gravity*.

I.—SOLUBLE MINERALS.

A. No Effervescence with muriatic acid.

a. *Not deflagrating on burning coals.*

	Sp. gr.		Sp. gr.
Glauber salt	1·4—1·5	Copperas,	2·0
Sal ammoniac,	1·5—1·6	White vitriol,	2·0—2·1
Epsom salt,	1·7—1·8	Blue vitriol,	2·2—2·3
Borax,	"	Common salt,	"
Alum,	"	White arsenic,	3·7

b. *Deflagrate on burning coals.*

Nit. of lime,	1·62	Nit. of soda,	2·0—3·0
Niter,	1·9—2·0		

B. Effervescing with muriatic acid.

Natron,	1·4—1·5

II.—INSOLUBLE MINERALS.

I. LUSTER UNMETALLIC.

A. Streak Uncolored.

a. No fumes before the blowpipe on charcoal.

1. *Wholly soluble in one or more of the acids, (cold or hot), usually with effervescence.*

* Infusible.

	Sp. gr.		Sp. gr.
Websterite,	1·6—1·7	Magnesite,	2·9—3·0
Brucite,	2·3—2·4	Mesitine spar,	3·3—3·7
Nemalite,	2·3—2·5	Diallogite,	3·5—3·6
Calc spar,	"	Oligon spar,	3·7—3·8
Hydromagnesite,	2·8	Yttrocerite,	
Aragonite,	2·8—3·0	Blende,	4·0—4·1
Dolomite,	2·8—2·9		

† Fusible with more or less difficulty.

Wavellite,	2·3—2·4	Triphyline,	3·4—3·6
Boracite,	2·9—3·0	Strontianite,	3·6—3·7
Apatite,	3·0—3·3	Spathic iron,	3·7—3·9
Fluor spar,	3·1—3·2	Witherite,	4·2—4·4
Cacoxene,	3·3—3·4	White lead ore,	6·1—6·5
Triplite,	3·4—3·8	Pyromorphite,	6·5—7·1

2. *Soluble in acids, excepting the silica, which separates as a jelly.*

* Infusible.

Allophane,	1·8—1·9	Halloylite,	1·8—2·1

† Fusible.

Philippsite,	2·0—2·2	Mesole,	2·3—2·4
Analcime,	2·0—2·3	Thomsonite,	"
Datholite,	"	Sodalite,	2·2—2·5
Natrolite,	2·1—2·3	Pectolite,	2·69
Scolecite,	2·2—2·3	Tabular spar,	2·7—2·9
Laumonite,	2·2—2·4	Calamine,	3·2—3·5
Dysclasite,	"		

3. *Not acted on by acids, or partially soluble without forming a jelly.*

* Infusible.

Chrysocolla,	2·3—2·4	Yenite,	2·4—5.2

	Sp. gr.		Sp. gr.
Opal,	"	Topaz,	3·4—3·6
Quartz,	2·6—2·8	Diamond,	3·4—3·7
Alum-stone,	"	Kyanite,	3·5—3·7
Talc,	2·7—2·9	Staurotide,	3·5—3·8
Pyrophyllite,	"	Chrysoberyl,	3·5—3·8
Mica,	2·8—3·0	Anatase,	3·8—3·9
Turquois,	"	Sapphire,	3·9—4·2
Nephrite,	2·9—3·1	Blende,	4·0—4·1
Andalusite,	2·9—3·2	Spinel,	3·5—4·6
Emerald nickel,	3·05	Zircon,	4·4—4·8
Clintonite,	3·0—3·1	Monazite,	4·8—5·1
Sillimanite,	3·0—3·4	Plumbo-resinite,	6·3—6·4
Bucholzite,	3·2—3·6	Tin ore,	6·5—7·1
Chrysolite,	3·3—3·6		

† Fusible with more or less difficulty.

Chabazite,	2·0—2·2	Prehnite,	2·8—3·0
Stilbite,	2·1—2·2	Boracite,	2·9—3·0
Heulandite,	2·2	Chrysolite,	"
Gypsum,	2·2—2·4	Euclase,	2·9—3·1
Apophyllite,	2·3—2·4	Hornblende,	2·9—3·4
Feldspar,	2·3—2·6	Lazulite,	3·0—3·1
Serpentine,	2·4—2·6	Tourmaline,	"
Obsidian,	2·2—2·8	Spodumene,	3·1—3·2
Harmotome,	2·3—2·5	Chondrodite,	"
Petalite,	2·4—2·5	Axinite,	3·2—3·3
Schiller spar,	2·5—2·7	Pyroxene,	3·1—3·5
Lapis Lazuli,	2·5—2·9	Sphene,	3·2—3·5
Albite,	2·6—2·7	Epidote,	"
Labradorite,	2·6—2·8	Idocrase,	3·3—3·4
Scapolite,	"	Manganese spar,	3·4—3·7
Iolite,	"	Garnet,	3·5—4·3
Beryl,	"	Celestine,	3·8—4·0
Chlorite,	2·6—2·9	Pyrochlore,	3·8—4·3
Chlorophyllite,	2·7—2·8	Heavy spar,	4·3—4·8
Talc,	2·7—2·9	Monazite,	4·8—5·1
Mica,	2·8—3·0	Tungstate of lime,	6·0—6·1
Anhydrite,	"	Anglesite,	6·2—6·3

b. Colored or odorous fumes before the blowpipe.

Scorodite,	3·1—3·3	Horn silver,	5·5—5·6
Blende,	4·0—4·1	Bismuth blende,	5·9—6·1
Calamine,	4·2—4·5	Mimetene,	6·4—6·5

B. Streak Colored.

a. *No fumes before the blowpipe.*

* Fusible.

	Sp. gr.		Sp. gr.
Vivianite,	2·6—2·7	Pyrochlore,	4·2—4·3
Uranite,	3·0—3·6	Minium,	4·6
Chondrodite,	3·1—3·3	Monazite,	4·8—5·1
Allanite,	3·2—4·1	Cupreous anglesite,	5·3—5·5
Triplite,	3·4—3·8	Red copper ore,	5·9—6·0
Azurite,	3·5—3·9	Chromate of lead,	6·0
Green malachite,	4·0—4·1	Pyromorphite,	6·8—7·1

† Infusible.

Sulphur,	2·07	Blende,	4·0—4·1
Copper mica,	2·55	Psilomelane,	4·0—4·4
Earthy cobalt,	2·2—2·3	Rutile,	4·2—4·3
Cobalt bloom,	2·9—3·0	Chromic iron,	4·3—4·5
Warwickite,	3·0—3·3	Atacamite,	4·4—4·5
Dioptase,	3·2—3·3	Red antimony,	4·4—4·6
Cacoxene,	3·3—3·4	Red zinc ore,	5·4—5·6
Orpiment,	3·4—3 5	Red silver ore,	5·4—5·9
Realgar,	3·3—3·7	Pitchblende,	6·47
Wad,	3·7	Tin ore,	6·5—7·1
Black copper,		Cinnabar,	8·0—8·1
Limonite,	3·9—4·1		

LUSTER METALLIC.

A. Streak Uncolored.

* No fumes before the blowpipe on charcoal.

Earthy cobalt,	2·2—2·3	Specular iron,	4·5—5·3
Wad,	3·7	Pyrolusite,	4·8—5·0
Yenite,	3·8—4·1	Franklinite,	4·8—5·1
Arkansite,	3·85	Magnetic iron ore,	5·0—5·1
Brown hematite,	3·9—4·0	Columbite,	5·9—6·1
Blende,	4·0—4·1	Pitchblende,	6·47
Psilomelane,	4·0—4·4	Wolfram,	7·1—7·4
Manganite,	4·3—4·4	Cinnabar,	8·0—8·1
Chromic iron,	4·3—4·5		

† Fumes before the blowpipe.

Copper pyrites,	4·0—4·2	Nickel glance,	6·0—6·2
Magnetic pyrites,	4·5—4·7	Mispickel,	6·1
White iron pyrites,	"	Cobaltine,	6·2—6·4
Iron pyrites,	4·8—5·1	Smaltine,	6·4—7·2
Variegated copper,	5·0—5·1	Leucopyrite,	7·2—7·4
Dark red silver,	5·7—5·9	Copper nickel,	7·3—7·7

B. Streak Metallic.

* Malleable.

	Sp. gr.		Sp. gr.
Native iron,	7·3—7·8	Native lead,	11—12
Native copper,	8·5—8·6	Native mercury,	13—14
Native silver,	10—11	Native platinum,	16—19
Native palladium,	10—12	Native gold,	12—20

† Not malleable: no fumes when heated.

Graphite,	2·21	Ilmenite,	4·4—4·8

‡ Not malleable: fumes when heated on charcoal.

Gray antimony,	4·5—4·7	Native antimony,	6·6—6·8
Molybdenite,	4·5—4·8	Fol. tellurium,	7·0—7·1
Gray copper,	4·7—5·1	White nickel,	7·1—7·2
Vitreous copper,	5·5—5·8	Vitreous silver,	7·1—7·4
Native arsenic,	5·6—5·8	Galena,	7·5—7·7
Native tellurium,	5·7—6·1	Native bismuth,	9·7—9·8
Brittle silver,	6·2—6·3	Amalgam,	10·5—11

TABLE III.—MINERALS ARRANGED ACCORDING TO THEIR CRYSTALLIZATION.

I.—CRYSTALS MONOMETRIC.

A. *Luster unmetallic.*

* Infusible.

		Hardness.	Sp. gr.	Cleavage.
Blende,	269	2·0—3·0	4·0—4·2	Dodecahedral.
Chromic iron,	241	5·5	4·3—4·5	Octahed. imperf.
Leucite,	175	5·5—6·0	2·4—2·5	Ncne.
Dysluite,	161	7·5—8·0	4·5—4·6	Oct. imp.
Spinel,	160	8·0	3·5—3·6	Oct. imp.
Diamond,	80	10·0		Oct. perfect.

† Fusible.

		Hardness.	Sp. gr.	Cleavage..
Alum,	127	1·5—2·0	1·7—1·8	Oct.
Common salt,	104	2·0	2·2—2·3	Cubic.
Red copper ore,	296	3·5—4·0	5·8—6·1	Oct. imperf.
Fluor spar,	121	4·0	3·0—3·3	Oct. perf.
Pyrochlore,	208	5·0—5·5	3·8—4·5	None.
Analcime,	168	"	2·0—2·3	Imperfect.
Lapis Lazuli,	196	5·5—6·0	2·5—2·9	Dodec. imperf.
Sodalite,	197	5·5—6·0	2·2—2·4	Dodec. imp.
Garnet,	184	6·5—7·5	3·5—4·3	Dod. oft. distinct
Boracite,	126	7·0	2·9—3·0	Oct. indistinct.

2.—*Luster metallic.*

* No fumes before the blowpipe on charcoal.

Native copper,	290	2·5—3·0	8·4— 8·8	None.
Native silver,	323	"	10·3—10·5	None.
Native gold,	312	"	12·0—20·0	None.
Blende,	269	3·5—4·0	4·0— 4·2	Dodec. perf!
Native platinum,	308	4·0—4·5	16·0—19·0	Cubic, indist.
Native iron,	230	4·5	5·1— 5·2	Oct. perfect.
Chromic iron,	241	5·0—5·5	4·3— 4·5	Oct. imp.
Franklinite,	240	5·5—6·5	4·8— 5·1	Oct. imp.
Magnetite,	235	"	5·0— 5·1	Oct. imp.

† Fumes before the blowpipe on charcoal.

Vitreous silver,	325	2·0—2·5	7·1—7·4	Dodec. imperf.
Native bismuth,	220	"	9·7—9·8	Oct. perf!
Native amalgam,	287	2·0—3·5	10·5—14	Dodec. imp.
Erubescite,	294	2·5—3·0	5·0—5·1	Oct. imp.
Galena,	277	"	7·5—7·7	Cubic perf!
Gray copper ore,	295	3·0—4·0	4·7—5·2	Indistinct.
Nickel glance,	253	5·0—5·5	6·0—6·2	Cubic perf!
Cobaltine,	266	"	6·1—6·3	Cubic perf.
Smaltine,	266	"	6·3—6·4	Oct. imp.
White nickel,	263	5·5	7·1—7·2	
Pyrites,	231	6·0—6·5	4·8—5·1	Cubic imp.

II.—CRYSTALS DIMETRIC.

1. *Luster unmetallic.*

* Infusible.

Anatase,	211	5·5—6·0	3·8—3·9	Oct. and basal.

		Hardness.	Sp. gr.	Cleavage.
Tin ore,	214	6·0—7·0	6·5—7·1	Indistinct.
Zircon,	200	7·5	4·4—4·8	Imperfect.

† Fusible.

Uranite,	228	2·0—2·5	3·0—3·6	Basal, perf !!
Apophyllite,	165	4·5—5·0	2·2—2·4	Basal, perf!
Scapolite,	180	5·0—6·0	2·5—2·8	Lat. distinct.
Idocrase,	184	6·0—6·5	3·3—3·5	Lat. indistinct !
Rutile,	210	"	4·1—4·3	Lat. imp.

2. *Luster metallic.*

Foliated tellurium,	280	1·0—1·5	7·0—7·2	Foliated !
Copper pyrites,	292	3·5—4·0	4·1—4·2	Indistinct.
Hausmannite,	261	5·0—5·5	4·7—4·8	Basal, distinct !
Braunite,	261	6·0—6·5	4·8—4·9	Oct. distinct.

III. CRYSTALS TRIMETRIC.

1. *Luster unmetallic.*

* Infusible.

Talc,	143	1·0—1·5	2·7—2·9	Basal, fol !!
Aragonite,	118	3·5—4·0	2·9—3·0	Lat. imp.
Red Zinc ore,	270	4·0—4·5	5·4—5·6	Basal, fol !!
Chrysolite,	156	6·5—7·0	3·3—3·5	Lat. imp.
Staurotide,	174	7·0—7·5	3·6—3·8	Indistinct.
Andalusite,	174	7·5	3·1—3·4	Indistinct.
Topaz,	194	8·0	3·4—3·6	Basal, perfect !
Chrysoberyl,	199	8·5	3·5—3·8	Imperfect.

† Fusible: gelatinize in acids.

Mesole,	167	3·5	2·3—2·4	One perfect.
Thomsonite,	167	4·5	2·2—2·4	Two rect. perf.
Phillipsite,	168	4·0—4·5	2·0—2·2	Imperfect.
Calamine,	272	4·5—5·0	3·3—3·5	Lat. perfect.
Natrolite,	166	4·5—5·5	2·1—2·3	Lat. perf.
Scolecite,	167	5·0—5·5	2·2—2·3	Imperfect.

‡ Fusible: not gelatizing; giving no odorous or colored fumes before the blowpipe.

Talc, (some var.,)	143	1·0—1·5	2·7—2·9	Foliated !!
Niter,	101	2·0	1·9—2·0	Imperfect.
Epsom salt,	124	2·0—2·5	1·7—1·8	One perfect.
Cryolite,	132	"	2·9—3·0	One prf; two

		Hardness.	Sp. gr.	Cleavage.
Mica,	193	"	2·8—3·1	Foliated !!
Anglesite,	281	2·5—3·0	6·2—6·3	Imperfect.
Heavy spar,	108	2·5—3·5	4·3—4·8	Imperfect.
Celestine,	110	"	3·9—4·0	Lat. distinct.
Anhydrite,	114	3·0—3·5	2·8—3·0	Three rect. dist.
White lead ore,	281	"	6·1—6·5	Lat. perf.
Witherite,	109	"	4·2—4·4	Imperfect.
Serpentine,	145	3·0—4·0	2·5—2·6	Sometimes fol.
Strontianite,	111	3·5—4·0	3·6—3·8	Lat. distinct.
Wavellite,	130	"	2·2—2·4	Two distinct.
Stilbite,	165	"	2·1—2·2	One perfect !
Harmotome,	168	4·0—4·5	2·4—2·5	Imperfect.
Wolfram,	244	5·0—5·5	7·1—7·4	One perfect.
Lazulite,	131	5·0—6·0	3·0—3·1	Indistinct.
Yenite,	245	5·5—6·0	3·8—4·1	Indistinct.
Prehnite,	170	6·0—7·0	2·8—3·0	Basal, distinct.
Iolite,	190	7·0—7·5	2·5—2·7	Indistinct.

§ Giving fumes before the blowpipe on charcoal.

Orpiment,	226	1·5—2·0	3·4—3·6	Foliated !
Sulphur,	98	1·5—2·5	2·0—2·1	Indistinct.
White vitriol,	271	2·0—2·5	2·0—2·1	One perfect.
White antimony,	224	2·5—3·0	5·5—5·6	Lat. perfect !!
Atacamite,	302	3·0—3·5	4·0—4·4	Basal, perfect.
Scorodite,	249	3·5—4·0	3·1—3·3	Imperfect.

2. *Luster metallic.*

* No fumes before the blowpipe on charcoal.

Pyrolusite,	259	2·0—2·5	4·8—5·0	Three imperfect.
Manganite,	261	4·0—4·5	4·3—4·4	One imperfect.
Wolfram,	244	5·0—5·5	7·1—7·4	One perfect.
Yenite,	245	5·5—6·0	3·8—4·1	Indistinct.
Columbite,	243	5·0—6·0	5·9—6·1	Indistinct.
Tantalite,	244	"	7·2—8·0	Imperfect.

† Fumes before the blowpipe on charcoal.

Gray antimony,	222	2·0	4·5—4·7	One perfect !
Brittle silver ore,	326	2·0—2·5	6·2—6·3	Imperfect.
Vitreous copper,	292	2·5—3·0	5·5—5·8	Lat. indistinct.
Leucopyrite,	235	5·0—5·5	7·2—7·4	One distinct.
Mispickel,	234	5·0—6·0	6·1—6·2	Lat imperfect.
White iron pyrites,	233	6·0—6·5	4·6—4·9	Lat imperfect.

IV.—CRYSTALS MONOCLINIC.

1. *Luster unmetallic.*

* Soluble.

		Hardness.	Sp. gr.	Cleavage.
Natron,	103	1·0—1·5	1·4—1·5	
Glauber salt,	102	1·5—2·0	1·5—2·0	
Copperas,	246	2·0	1·8—1·9	One perfect.
Borax,	107	2·0—2·5	1·7	Lat. perfect.

† Insoluble: no fumes before the blowpipe on charcoal.

Vivianite,	248	1·5—2·0	2·6—2·7	Basal, perfect !
Gypsum,	112	2·0	2·3—2·4	Foliated !
Mica,	191	2·0—2·5	2·8—3·0	Foliated !!
Heulandite,	164	3·5—4·0	2·1—2·2	Foliated.
Laumonite,	166	3·5—4·0	2·3	One distinct.
Green malachite,	298	"	4·0—4·1	Basal, perfect.
Azurite,	300	3·5—4·5	3·5—3·9	Lateral.
Clintonite,	148	4·0—5·0	3·0—3·1	Foliated.
Monazite,	206	5·0	4·8—5·1	Basalt, perfect !
Datholite,	142	5·5—6·0	2·9—3·0	Indistinct.
Sphene,	211	"	3·2—3·5	Indistinct.
Hornblende,	152	5·0—6·0	2·9—3·4	Lat. perfect.
Pyroxene,	150	"	3·2—3·5	Lat. distinct.
Allanite,	207	"	3·3—3·8	Indistinct.
Feldspar,	176	6·0	2·3—2·6	One prf; one imp.
Chondrodite,	157	6·0—6·5	3·1—3·2	Indistinct.
Epidote,	182	6·0—7·0	3·2—3·5	Lat. imperf.
Spodumene,	156	6·5—7·0	3·1—3·2	Lat. perfect.
Euclase,	199	7·5	2·9—3·1	Basal, perfect.

‡ Fumes before the blowpipe.

Cobalt bloom,	267	1·5—2·0	2·9—3·0	Basal, perfect !
Realgar,	226	"	3·3—3·6	Imperfect.
Pharmacolite,	226	2·0—2·5	2·6—2·8	Basal, perfect !!
Miargyrite,	327	"	5·2—5·4	Lat. imperfect.

2. *Luster metallic.*

Miargyrite,	327	2·0—2·5	5·2—5·4	Lat. imperfect.
Wolfram,	244	5·0—5·5	7·1—7·4	One perfect.
Warwickite,		5·5—6·0	3·0—3·3	One perfect.
Allanite,	207	"	3·3—3·8	Imperfect.

V.—CRYSTALS TRICLINIC.

* Soluble.

Blue vitriol,	97	2·5	2·2—2·3	Imperfect.

† Insoluble: fusible.

		Hardness.	Sp. gr.	Cleavage.
Albite,	177	6·0	2·6—2·7	One perf.; two imperfect.
Labradorite,	178	"	2·6—2·8	One perf.; one imperfect.
Manganese spar,	258	6·0—7·0	3·4—3·7	One perfect.
Axinite,	190	6·5—7·0	3·2—3·3	Imperfect.

‡ Infusible.

Kyanite,	173	5·0—7·0	3·5—3·7	Lat. distinct.
Sillimanite,	172	7·0—7·5	3·2—3·3	Diagonal perf.!!

6. CRYSTALS HEXAGONAL OR RHOMBOHEDRAL.

1. *Luster unmetallic.*

* Soluble.

Nitrate of soda,	103	1·5—2·0	2·0—2·1	Rhomb. perf.
Coquimbite,	256	"		Hexag. imperf.

† Insoluble: infusible.

Brucite,	126	1·5	2·35	Foliated!
Mica, (hexagonal)	193	2·0—2·5	2·8—3·1	Foliated!!
Calc spar,	115	2·5—3·5	2·5—2·8	Rhomb. perf!
Diallogite,	261	3·5	3·5—3·6	Rhombohedral.
Magnesite,	124	3·0—4·0	2·8—3·0	Rhomb. perf.
Ankerite,	120	"	2·9—3·2	Rhomb. perf.
Dolomite,	118	3·5—4·0	3·5—4·0	Rhomb. perf.
Spathic iron,	247	"	3·7—3·9	Rhomb. perf.
Alum stone,	129	5·0	2·6—2·8	Basal, near perf.
Dioptase,	301	"	3·2—3·3	Rhombohedral.
Quartz,	132	7·0	2·6—2·7	Imperfect.
Sapphire,	158	9·0	3·9	Basal, perf.

‡ Insoluble: fusible, without fumes.

Chlorite,	145	1·5—2·0	2·6—2·9	Foliated!
Chabazite,	169	4·0—4·5	2·0—2·2	Rhombohed. ma.
Apatite,	120	5·0	3·0—3·3	Indistinct.
Nepheline,	179	5·5—6·0	2·4—2·7	Imperfect.
Tourmaline,	187	7·0—8·0	3·0—3·1	Indistinct.
Beryl,	197	7·5—8·0	2·6—2·8	Basal, indistinct.

§ Insoluble: fumes before the blowpipe on charcoal.

		Hardness.	Sp. gr.	Cleavage.
Red silver ore,	327	2·0—3·0	5·4—5·9	Imperfect.
Cinnabar,	287	2·0—2·5	7·8—8·1	Hexag. perfect
Smithsonite,	272	5·0	4·3—4·5	Rhomb. perf.

2. *Luster metallic.*

* No fumes before the blowpipe on charcoal.

Graphite,	91	1·0—2·0	2·0—2·1	Foliated!
Ilmenite,	241	5·0—6·0	4·4—5·0	Indistinct.
Specular iron,	237	5·5—6·5	5·0—5·3	Indistinct.

† Fumes before the blowpipe on charcoal.

Molybdenite,	217	1·0—1·5	4·5—4·8	Foliated !!
Native tellurium,	219	2·0—2·5	5·7—6·1	Imperfect.
Dark red silver,	327	2·5	5·7—5·9	Imperfect.
Cinnabar,	287	"	7·8—8·1	Hexag. perfect.
Native antimony,	222	3·0—3·5	6·6—6·8	Basal, perfect; rhombohed. dist.
Native arsenic,	225	3·5	5·6—6·0	Imperfect.
Magnetic pyrites,	233	3·5—4·5	4·6—4·7	Basal hexag. prf.
Copper nickel,	263	5·0—5·5	7·3—7·7	

INDEX.

A.

B.

D.

E.

F.

G.

H.

I.

O.

P.

S.

Y.

Z.

ERRATUM.—On page 156, 15th line from the bottom, read "alumina, 29·3," instead of "alumina, 2·3."

www.ingramcontent.com/pod-product-compliance
Lightning Source LLC
LaVergne TN
LVHW021123110826
845150LV00005B/931

* 9 7 8 1 4 2 5 5 5 0 6 9 1 *